ANTOINE-LAURENT LAVOISIER

Elements of Chemistry

*in a new systematic order, containing
all the modern discoveries*

TRANSLATED BY ROBERT KERR

WITH A NEW INTRODUCTION BY

DOUGLAS McKIE, D.Sc., Ph.D., F.R.S.E.

PROFESSOR OF THE HISTORY OF SCIENCE
UNIVERSITY COLLEGE LONDON

DOVER PUBLICATIONS, INC
New York

This Dover edition, first published in 1965, is an unabridged republication of the English translation first published by William Creech in 1790, to which has been added a new introduction by Douglas McKie, D. Sc., Ph. D., F.R.S.E., Professor of the History of Science, University College London.

International Standard Book Number: 0-486-64624-6

Library of Congress Catalog Card Number 65-15513

Manufactured in the United States of America

Dover Publications, Inc.
180 Varick Street
New York, N.Y. 10014

Introduction to Dover Edition

The *Elements of Chemistry, in a new systematic order, containing all the modern discoveries* (Edinburgh, 1790), here reprinted in facsimile, is a translation by Robert Kerr (1755–1813) of the classic *Traité élémentaire de Chimie, présenté dans un ordre nouveau et d'après les découvertes modernes* (Paris, 1789) of Antoine-Laurent Lavoisier (1743–1794), the founder of modern chemistry. The *Traité* was first published in Paris in March 1789,[1] the *Elements* in Edinburgh in November 1790 in readiness for the opening of the university session of 1790–1791. That the English translation was published in Edinburgh was most appropriate; for Lavoisier's great forerunner in these matters, Joseph Black (1728–1799), was still in active occupation of the chair of chemistry in the University of Edinburgh, lecturing to large classes in what was then the leading school of chemistry in the world, and had declared, with considerable effect on his colleagues and on his students, his acceptance of the revolutionary changes introduced by Lavoisier. Something, however, must be credited to the initiative of William Creech (1745–1815), who was then Scotland's foremost publisher and from whose presses there poured forth a stream of literary as well as scientific works, including translations from several languages.

Robert Kerr, the translator, was a man of some reputation in Edinburgh. Born in Bughtridge, Roxburghshire, in 1755, the son of James Kerr, who had been Member of Parliament for Edinburgh from 1747 to 1754, he was related on his father's side to the Royalist, Robert, first Earl of Ancrum (1578–1654); and his mother, Elizabeth Kerr, was a granddaughter of Robert, first Marquis of Lothian (1636–1703), a supporter of "the glorious revolution" of 1688 and a Privy Councillor to King

[1] It is sometimes stated that there were two issues of the first edition, but this does not appear to be correct. See D. McKie, "On some pre-publication copies of Lavoisier's *Traité* (1789)," *Ambix*, 1961, Vol. 9, pp. 37–46.

William III. He was educated in Edinburgh, first at the High School, and then at the University where he studied medicine, attending the lectures of Joseph Black on chemistry in the sessions 1774–75, 1775–76, and 1776–77. Afterwards he became surgeon to the Orphan Hospital in Edinburgh. He was a man of broad scholarly gifts and his translations included not only the classic chemical treatise of Lavoisier, but also the *Essay on the New Method of Bleaching* (by means of chlorine) (Dublin [1790]; 2nd ed., Edinburgh, 1791) from the French of C.-L. Berthollet, *The Animal Kingdom* . . . *Mammalia* (London, 1792) from Part I of the Latin *Systema Naturae* of Linnaeus, *The Natural History of Oviparous Quadrupeds and Serpents* (Edinburgh, 1802) from the French of two volumes edited by Lacépède as posthumous additions to Buffon's encyclopaedic *Histoire Naturelle,* and an *Essay on the Theory of the Earth* (Edinburgh, 1813, with four further editions in 1815, 1817, 1822, and 1825) from the French of Cuvier with additional notes by Robert Jameson (1774–1854), the distinguished mineralogist and Regius Professor of Natural History in the University of Edinburgh. Kerr compiled part of a series of volumes of *A General History and Collection of Voyages and Travels . . . forming a complete history of the origin and progress of navigation, discovery and commerce, by sea and land, from the earliest to the present time* (Edinburgh, 18 vols., 1811–1824); he wrote a *General View of the Agriculture of the County of Berwick* (London, 1809 and 1813) for the Board of Agriculture, and a *History of Scotland during the reign of Robert I, surnamed the Bruce* (Edinburgh, 2 vols., 1811); and he edited the *Memoirs* (Edinburgh, 2 vols., 1811) of William Smellie (1740–1795), printer, naturalist, antiquary, and first editor of the *Encyclopædia Britannica.* He was elected a Fellow of the Royal Society of Edinburgh in 1788 and of the Society of Antiquaries of Scotland. Kerr's great abilities attracted him to science, and especially to chemistry, in which he had been instructed by Joseph Black, the pre-eminent teacher of his time, and he was well equipped to translate Lavoisier's *Traité.*

Antoine-Laurent Lavoisier was born in Paris on Monday, 26 August 1743. His father, Jean-Antoine, was a lawyer, and so was his grandfather, but his earliest known ancestor was Antoine Lavoisier, who died in 1620, and who was a postillion in the

King's service at Villers-Cotterets, a small country town some fifty miles northeast of Paris; his mother was Émilie Punctis, daughter of Clément Punctis, advocate to the Parliament of Paris. There was wealth on both sides of the family. Lavoisier's mother died when he was five years old, and he and his young sister, born in 1745, together with their father, went to live at the house of their widowed grandmother, Mme Punctis, where they were devotedly cared for by their aunt, their mother's younger sister, Constance Punctis, then aged twenty-two. Lavoisier's sister died in 1760 at the age of fifteen.

Lavoisier was educated in Paris at the Collège Mazarin, his father's old school, from 1754 to 1760, and here he made his first acquaintance with science. But on leaving school he followed the family tradition and entered on the study of law, qualifying as Bachelor in 1763 and Licentiate in 1764. In these years of legal studies, however, he maintained and developed his interest in science, first aroused at school, learning geology from Jean-Étienne Guettard (1715–1786), Member of the Académie Royale des Sciences, the most eminent geologist in France, pioneer of geological surveys and of geological cartography and an intimate friend in the Punctis–Lavoisier home, and attending the inspiring lectures on chemistry given by another Academician, Guillaume-François Rouelle (1703–1770), at the Jardin du Roi. At the same time he studied astronomy and mathematics under Nicolas-Louis de Lacaille (1713–1762), botany under Bernard de Jussieu (1699–1777), both of them Academicians, and also anatomy; and he was much attracted to meteorology, recording the readings of the barometer in his home several times a day, an interest that later led him to organize such observations in other parts of France and in other countries with the object of discovering the laws relating to the movements of the atmosphere and thence devising rules for forecasting the weather.

But it was significantly to chemistry that Lavoisier turned with, however, some leanings towards geology and mineralogy through the influence of Guettard. The chemistry that he learned in the Jardin du Roi was very different from chemistry as he left it thirty years later; and to appreciate his reforms we need to consider, as briefly as we can, the state of that science in the mid-eighteenth century.

The long theoretical inheritance of chemistry was the four-element theory of the philosophers of Greek antiquity, according to which all substances were composed of the four elements, earth and air, fire and water, combined in an infinite variety of proportions. Each element had two qualities: earth was cold and dry, water was cold and wet, fire was hot and dry, and air was hot and wet. By adjustment of the qualities one element could be converted into another, according to this purely speculative idea; and, similarly, all substances were transmutable one into another by suitably adjusting the proportions of the universal four elements contained in each and all of them. Matter was a unity and substances were transmutable. The theory was not dead in Lavoisier's time—in Edinburgh Black was teaching his classes that water was transmutable into earth.

Another theory of the composition of matter was put forward in the seventeenth century by Johann Baptista van Helmont (*c.* 1580–1644) in his posthumous *Ortus medicinae* (Amsterdam, 1648). He argued that all substances, except air, were ultimately derived from water. To demonstrate this he made his famous quantitative experiment with a small willow tree, an experiment that took five years, and he concluded that the tree had grown entirely from the water that he had supplied to it during this long period. His theory had one very great patron, Isaac Newton, who accepted it and referred to it in the *Principia* (London, 1687). His most significant work was, however, his recognition of the material nature of what he called *gas*, a generic name that he used for those products of chemical reactions that had been previously regarded as merely spirituous and immaterial; he explained to chemists that the many familiar and destructive explosions that shattered their glass apparatus when they experimented on reactions in sealed or closed vessels were due to the release of "a wild spirit" or "gas." In a simple way he observed differences between gas from various sources but, as he did not isolate any gas, his distinctions were not precise, and he sometimes confused one gas with another. He had, however, advanced the chemistry of his time by demonstrating that these substances were material.

The more recent theory, the phlogiston theory, founded by Johann Joachim Becher (1635–1682) and developed by Georg Ernst Stahl (1660–1734), was to some extent derived from the old belief that there was a fire element; it applied to metals, minerals and combustible substances in general; and it supposed that all combustible bodies contained a common principle, that is, an element, which was named phlogiston and which was released in the process of combustion in the form of fire, flame and, sometimes, light. That combustion could proceed only to a limited degree in a confined volume of air, a fact long familiar to generations of observers, was accordingly explained by asserting that a limited amount of air could take up only a limited amount of the phlogiston released from a substance subjected to combustion; a similar explanation was advanced for the corresponding and well-known fact that respiration ceased and animal life was likewise extinguished in a limited amount of air, respiration being considered as resembling combustion and releasing phlogiston. The proportion of phlogiston in different combustibles varied widely. Bodies that resisted the action of fire and those that would not burn contained very little or none at all. On the other hand, charcoal, oils, fats, spirit of wine, wheat, flour, must be composed entirely or almost entirely of phlogiston, since it was observed that on burning they either completely disappeared or left negligible residues.

The application of the phlogiston theory to metals had important consequences. It had long been known that when metals were heated in air they lost their metallic properties and were changed into powders or, as they were called, *calces* (literally, "ashes"); and it was equally well known, long before the phlogiston theory was formulated, that the calces of metals on heating with charcoal were reconverted into the original metals, recovering all their metallic properties, their form, and their lustre. Therefore, it appeared, the restoration of phlogiston to the calx reconstituted the metal from which it had been formed, and the metals, like combustible bodies, contained this same common constituent, phlogiston, which, it will be seen, could now be demonstrably transferred from charcoal to calx, that is, from one substance to another. Phlogiston could not, however, be isolated, put in a bottle and labelled, but neither could electricity; yet the existence of phlogiston was satisfactorily established in the minds of the chemists of that age by these and by many other observations. The theory was applied to explain both the chemical properties of substances and chemical changes in general. Chemistry was, indeed, systematized in terms of this theory; and, erroneous as it proved to be, it brought about a great change in chemical thought, which is too infrequently emphasized; for, whereas the old theory of the four elements or even its variant, the three principles of Paracelsus (1493–1541), which we have not space to consider in detail here, was merely a theory of the elementary chemical composition of matter, the phlogiston theory could be applied to explain chemical changes, not merely chemical composition, and it could be and was applied to explain and even to systematize the vast, but hitherto disorderly, and ever-increasing knowledge of chemical reactions and chemical processes. The phlogiston theory was thus the first great and systematizing generalization in chemistry. It was wrong; but it was accepted almost without exception by all the leading chemists of eighteenth-century Europe, whose distinguished names are too many to enumerate here; and, if there is one thing more than others to be said in favour of scientists, it is that their history shows that they learn from their mistakes.

The phlogiston theory was, however, in spite of its success and its acceptance, being slowly and imperceptibly undermined even before Lavoisier turned the synthetic powers of his genius to the reconsideration of chemical phenomena. Already a century earlier, Robert Boyle (1627–1691), in his *Sceptical Chymist* (London, 1661), had destructively criticized the four-element theory together with the Paracelsian theory of the three principles on the simple basis that neither theory was in agreement with the facts of chemistry. Chemists, of course, considered that fire resolved all substances into their elements, but Boyle pointed out that the number of products varied according to the means by which fire was applied, either on an open hearth or by distillation in a retort. He mentioned as a notable instance that wood heated in the open gave ashes and soot, but fired in a retort it yielded oil, spirit, vinegar, water, and charcoal, and therefore, according to the method of heating, the same substance gave either two or five products, which was either one less than the three principles of the recent Paracelsus or one more than the four elements of the ancient Greeks. Moreover, if one took a substance prepared from known ingredients, such ¡ made from fat and alkali, treatment with fire did not reproduce those ingredients, but two very different products which were, Boyle pointedly added, quite useless for making soap. Fire, he said, did not decompose substances into their elements, as the facts of chemistry plainly showed, but, on the contrary, re-arranged their component particles to form different compounds. Chemists, he urged, should reconsider the many experiments they had made and also devise new ones; and, to help towards a solution of the problem of what constituted a chemical element, he formulated an entirely new definition, which, stripped of its cumbersome seventeenth-century phraseology, defined an element as "a substance that cannot be decomposed into any simpler substance." It was this definition that Lavoisier eventually revived more than a hundred years later. In the meantime, however, while Boyle's criticism of the accepted theory of chemical elements had been destructive, chemists continued to make use of either the four elements or the three principles or even of both, for want of a better theory; and they did not treat the problem very seriously. Often dismissing it briefly in their books, they were more concerned with describing the preparation and properties of substances. Boyle also maintained that it was too simple to assume that the great book of Nature, still to be regarded as written in cipher, could be decoded on the supposition that it was written in four characters only; there might well be neither three nor four but, more probably, a far greater number of chemical elements.

In another direction, Boyle, in his researches on air, opened a new field of investigation that eventually led by very indirect ways to the closer chemical study of a substance, namely, air, long considered to be unquestionably an element. In the air pump constructed for him by Robert Hooke (1635–1703) he carried out experiments both on combustion and on respiration, reported in his *New Experiments Physico-Mechanicall* (Oxford, 1660), finding that both fire and life were extinguished much more rapidly when the air was pumped out of the receiver of the pump into which various burning substances and small animals had been placed in turn; while he knew that in the open air both processes persisted, he felt that he could conclude only that air supported fire and life in some way not understood. His contemporary, Hooke, went, however, much further, arguing in his *Micrographia* (London, 1665) that since certain combustible mixtures containing saltpetre would burn under water, that is, in the absence of air, it followed that the common air and the substance saltpetre each contained a component that supported combustion. Both Boyle and Hooke returned to this problem on a number of occasions, far too many to recount here. Boyle in 1670 reported to the Royal Society that air was not diminished in respiration, since he could detect no reduction in pressure when he placed a mouse and then a bird in the air in a sealed glass vessel together with a mercury pressure-gauge; and in some *Tracts* (London, 1674) he claimed, after experiments in a glass vessel with the flame of lamps burning either oil or spirit of wine, that combusion similarly did not diminish air.[2] Hooke in the previous year, 1673, had reported to the Royal Society an experiment that showed air to be decreased by combustion, but he was unable to repeat this experiment before the Society. Boyle went only so far as to conclude, on the evidence before him, that common air contained a minute proportion of some "vital substance" that supported fire and life. At this same time John Mayow (1641–1679), in his *Tractatus quinque* (Oxford, 1674), changed the current experimental technique, putting burning combustibles (such as a lighted candle) and small animals (mice), not in air contained in sealed glass vessels, but in air confined over water, and thereby showed that air was decreased both by combustion and by respiration.[3] His results, however, appear to have had no effect on the further exploration of the problem by his contemporaries. So ended many years of experiment and discussion; but it is to be noted that all three of these able experimenters had concluded that air contained something that supported combustion and respiration, although to Boyle this

[2] There would be no change in pressure in these experiments of 1670 and 1674 as made by Boyle, because oxygen, as we now know, is converted into an equal volume of carbon dioxide.

[3] Carbon dioxide, produced, as we now know, by the combustion of the candle and the respiration of the mice, had dissolved in the water and thereby brought about a decrease in the volume of air, a result that could not be revealed by the experimental technique used by Boyle.

substance was present in only a minute proportion.[4] The study of the phenomena of the calcination of metals, by which the problem of the composition of the air was finally resolved by Lavoisier a century later, played no significant part in these seventeenth-century studies on combustion, although Mayow had asserted that in calcination metals took up "nitro-aereal particles" from the air; for, while Boyle had in his *Essays of Effluviums* (London, 1673) established once and for all in a long series of experiments the important experimental result that metals gained in weight on calcination, he had ascribed this increase of weight to the addition of material particles of fire that had become "corporified" with his metals.[5]

Boyle had also made another great change in chemical thought by applying the ancient Greek atomic theory of Democritus to chemical changes. According to this theory the material world consisted of atoms and void. The atoms were minute, indestructible, eternal; they were all qualitatively the same; and their arrangements in different numbers and in different configurations produced the confusing variety of different substances in the world. This mechanical philosophy, as it was termed, had recently been brought into greater prominence by the work of Pierre Gassend (1592–1655) and Boyle had applied it throughout his chemical studies. If matter had such a structure, one substance could, therefore, be changed into another by the mere reordering of its particles into a new configuration. Thus the unity and the transmutability of matter were alike implicit in the atomic theory and in the four-element theory. But Boyle and his like-minded contemporaries are not on that account to be inconsiderately dubbed alchemists; for the objective of the alchemists was to adjust the proportions of the four qualities and the four elements, not to rearrange atoms in new configurations. Isaac Newton (1642–1727), having followed Boyle in accepting the old atomic theory, and accepting it even more completely

[4] For the detailed history of this problem, see D. McKie, "Fire and the Flamma Vitalis" in *Science, Medicine and History* (Essays on the Evolution of Scientific Thought and Medical Practice written in honour of Charles Singer, ed. E. A. Underwood), London, 1953, Vol. I, pp. 469–488.

[5] For a detailed study of these *Essays*, see D. McKie, *Science Progress*, 1935, Vol. 29, pp. 253–265. See also *Science Progress*, 1936, Vol. 31, pp. 55–67, for a contemporary criticism of Boyle's experiments.

than Boyle, wrote in his *Opticks* (London, 1717, 2nd ed., Query 30) that dense, that is, solid, bodies could be rarefied into air, and this air reconverted into a solid. Newton's comment was taken up later by Stephen Hales (1677–1761) in his *Vegetable Staticks* (London, 1727). Hales heated weighed amounts of various substances and collected the air that was released from them in the process; unlike Newton, however, he supposed that the air existed as air in the bodies from which he obtained it, but he had demonstrated that air was widely contained in many different bodies or, as he said, "fixed" in them. For his experiments Hales devised an early form of gas-collecting apparatus, an apparatus in which the receiver for collecting the product was separated from the part where the "air" was produced, the two components being connected by a delivery tube. This was an important practical improvement in apparatus. Boyle had earlier described in his *New Experiments* of 1660, referred to above, a simple method of collecting "air" produced by the action of acids on metals. It consisted merely of a phial, filled with dilute acid and inverted in a bowl containing water, the metal being inserted into the phial in small pieces just before inversion. Boyle did not give an illustration of this apparatus, but it was subsequently illustrated in Mayow's *Tractatus* of 1674. Both Boyle and Mayow obtained hydrogen and nitric oxide in this way and each supposed that he had obtained "air," "air regenerated *de novo*." Neither with Hales nor with his predecessors was there any suspicion that there might exist different "airs," or, as we would now say, "gases." Hales had, however, made an interesting observation, namely, that burning phosphorus and also burning sulphur absorbed a considerable quantity of air.

In the second half of the eighteenth century, further significant advances came somewhat less slowly. In 1754 Joseph Black presented for the degree of M.D. in the University of Edinburgh a dissertation[6] in the customary Latin on the acid humour arising from food and on *magnesia alba*, a new drug recently introduced into medicine; he read an enlarged account of this a year later to a society that subsequently became in 1783 the Royal

---

[6] *De humore acido a cibis orto, et magnesia alba*, Edinburgh, 1754.

Society of Edinburgh, and this was published in 1756.[7] In this extension of his medical dissertation Black concluded from his experiments that magnesia alba and other alkaline substances in the mild state contained what he called, following the example of Hales, "fixed air," whereas in the caustic state they did not contain this air. Moreover, he carried chemistry forward at one stride by showing that "fixed air" was chemically different from common air. Black had traced by means of the balance the changes in the mild alkalis produced by the action of heat in converting them into their caustic form. He did not isolate "fixed air" (our modern carbon dioxide), but he showed later in his lectures that it was produced in the combustion of charcoal, in the burning of a candle, in respiration, and in fermentation. Thus for the first time an "air" had been chemically differentiated from the common air of the atmosphere. Ten years later, in 1766, Henry Cavendish (1731–1810) isolated "inflammable air" (hydrogen) and studied its chemical and physical properties.[8] Then in 1772 Daniel Rutherford (1749–1819) showed that, after the removal of "fixed air" from air vitiated by the respiration of mice or by the combustion of coals, there was a residue of another mephitic air, which he called "noxious air" (nitrogen), and which was also obtained, he added, as a residue from air after the burning of sulphur or phosphorus and by the calcination of metals.[9] Between 1772 and 1777 Joseph Priestley (1733–1804) greatly enlarged this new field of research by the isolation and recognition of seven other new "airs," namely, "nitrous air" (nitric oxide), "acid air" or "muriatic acid air" (hydrogen chloride), "alkaline air" (ammonia), "diminished nitrous air" (nitrous oxide), "vitriolic acid air" (sulphur dioxide), "dephlogisticated air" (oxygen) and "nitrous acid vapour" (nitrogen dioxide).[10]

One of Priestley's discoveries was of particular significance at this time and it needs some detailed notice here. On 1 August

[7] "Experiments upon Magnesia alba, Quicklime, and some other Alcaline Substances" (*Essays and Observations, Physical and Literary, Read before a Society in Edinburgh*, Edinburgh, 1756, Vol. II, pp. 157–225).

[8] *Phil. Trans.*, 1766, Vol. 56, pp. 141–159.

[9] *De aere fixo dicto, aut mephitico*, Edinburgh, 1772.

[10] *Phil. Trans.*, 1772, Vol. 62, pp. 147–264; *Experiments and Observations on Different Kinds of Air*, London, 3 vols., 1774, 1775, 1777.

1774 he obtained a new "air" by heating *mercurius calcinatus per se* (mercuric oxide, prepared by heating mercury). It had unusual properties; while it was not soluble in water, it greatly enhanced the flame of a candle. His experiments were interrupted by a journey to the Continent. In Paris in November 1774 he met Lavoisier and other men of science associated with him and he told Lavoisier of his most recent discovery. Priestley's reputation was already high in Paris; he was a Fellow of the Royal Society of London; and a year earlier, on 30 November 1773, he had been awarded the Copley Medal, the highest award in the Society's gift, for the paper on "airs" that he had published in 1772 in the *Philosophical Transactions*.[11] After his return to England he resumed his experiments and in March 1775 he discovered that his new "air" was respirable, even more respirable than atmospheric air since it was far less rapidly and far less extensively vitiated by respiration; likewise it was a far better supporter of combustion. Thinking on these things, Priestley named it "dephlogisticated air," because, since it was such a remarkable supporter of combustion and of respiration, it must be free of the phlogiston that was always present in considerable proportions in the common air of the atmosphere into which it was being uninterruptedly poured from fires, from breathing animals, and from fermentations.[12] Thus, the new "air" was common air freed from the phlogiston that it normally contained. This information was of the greatest interest to Lavoisier, to whose early work we now turn.

Before doing so, however, we may remind ourselves that when Lavoisier embarked on his first researches in chemistry the four-element theory was generally, though often tacitly, accepted or rather not rejected for want of a better one; that Boyle's definition of the chemical element had fallen dead from his hands over a century earlier; that both combustion and the calcination of metals were regarded as processes of decomposition; that the chemical compositions of air and of water were unknown and even unsuspected, as also were those of the vast array of other less common and less familiar substances; and that

[11] See D. McKie, "Joseph Priestley and the Copley Medal," *Ambix*, 1960, Vol. 9, pp. 1–22.

[12] *Phil. Trans.*, 1775, Vol. 65, pp. 384–390.

the phlogiston theory dominated chemistry. Within twenty years Lavoisier changed all this.

After his studies under Guettard, Rouelle, and the other Academicians whom we have mentioned, Lavoisier worked for three years with Guettard on the collection of details for the projected geological map of France, and in the summer of 1767 he accompanied Guettard on a geological survey of Alsace and Lorraine. In 1764 he submitted a memoir to the Académie Royale des Sciences on the properties of gypsum and the setting of plaster of Paris; it was his first contribution to chemistry and it was a careful and exact research. Other memoirs followed. In 1765 he submitted an essay in a competition organized by the Academy on the problem of lighting the streets of cities and large towns at night. He did not win the prize; but he was specially awarded a gold medal by the King for this remarkable study, which included not only many scientific experiments but also an economic analysis of costs. He continued his geological field work and at this time he seems to have finally decided to abandon the profession of law for the pursuit of science. In 1768, at the early age of twenty-five, he was elected a member of the Academy of Sciences in the junior grade of assistant. Earlier in that year he entered the Tax-Farm as an Assistant Farmer-General, in its dismal consequences the most ill-fated action in his life; he had recently inherited the fortune left to him by his mother and he wished to invest it and use the income to enable him to devote as much of his time as possible to scientific research. In 1771 he married Marie-Anne-Pierrette, daughter of the Farmer-General Jacques Paulze. Madame Lavoisier was an able and well-educated woman. She learned English to help her husband in his reading of the growing mass of scientific literature, for which he lacked adequate time on account of his heavy administrative duties in the Tax-Farm and his many prolonged absences from Paris on journeys of supervision and inspection in the provinces; she also helped him in his experiments, recording observations and results in his laboratory notebooks and drawing sketches of his apparatus; and she drew the diagrams for the thirteen folding plates of his *Traité*.

As Lavoisier entered more actively into the scientific life of Paris, there was much discussion about the improvement of the supply of drinking water of good quality to the city. During his field work with Guettard he had determined with hydrometers and recorded the densities of the waters of rivers and springs and even of the water supplied at the inns at which they stayed. He used this physical means because there was no chemical method for ascertaining the purity of water. Moreover, as it was still generally accepted that water on heating was converted, at least in part, into earth, the solution of the problem of the analysis of water did not seem at all promising since, if this belief were well founded, the water itself would be slowly and constantly changing into earth during the investigation, which would inevitably be confusing. After studying all that had been published on the conversion of water into earth, Lavoisier concluded that what had been done was not satisfactory and decided that further experiment was necessary. Influenced by Boyle's method of studying the calcination of metals in sealed vessels, from 24 October 1768 to 1 February 1769 he heated a weighed amount of water, as pure as could be obtained by repeated distillation, in a weighed sealed glass vessel, the alchemist's "pelican," in which a liquid could be continuously distilled on itself. At the end of this long experiment, the total weight of the unopened vessel and its contents was the same as it was at the beginning of the experiment over a hundred days previously. That he had Boyle's experiments in mind is clear, because his immediate conclusion at this stage of the research was that no material particles of fire or of any other external matter had penetrated the glass walls of the "pelican." He then weighed the "pelican" after opening it and pouring the contents into another glass vessel. Some earth had been formed and particles of it were visible in the water. However, he found that the weight of the "pelican" had decreased by an amount nearly equal to the weight of the earth obtained, and therefore the earth had been produced by the disintegrating erosive action of the water on the glass fabric of the "pelican," not by the conversion of water into earth. In the chemical context of the time, this was a most striking result; and a theory held for twenty centuries and still accepted by many of his contemporaries had been refuted by this patient and difficult research, well designed and skilfully performed.

Early in 1772 Lavoisier collaborated with some of his colleagues in the Academy in a series of experiments with the great burning-lens of Tschirnhausen, owned by the Academy, on the question of whether the diamond was combustible. The results showed that the diamond was unaffected by heat if protected from access to the air. Further experiments with another of Tschirnhausen's burning-lenses showed that the diamond was combustible. Later in 1772 he experimented on the combustion of phosphorus and of sulphur. On 20 October he sent a note to the Academy stating that phosphorus absorbed air in burning, that it combined with air to produce "acid spirit of phosphorus" (phosphoric acid), and that it increased in weight by this combination with air. Twelve days later, on 1 November, he deposited a sealed note with the Secretary of the Academy, which was not to be opened until he so desired. In this historic document he stated that he had discovered that both sulphur and phosphorus did not lose but, on the contrary, gained weight on burning; that these increases in weight arose from the combination of sulphur and of phosphorus with "a prodigious quantity of air"; that what he had discovered with regard to the combustion of sulphur and of phosphorus might well occur in all substances that gained in weight in combustion and in calcination; and that, by the reduction of the calx of lead with charcoal, he had found that a large quantity of air was liberated, a thousand times greater in volume than the quantity of calx that he had used, which result, he claimed, completely confirmed his conjecture. Hales, it will be recalled, had observed that both burning phosphorus and burning sulphur absorbed a considerable amount of air, an observation slight in importance and in significant detail as compared with that now reported by Lavoisier.

Three months later Lavoisier wrote in his laboratory notebook on 20 February 1773 that he intended to make many experiments on the air that combined with substances or was liberated from them, "an immense series of experiments," he wrote, "destined to bring about a revolution in physics and in chemistry." As a beginning of this plan, he repeated the experiments made by Black and others and published his results in a volume entitled *Opuscules physiques et chymiques* (Paris, 1774). The book was published in January 1774. At the conclusion of this long research, Lavoisier was not quite sure whether it was Black's "fixed air" that combined with the metals in calcination, but he was more inclined to believe that it was common air or some "elastic fluid" (a gas, as we would now say) contained in common air; his reasons for this latter conclusion were that metals could not be calcined in vessels exhausted of air, that the calcination was greater when a greater surface of metal was exposed to the air, and that an "elastic fluid" was released when metallic calces were reduced to metals by heating with charcoal. Further, he found that the "elastic fluid" produced by heating the calces with charcoal was identical with Black's "fixed air," which was probably the most significant result in this series of experiments; that 6 to 7 grains of phosphorus ignited in a confined volume of air gained in weight through absorption of air by 10 to 12 grains while, in another experiment, 154 grains of phosphorus absorbed during its combustion 89 grains either of air itself or "of some other elastic fluid contained, in a certain proportion, in the air that we breathe"; and that burning phosphorus reduced the volume of air in which it was confined between one-sixth and one-fifth, the maximum reduction that he observed being nearly one-fifth. At this point in his researches, Lavoisier had satisfied himself that both combustion and the calcination of metals involved combination with some sort of air and concurrent gain in weight.

In the early months of 1774 Lavoisier continued his researches on the calcination of metals and on 14 April he reported his results in a memoir that was read on 12 November at a public meeting of the Academy. Once again he turned to the method used by Boyle in his _Essays of Effluviums_, namely, heating weighed amounts of the metals in sealed vessels, the amounts being weighed before and after heating. But, he argued, Boyle's method was wrong in an important detail; for Boyle had omitted weighing the closed vessels before he opened them after the calcination of their contents, merely determining the gain in weight by removing the calcined residues from the vessels and comparing their weights with the weights originally taken, whereas he ought to have weighed the sealed vessels before opening them in order to justify his conclusion that material particles of fire had passed through the glass and had combined with the metals. Lavoisier therefore repeated these experiments with this necessary change in procedure, taking weighed amounts of lead and of tin. He found that after calcination and before opening the total weights of the vessels with their contents had remained unchanged and therefore that nothing had entered through the glass; that on opening the vessels by breaking their seals a whistling noise could be heard as air rushed in; that the total weights of the vessels after opening were increased by amounts equal to the increases in weight found in the calcined metals when they were removed from the vessels and weighed separately; that calcination proceeded to a greater degree in large vessels than in small ones; that the gain in weight of metals on calcination was therefore due to combination with the air contained in the vessels in which they were calcined; and that these experiments provided a means of analysing the common air of the atmosphere, which, hitherto supposed to be an element, now almost certainly appeared to be at least a mixture if not a compound. Full details could not be given in the short time available at a public meeting of the Academy, but the memoir as read was published almost immediately in the next monthly issue (December 1774) of the new journal edited by the Abbé Rozier.[13]

Some weeks before reading this memoir at the Academy, Lavoisier received Priestley in Paris and heard of the new "air" that Priestley had isolated in the preceding August from *mercurius calcinatus per se*, a calx of mercury with the unusual property of being reduced to its metallic state on heating without the addition of charcoal. Lavoisier resumed his experiments on calcination and at a public meeting of the Academy on 26 April 1775 he read another memoir, in which he announced his conclusion that in calcination the metals combined, not with any part of the air, but with pure air or what he called "the purest part of the air"; that this pure air, when recovered by heating the calx of mercury, supported combustion and respiration much better than common air; and that mercury calx, when heated with charcoal, gave "fixed air." Priestley's otherwise valuable and highly significant communication had evidently confused him and he momentarily

[13] *Observations sur la Physique*, 1774, Vol. 4, pp. 448–451.

abandoned the opinion that he had expressed in the preceding November that air was a mixture, perhaps even a compound, and that the metals on calcination combined with one of its components. The memoir was published at once in the next monthly issue (May 1775) of Rozier's journal.[14]

As the annual volumes of the *Mémoires* of the Academy were published in arrears, sometimes of several years, Lavoisier was able to revise these two memoirs as read in 1774 and 1775, and the recensions appeared in the volumes for those years, which were both published in 1778. In revising the first memoir Lavoisier reverted to his conclusion of 1774 that it was only a part of the air that combined with the metals, now named the "salubrious part," and that air was a mixture of two very different "elastic fluids," one that supported combustion and respiration and another incapable of either, the former, he thought from other considerations, being a little heavier and the latter a little lighter than common air.[15] In the new version of the second memoir, he referred to the active part of the air as "eminently respirable air" and he showed that "fixed air" was a compound of charcoal with this part of the common air.[16] There is no space here to deal with his experiments, but it may be pointed out that when he read the recension of the second memoir to the Academy on 8 August 1778 he exhibited the apparatus with which he had performed his classic experiment on the composition of air (first illustrated in Mme Lavoisier's drawing, engraved as Fig. 2 of Plate IV of the *Traité*).

The problem of the chemical composition of common air was now solved and in a long series of memoirs Lavoisier applied this result to show that "eminently respirable air" was contained in phosphoric and vitriolic acids, and also in "nitrous" (nitric) acid, that it was converted into "fixed air" or "aerial acid" by respiration and by the burning of candles, and that it combined with nonmetallic substances to form acids, for which reason in 1779 he renamed it *principe acidifiant* or *principe oxygine* (from the Greek, ὀξύς, acid, and γείνομαι, I beget). He had already used the term *mofette* for that part of the air that was

[14] *Observations sur la Physique*, 1775, Vol. 5, pp. 429–434.
[15] *Mém. Acad. R. Sci.*, 1774 (published 1778), pp. 351–367.
[16] *Mém. Acad. R. Sci.*, 1775 (published 1778), pp. 520–526.

inactive in combustion and in respiration. He applied Black's discovery of latent heat to explain the general constitution of "elastic fluids" (gases) as compounds of evaporable liquids or volatile solids with the "matter of fire" or material heat. He read the French version of the recent book on air and fire by Carl Wilhelm Scheele (1742–1786)[17] and, although Scheele's phlogistic explanations were very confused, Lavoisier found much experimental evidence in this treatise to support the theory that he himself had put forward. He had sent Scheele a copy of his *Opuscules* in 1774, and Scheele had replied with the suggestion that Lavoisier should heat the precipitate obtained by adding "alkali of tartar" to the solution of silver in "acid of nitre," remove the "fixed air" with quicklime and test whether the residue would allow a candle to burn and animals to breathe in it. It is not known whether Lavoisier made this experiment.

During the winter of 1782–1783 Lavoisier in collaboration with his friend, the famous mathematician, Pierre-Simon de Laplace (1749–1827), again applied Black's discovery of latent heat and devised the first ice-calorimeter (see *Traité*, Plate VI), with which the heat evolved in various chemical changes was measured in a series of experiments marking the foundation of thermochemistry. From this research Lavoisier and Laplace concluded that respiration is a kind of combustion and that the preservation of the temperature of the animal body at a constant degree above the temperature of its surroundings, while it is constantly losing heat in that direction, is due to the heat given out, in the lungs as they supposed, by the conversion of the inhaled oxygen into "fixed air."[18]

Lavoisier now began to attack the phlogiston theory, whereas so far he had gone no further than suggesting that his new theory gave a better explanation of many chemical phenomena more in accordance with experimental facts.[19] In this magnificent criticism, he claimed that phlogiston was imaginary; that its existence in metals, in sulphur, in phosphorus and in all combustible

---

[17] *Chemische Abhandlung von der Luft und dem Feuer*, Upsala and Leipzig, 1777; French translation by Baron de Dietrich, *Traité chimique de l'air et du feu*, Paris, 1781.
[18] *Mém. Acad. R. Sci.*, 1780 (published 1784), pp. 355–408.
[19] *Mém. Acad. R. Sci.*, 1783 (published 1786), pp. 505–538.

bodies was a baseless supposition; and that all the facts of combustion and calcination were explained in a much simpler and a much easier way without phlogiston than with it. Like Boyle in his *Sceptical Chymist* of 1661, he destructively criticized the current chemical theory; but, unlike Boyle, he put a better theory in its place.

In June 1783 Lavoisier heard from Dr. Charles Blagden, then visiting Paris, that Cavendish had obtained water by burning "inflammable air" (hydrogen) and in the presence of Blagden and others he made a hurried and improvised experiment to verify Cavendish's discovery. At a public meeting of the Academy on 12 November of that year, he read a memoir reporting further experiments (including the decomposition of water by iron), the receipt of more information from Blagden, and some new experiments by Monge, and he put forward the explanation that water was composed of "dephlogisticated air" (curiously using this name instead of his own term "oxygen") and "inflammable air." Water, therefore, was not an element. Lavoisier had realized the importance of Cavendish's discovery and his memoir was published at once in the next monthly issue (December 1783) of Rozier's journal.[20] On the history of this phase of Lavoisier's work, much has been written about the vexed question of priority in the so-called "Water Controversy." We shall do no more here than recall that while Cavendish discovered the formation of water in the combustion of "inflammable air" either in common air or in "dephlogisticated air" he supposed that the water formed was deposited from both airs in each of which it was originally present as a constituent; that Lavoisier concluded that water was a compound of "inflammable air" and "dephlogisticated air"; and that Cavendish rejected this explanation in the memoir in which he published the results of his experiments.[21] The solution of the problem of the chemical composition of water may be regarded as completing Lavoisier's researches on combustion.

[20] *Observations sur la Physique*, 1783, Vol. 23, pp. 452–455. An extended version of this memoir was published later in *Mém. Acad. R. Sci.*, 1781 (published 1784), pp. 468–494.

[21] *Phil. Trans.*, 1784, Vol. 74, pp. 119–153.

Lavoisier's new theory of combustion and calcination gained adherents only slowly. Black was teaching it to his students in Edinburgh at some date before 1784. Laplace had accepted it probably before that date, Berthollet in 1785, Guyton de Morveau in 1786, Fourcroy in 1787, and Monge, Meusnier, and Chaptal later. In the meantime Guyton had been considering the reform of chemical nomenclature, and from 1782 onwards Lavoisier, Berthollet, and Fourcroy collaborated with him in devising a new system. Their proposals were read to the Academy on 18 April 1787. Briefly, we may say that the reformers adopted Boyle's definition that substances that could not be decomposed into simpler substances were to be regarded, provisionally at any rate, as elements; but the other aim of their system was to give names to substances that corresponded to their chemical composition in place of such irrational terms as oil of tartar, butter of antimony, flowers of zinc, and similar denominations. The elements having been defined, they proceeded to deal with substances containing two elements: the acids, regarded as compounds of oxygen with various other elements, received names indicating their composition, the specific name indicating the element with which oxygen was combined, and the class name (acid) indicating to what group of two-component substances they belonged, as, for example, phosphoric acid and sulphuric acid; and, where there were two acids formed from the same two elements, the one with the greater proportion of oxygen was named, for instance, sulphuric acid, while that with the lower proportion was called sulphurous acid. Another group in the two-component class consisted of the substances hitherto known as calces, but now shown to be compounds of metals with oxygen. The general name proposed for this group was oxides, the different oxides being distinguished by the name of the metals from which they were separately formed, as, for example, zinc oxide. Substances composed of three simple substances presented greater difficulties, but the salts, for example, were renamed to indicate the base and the acid composing them, and thus "vitriol of Venus" became "copper sulphate," and so on. It is enough to say that the new system became the basis of modern chemical nomenclature and that it proved adaptable to new discoveries. These proposals were published under the title of *Méthode de Nomenclature chimique* (Paris, 1787), and the volume included an alphabetic synonymy of the old and the new names with a dictionary of the new nomenclature. The list of elements included light, matter of heat, oxygen, hydrogen (formerly "inflammable air"), azote (formerly "phlogisticated air," our modern nitrogen), carbon, sulphur, phosphorus, the unknown "radicals" of muriatic (hydrochloric) acid and also of boracic acid and "fluoric" acid, the metals (seventeen in all), the earths (five), and the alkalis (potash, soda, and ammonia), together with the unknown radicals of nineteen organic acids, these last three groups being the least satisfactory part of a table that recognized fifty-five elements, and it is for this reason that it cannot be regarded as the first modern table of the chemical elements. Such a table appeared first in the *Traité* after Lavoisier had rightly removed the radicals of the organic acids and also the alkalis.

With the reconstruction of its language in terms related to the new theory of combustion, and with the compositions of air and of water revealed, chemistry at last set out on its modern road; and Lavoisier turned to a very different problem. As Maurice Daumas has shown in a recent and most scholarly study, Lavoisier had long intended, perhaps for more than twelve years, to write a book on chemistry; he had in mind a book for beginners, one of his objects in compiling such a work being to show how chemistry should be taught to those who were embarking on its study for the first time.[22] After revising his various projects, his plans took shape in 1788, and the *Traité* was published in March 1789 in Paris. It was republished there six times, in a *nouvelle édition* in 1789, the year of its first publication, in a second edition in 1793, in a third edition in 1801, and in another *nouvelle édition* in 1805, together with two pirated editions in 1793; and there was one provincial French edition (Avignon, 1804). Four further editions of Kerr's translation were published in Edinburgh, the second in 1793, the third in 1796, the fourth in 1799, and the fifth (in two volumes) in 1802. There were also translations into other languages, including three editions in Italian (Venice, 1791, 1792, and 1796), two in German (Berlin,

[22] Maurice Daumas, *Lavoisier, théoricien et expérimentateur*, Paris, 1955. For the history of the writing of the *Traité*, see especially Chap. IV, pp. 91–112.

1792 and 1803), one in Spanish (Madrid, 1798), and one in Dutch (Utrecht, 1800), together with three in America (Philadelphia, 1799; New York, 1801 and 1806) reprinted from Kerr's translations. Thus there were altogether twenty-three editions of this classic of science published in seven countries within seventeen years of its first appearance; and this does not include summaries in German (Greifswald, 1794) and in Dutch (Amsterdam, 1791) and an incomplete Spanish version of the first volume only (Mexico, 1797).[23] It is abundantly clear that the book had a wide circulation in its time, a result that scarcely needs any comment, since Lavoisier had done for chemistry what Newton had done for mechanics a century earlier in his immortal *Principia*.

The first edition of the *Traité* was published in two volumes continuously paginated and it was divided into three parts: first, on the formation and decomposition of aeriform fluids (gases), on the combustion of simple bodies (elements), and on the formation of acids; second, on the combination of acids with salifiable bases, and on the formation of neutral salts; and third, on the instruments and operations of chemistry. Appended were eight tables for the use of chemists, including conversion tables for various weights, volumes, lengths, and pressures, and others giving the densities of gases and the specific gravities of many mineral substances; Kerr omitted some of these from his translation and added others. There were thirteen folding plates to illustrate the book, and these were engraved from drawings made by Mme Lavoisier, but they were redrawn for Kerr's translation and in some of them the different pieces of apparatus in a particular plate were rearranged. In the opening pages of the *Traité* Lavoisier inserted a *Discours préliminaire* which appeared in Kerr's translation as "Preface by the Author" and which is one of Lavoisier's more notable compositions, written in that clear prose of which he was a master; here he explained his aims in the writing of such a book and his principles of regarding it as a law "never to advance but from what is known

[23] See Denis I. Duveen and Herbert S. Klickstein, *A Bibliography of the Works of Antoine Laurent Lavoisier 1743–1794*, London, 1954, pp. 154–199, a most valuable work of reference to which every student of Lavoisier is deeply indebted.

to what is unknown" and of "forming no conclusions which are not fully warranted by experiment." In its presentation as an elementary textbook of chemistry the *Traité* broke with a long-established tradition in the composition of such works in that, as the titles of its three parts indicate, it was not a series of descriptions of the laboratory methods of preparing a great range of substances and that it gave much space to the description of the apparatus and instruments of chemistry with full details of their uses. It was an entirely new type of textbook of chemistry.

There is one of several interesting details about Kerr's translation that is appropriate to mention on this occasion. In the *Elements* Kerr opened his translation with an "Advertisement of the Translator," in which he said (p. vi) that the French copy did not reach him before the middle of September and that the publisher judged it necessary that the translation should be ready for the beginning of the university session at the end of October. It has often been stated that the book was translated and published in this short period of about seven weeks. However, in a letter to Lavoisier dated 21 January 1791, Kerr wrote that he had had the good fortune to procure a copy "soon after the publication"; [24] and it therefore appears, and is certainly more probable, that the September that Kerr was referring to in his "Advertisement" was September 1789, not September 1790, which latter would not be soon after publication. The date of 23 October 1789 subscribed to his "Advertisement" (p. xii) of a book published in November 1790 should not be given too much significance.

The first part of the *Traité* applied Black's discovery of latent heat to explain that gases were compounds of the matter of heat (caloric) with the bases of the various gases, oxygen gas being thus composed of oxygen and caloric. It also described the formation and composition of the atmosphere, with its analysis in the classic apparatus now illustrated (Plate IV, Fig. 2) for the first time, the nomenclature of its components, and its decomposition by sulphur, phosphorus, and charcoal with the

---

[24] See D. McKie, *Notes and Records of the Royal Society of London*, 1949–50, Vol. 7, pp. 13. Kerr repeated this statement in the fifth edition of the *Elements* (1802).

formation of acids and by metals with the formation of oxides. Then followed a chapter on the decomposition and composition of water, showing that water was a compound of hydrogen and oxygen. Experiments were described giving measurements of the quantities of caloric released in various combustions. The bases of the vegetable acids were shown to consist either of hydrogen and carbon or of hydrogen, carbon, and phosphorus, while animal acids were more compound, their bases generally consisting of combinations of carbon, phosphorus, hydrogen, and azote (nitrogen). In a chapter on vinous fermentation, a quantitative experiment was reported showing that the total weight of the products in this chemical change was equal to the total weight of the reactants, which showed that "in all the operations of art and nature, nothing is created; an equal quantity of matter exists both before and after the experiment." This was the first explicit statement in the history of chemistry of the law of the conservation of matter in chemical change.

The second part of the *Traité* dealt with the compounds of the acids with various bases, giving extensive tables of these compounds. But its most interesting and revolutionary feature was the table of simple substances or elements given in the first section. This was the first modern list of the chemical elements. Lavoisier had removed from the table given in 1787 in the new nomenclature the radicals of the organic acids, because they had been shown to be compounds of hydrogen and carbon, but he retained the earths, although he suspected that they were compounds, probably oxides, and he omitted the alkalis because they were evidently compounds. The list therefore contained thirty-three elements, including light and caloric.

The third part, almost a half of the book, described in excellent detail the apparatus and instruments of chemistry and their uses.

With the publication of the *Traité* the revolution that Lavoisier had seen to be possible in chemistry when he composed the memorandum in his laboratory notebook in 1773 was at last achieved, and in 1791 he wrote: "All young chemists adopt the theory and from that I conclude that the revolution in chemistry has come to pass."

When the political revolution began in 1789, Lavoisier was engaged on some further well-known researches on respiration and transpiration, but he was soon involved in other more desperate matters. The Academy was suppressed in 1793 in spite of his heroic efforts maintained to the last minute to save it from extinction. By a decree passed by the Revolutionary Convention on 24 November 1793, he was arrested with his former colleagues in the suppressed Tax-Farm on the pretext that the final presentation of the Farm's accounts had been delayed. The twenty-eight prisoners were brought before the Revolutionary Tribunal in Paris on 8 May 1794, convicted on a charge, on which they had not been accused or tried, of plotting with the enemies of France, and executed on the same day. Thus perished the most illustrious victim of the Revolution.

In conclusion we may compare Lavoisier with Newton. Each had rejected his immediate scientific inheritance. Newton had overthrown Cartesianism and rebuilt the science of mechanics, terrestrial and celestial, on the laws of motion and the inverse-square law of gravitational attraction; and Lavoisier had refuted the phlogiston theory and reconstructed chemistry by the discovery of the compositions of air and of water and by the recognition of the chemical elements as the ultimate residues of chemical analysis. Each built on the work of others, but each added his own great contribution; and, while this indebtedness to others is to be remembered, it should not be forgotten that both the *Principia* and the *Traité* are each the work of one supreme mind. Newton and Lavoisier were not altogether different in their characters or in their occupations; both were reserved and, in different degrees, distant; both had acquaintances and few friends; both held high and responsible offices in the administration of their countries, Newton as Warden and later Master of the Royal Mint, Lavoisier as Farmer-General of Taxes, Commissioner for Gunpowder and Saltpetre, and Commissioner of the National Treasury; Newton was President of the Royal Society of London, Lavoisier Director and Treasurer of the Académie Royale des Sciences of Paris. Their lives ended very differently. Newton lived to the age of eighty-five, honoured and respected; he was buried in Westminster Abbey. Lavoisier was murdered in the prime of his life at the age of fifty-one; his memorial is his great place in the history of the science to which he devoted his genius for a mere twenty years, while heavily engaged in many official duties.[25]

University College London   Douglas McKie
15 July 1964

[25] Readers interested in a wider and more detailed study of Lavoisier's life and work may consult D. McKie, *Antoine Lavoisier, Scientist, Economist and Social Reformer*, London and New York, 1952.

ELEMENTS

of

CHEMISTRY.

ELEMENTS

O F

CHEMISTRY,

IN A

NEW SYSTEMATIC ORDER,

CONTAINING ALL THE

MODERN DISCOVERIES.

ILLUSTRATED WITH THIRTEEN COPPERPLATES.

By Mr LAVOISIER,

Member of the Academy of Sciences, Royal Society of Medicine, and Agricultural Society of Paris, of the Royal
Society of London, and Philofophical Societies
of Orleans, Bologna, Bafil, Philadelphia,
Haerlem, Manchefter, &c. &c.

TRANSLATED FROM THE FRENCH,

By ROBERT KERR, F.R. & A.SS.E.

Member of the Royal College of Surgeons, and Surgeon
to the Orphan Hofpital of Edinburgh.

EDINBURGH:

PRINTED FOR WILLIAM CREECH, AND SOLD IN
LONDON BY G. G. AND J. J. ROBINSONS.

MDCCXC.

ADVERTISEMENT

OF THE

TRANSLATOR.

THE very high character of Mr Lavoifier as a chemical philofopher, and the great revolution which, in the opinion of many excellent chemifts, he has effected in the theory of chemiftry, has long made it much defired to have a connected account of his difcoveries, and of the new theory he has founded upon the modern experiments written by himfelf. This is now accomplifhed by the publication of his Elements of Chemiftry; therefore no excufe can be at all neceffary for giving the following work to the public in an Englifh drefs; and the only hefitation of the Tranflator is with regard to his own abilities for the tafk. He is moft ready to confefs, that his knowledge of the compofition of language fit for publication is far

inferior

inferior to his attachment to the fubject, and to his defire of appearing decently before the judgment of the world.

He has earneftly endeavoured to give the meaning of the Author with the moft fcrupulous fidelity, having paid infinitely greater attention to accuracy of tranflation than to elegance of ftile. This laft indeed, had he even, by proper labour, been capable of attaining, he has been obliged, for very obvious reafons, to neglect, far more than accorded with his wifhes. The French copy did not reach his hands before the middle of September ; and it was judged neceffary by the Publifher that the Tranflation fhould be ready by the commencement of the Univerfity Seffion at the end of October.

He at firft intended to have changed all the weights and meafures ufed by Mr Lavoifier into their correfpondent Englifh denominations, but, upon trial, the tafk was found infinitely too great for the time allowed ; and to have executed this part of the work inaccurately, muft have been both ufelefs and mifleading to the reader. All that has been attempted in this way is adding, between brackets ( ), the degrees of Fahrenheit's

hrenheit's fcale correfponding with thofe of
Reaumeur's thermometer, which is ufed by the
Author. Rules are added, however, in the
Appendix, for converting the French weights
and meafures into Englifh, by which means the
reader may at any time calculate fuch quantities
as occur, when defirous of comparing Mr Lavoifier's experiments with thofe of Britifh authors.

By an overfight, the firft part of the tranflation went to prefs without any diftinction being
preferved between charcoal and its fimple elementary part, which enters into chemical combinations, efpecially with oxygen or the acidifying principle, forming carbonic acid. This
pure element, which exifts in great plenty in
well made charcoal, is named by Mr Lavoifier
carbone, and ought to have been fo in the tranflation ; but the attentive reader can very eafily
rectify the miftake. There is an error in Plate
XI. which the engraver copied ftrictly from the
original, and which was not difcovered until the
plate was worked off at prefs, when that part of
the Elements which treats of the apparatus there
reprefented came to be tranflated. The two
tubes 21. and 24. by which the gas is conveyed
into

into the bottles of alkaline folution 22. 25. fhould have been made to dip into the liquor, while the other tubes 23. and 26. which carry off the gas, ought to have been cut off fome way above the furface of the liquor in the bottles.

A few explanatory notes are added; and indeed, from the perfpicuity of the Author, very few were found neceffary. In a very fmall number of places, the liberty has been taken of throwing to the bottom of the page, in notes, fome parenthetical expreffions, only relative to the fubject, which, in their original place, tended to confufe the fenfe. Thefe, and the original notes of the Author, are diftinguifhed by the letter A, and to the few which the Tranflator has ventured to add, the letter E is fubjoined.

Mr Lavoifier has added, in an Appendix, feveral very ufeful Tables for facilitating the calculations now neceffary in the advanced ftate of modern chemiftry, wherein the moft fcrupulous accuracy is required. It is proper to give fome account of thefe, and of the reafons for omitting feveral of them.

No.

No. I. of the French Appendix is a Table for converting ounces, gros, and grains, into the decimal fractions of the French pound; and No. II. for reducing thefe decimal fractions again into the vulgar fubdivifions. No. III. contains the number of French cubical inches and decimals which correfpond to a determinate weight of water.

The Tranflator would moft readily have converted thefe Tables into Englifh weights and meafures; but the neceffary calculations muft have occupied a great deal more time than could have been fpared in the period limited for publication. They are therefore omitted, as altogether ufelefs, in their prefent ftate, to the Britifh chemift.

No. IV. is a Table for converting lines or twelfth parts of the inch, and twelfth parts of lines, into decimal fractions, chiefly for the purpofe of making the neceffary corrections upon the quantities of gaffes according to their barometrical preffure. This can hardly be at all ufeful or neceffary, as the barometers ufed in Britain are graduated in decimal fractions of the inch, but, being referred to by the Author in

the

the text, it has been retained, and is No. 1. of the Appendix to this Tranſlation.

No. V. Is a Table for converting the obſerved heights of water within the jars uſed in pneumato-chemical experiments into correſpondent heights of mercury for correcting the volume of gaſſes. This, in Mr Lavoiſier's Work, is expreſſed for the water in lines, and for the mercury in decimals of the inch, and conſequently, for the reaſons given reſpecting the Fourth Table, muſt have been of no uſe. The Tranſlator has therefore calculated a Table for this correction, in which the water is expreſſed in decimals, as well as the mercury. This Table is No. II. of the Engliſh Appendix.

No. VI. contains the number of French cubical inches and decimals contained in the correſponding ounce-meaſures uſed in the experiments of our celebrated countryman Dr Prieſtley. This Table, which forms No. III. of the Engliſh Appendix, is retained, with the addition of a column, in which the correſponding Engliſh cubical inches and decimals are expreſſed.

No.

No. VII. Is a Table of the weights of a cubical foot and inch, French meafure, of the different gaffes expreffed in French ounces, gros, grains, and decimals. This, which forms No. VI. of the Englifh Appendix, has been, with confiderable labour, calculated into Englifh weight and meafure.

No. VIII. Gives the fpecific gravities of a great number of bodies, with columns, containing the weights of a cubical foot and inch, French meafure, of all the fubftances. The fpecific gravities of this Table, which is No. VII. of the Englifh Appendix, are retained, but the additional columns, as ufelefs to the Britifh philofopher, are omitted ; and to have converted thefe into Englifh denominations muft have required very long and painful calculations.

Rules are fubjoined, in the Appendix to this tranflation, for converting all the weights and meafures ufed by Mr Lavoifier into correfponding Englifh denominations ; and the Tranflator is proud to acknowledge his obligation to the learned Profeffor of Natural Philofophy in the Univerfity of Edinburgh, who kindly fupplied him with the neceffary information for this purpofe. A Table is likewife added, No. IV. of
the

the Englifh Appendix, for converting the degrees of Reaumeur's fcale ufed by Mr Lavoifier
into the correfponding degrees of Fahrenheit,
which is univerfally employed in Britain *.

This Tranflation is fent into the world with
the utmoft diffidence, tempered, however, with
this confolation, that, though it muft fall greatly
fhort of the elegance, or even propriety of language, which every writer ought to endeavour to
attain, it cannot fail of advancing the interefts of
true chemical fcience, by diffeminating the accurate mode of analyfis adopted by its juftly celebrated Author. Should the public call for a fecond
edition, every care fhall be taken to correct the
forced imperfections of the prefent tranflation,
and to improve the work by valuable additional
matter from other authors of reputation in the
feveral fubjects treated of.

EDINBURGH, }
Oct. 23. 1789. }

* The Tranflator has fince been enabled, by the kind
affiftance of the gentleman above alluded to, to give
Tables, of the fame nature with thofe of Mr Lavoifier,
for facilitating the calculations of the refults of chemical experiments.

P R E F A C E

A U T H O R

———————

WHEN I began the following Work, my only object was to extend and explain more fully the Memoir which I read at the public meeting of the Academy of Sciences in the month of April 1787, on the neceſſity of reforming and completing the Nomenclature of Chemiſtry. While engaged in this employment, I perceived, better than I had ever done before, the juſtice of the following maxims of the Abbé de Condillac, in his Syſtem of Logic, and ſome other of his works.

" We think only through the medium of " words.—Languages are true analytical me- " thods.

" thods.—Algebra, which is adapted to its pur-
" pofe in every fpecies of expreffion, in the
" moft fimple, moft exact, and beft manner
" poffible, is at the fame time a language and
" an analytical method.—The art of reafoning
" is nothing more than a language well aran-
" ged."

Thus, while I thought myfelf employed only
in forming a Nomenclature, and while I propofed to myfelf nothing more than to improve the
chemical language, my work transformed itfelf
by degrees, without my being able to prevent
it, into a treatife upon the Elements of Chemiftry.

The impoffibility of feparating the nomenclature of a fcience from the fcience itfelf, is
owing to this, that every branch of phyfical fcience muft confift of three things; the feries of
facts which are the objects of the fcience, the
ideas which reprefent thefe facts, and the words
by which thefe ideas are expreffed. Like three
impreffions of the fame feal, the word ought to
produce the idea, and the idea to be a picture of
the fact. And, as ideas are preferved and communicated by means of words, it neceffarily follows

lows that we cannot improve the language of any ſcience without at the ſame time improving the ſcience itſelf; neither can we, on the other hand, improve a ſcience, without improving the language or nomenclature which belongs to it. However certain the facts of any ſcience may be, and, however juſt the ideas we may have formed of theſe facts, we can only communicate falſe impreſſions to others, while we want words by which theſe may be properly expreſſed.

To thoſe who will conſider it with attention, the firſt part of this treatiſe will afford frequent proofs of the truth of the above obſervations. But as, in the conduct of my work, I have been obliged to obſerve an order of arrangement eſſentially differing from what has been adopted in any other chemical work yet publiſhed, it is proper that I ſhould explain the motives which have led me to do ſo.

It is a maxim univerſally admitted in geometry, and indeed in every branch of knowledge, that, in the progreſs of inveſtigation, we ſhould proceed from known facts to what is unknown. In early infancy, our ideas ſpring from our wants; the ſenſation of want excites the idea of

the

the object by which it is to be gratified. In this manner, from a feries of fenfations, obfervations, and analyfes, a fucceffive train of ideas arifes, fo linked together, that an attentive obferver may trace back to a certain point the order and connection of the whole fum of human knowledge.

When we begin the ftudy of any fcience, we are in a fituation, refpecting that fcience, fimilar to that of children; and the courfe by which we have to advance is precifely the fame which Nature follows in the formation of their ideas. In a child, the idea is merely an effect produced by a fenfation; and, in the fame manner, in commencing the ftudy of a phyfical fcience, we ought to form no idea but what is a neceffary confequence, and immediate effect, of an experiment or obfervation. Befides, he that enters upon the career of fcience, is in a lefs advantageous fituation than a child who is acquiring his firft ideas. To the child, Nature gives various means of rectifying any miftakes he may commit refpecting the falutary or hurtful qualities of the objects which furround him. On every occafion his judgments are corrected by experience; want and pain are the neceffary

con-

confequences arifing from falfe judgment ; gra‹
tification and pleafure are produced by judging
aright. Under fuch mafters, we cannot fail to
become well informed ; and we foon learn to
reafon juftly, when want and pain are the ne‹
ceffary confequences of a contrary conduct.

In the ftudy and practice of the fciences it is
quite different ; the falfe judgments we form
neither affect our exiftence nor our welfare ; and
we are not forced by any phyfical neceffity to
correct them. Imagination, on the contrary,
which is ever wandering beyond the bounds of
truth, joined to felf-love and that felf-confidence
we are fo apt to indulge, prompt us to draw
conclufions which are not immediately derived
from facts ; fo that we become in fome meafure
interefted in deceiving ourfelves. Hence it is
by no means to be wondered, that, in the fcience
of phyfics in general, men have often made fuppofitions, inftead of forming conclufions. Thefe
fuppofitions, handed down from one age to another, acquire additional weight from the autho-
rities by which they are fupported, till at laft
they are received, even by men of genius, as
fundamental truths.

The

The only method of preventing fuch errors from taking place, and of correcting them when formed, is to reftrain and fimplify our reafoning as much as poffible. This depends entirely upon ourfelves, and the neglect of it is the only fource of our miftakes. We muft truft to nothing but facts : Thefe are prefented to us by Nature, and cannot deceive. We ought, in every inftance, to fubmit our reafoning to the teft of experiment, and never to fearch for truth but by the natural road of experiment and obfervation. Thus mathematicians obtain the folution of a problem by the mere arrangement of data, and by reducing their reafoning to fuch fimple fteps, to conclufions fo very obvious, as never to lofe fight of the evidence which guides them.

Thoroughly convinced of thefe truths, I have impofed upon myfelf, as a law, never to advance but from what is known to what is unknown ; never to form any conclufion which is not an immediate confequence neceffarily flowing from obfervation and experiment ; and always to arrange the facts, and the conclufions which are drawn from them, in fuch an order as fhall render it moft eafy for beginners in the

ftudy

ftudy of chemiftry thoroughly to underftand
them. Hence I have been obliged to depart
from the ufual order of courfes of lectures and
of treatifes upon chemiftry, which always affume the firft principles of the fcience, as known,
when the pupil or the reader fhould never be
fuppofed to know them till they have been explained in fubfequent leffons. In almoft every
inftance, thefe begin by treating of the elements
of matter, and by explaining the table of affinities, without confidering, that, in fo doing, they
muft bring the principal phenomena of chemiftry
into view at the very outfet :- They make ufe of
terms which have not been defined, and fuppofe
the fcience to be underftood by the very perfons
they are only beginning to teach. It ought
likewife to be confidered, that very little of chemiftry can be learned in a firft courfe, which is
hardly fufficient to make the language of the
fcience familiar to the ears, or the apparatus
familiar to the eyes. It is almoft impoffible to
become a chemift in lefs than three or four years
of conftant application.

Thefe inconveniencies are occafioned not fo
much by the nature of the fubject, as by the
method of teaching it ; and, to avoid them, I
was

was chiefly induced to adopt a new arrangement
of chemiſtry, which appeared to me more con-
ſonant to the order of Nature. I acknowledge,
however, that in thus endeavouring to avoid
difficulties of one kind, I have found myſelf involved in others of a different ſpecies, ſome of
which I have not been able to remove ; but I
am perſuaded, that ſuch as remain do not ariſe
from the nature of the order I have adopted,
but are rather conſequences of the imperfection
under which chemiſtry ſtill labours. This ſcience
ſtill has many chaſms, which interrupt the ſeries
of facts, and often render it extremely difficult
to reconcile them with each other : It has not,
like the elements of geometry, the advantage of
being a complete ſcience, the parts of which are
all cloſely connected together : Its actual progreſs, however, is ſo rapid, and the facts, under
the modern doctrine, have aſſumed ſo happy an
arrangement, that we have ground to hope, even
in our own times, to ſee it approach near to the
higheſt ſtate of perfection of which it is ſuſceptible.

The rigorous law from which I have never
deviated, of forming no concluſions which are
not fully warranted by experiment, and of never
ſupplying

supplying the abfence of facts, has prevented
me from comprehending in this work the branch
of chemiftry which treats of affinities, although
it is perhaps the beft calculated of any part of
chemiftry for being reduced into a completely
fyftematic body. Meffrs Geoffroy, Gellert, Bergman, Scheele, De Morveau, Kirwan, and many
others, have collected a number of particular
facts upon this fubject, which only wait for a
proper arrangement ; but the principal data are
ftill wanting, or, at leaft, thofe we have are either
not fufficiently defined, or not fufficiently proved, to become the foundation upon which to
build fo very important a branch of chemiftry.
This fcience of affinities, or elective attractions,
holds the fame place with regard to the other
branches of chemiftry, as the higher or tranfcen-
dental geometry does with refpect to the fimpler
and elementary part; and I thought it improper
to involve thofe fimple and plain elements, which
I flatter myfelf the greateft part of my readers
will eafily underftand, in the obfcurities and
difficulties which ftill attend that other very ufeful and neceffary branch of chemical fcience.

Perhaps a fentiment of felf-love may, without
my perceiving it, have given additional force to
thefe

thefe reflections. Mr de Morveau is at prefent engaged in publifhing the article *Affinity* in the Methodical Encyclopædia; and I had more reafons than one to decline entering upon a work in which he is employed.

It will, no doubt, be a matter of furprife, that in a treatife upon the elements of chemiftry, there fhould be no chapter on the conftituent and elementary parts of matter; but I fhall take occafion, in this place, to remark, that the fondnefs for reducing all the bodies in nature to three or four elements, proceeds from a prejudice which has defcended to us from the Greek Philofophers. The notion of four elements, which, by the variety of their proportions, compofe all the known fubftances in nature, is a mere hypothefis, affumed long before the firft principles of experimental philofophy or of chemiftry had any exiftence. In thofe days, without poffeffing facts, they framed fyftems; while we, who have collected facts, feem determined to reject them, when they do not agree with our prejudices. The authority of thefe fathers of human philofophy ftill carry great weight, and there is reafon to fear that it will even bear hard upon generations yet to come.

It

It is very remarkable, that, notwithstanding of the number of philosophical chemists who have supported the doctrine of the four elements, there is not one who has not been led by the evidence of facts to admit a greater number of elements into their theory. The first chemists that wrote after the revival of letters, considered sulphur and salt as elementary substances entering into the composition of a great number of substances ; hence, instead of four, they admitted the existence of six elements. Beccher assumes the existence of three kinds of earth, from the combination of which, in different proportions, he supposed all the varieties of metallic substances to be produced. Stahl gave a new modification to this system ; and succeeding chemists have taken the liberty to make or to imagine changes and additions of a similar nature. All these chemists were carried along by the influence of the genius of the age in which they lived, which contented itself with assertions without proofs ; or, at least, often admitted as proofs the slightest degrees of probability, unsupported by that strictly rigorous analysis required by modern philosophy.

All

All that can be faid upon the number and nature of elements is, in my opinion, confined to difcuffions entirely of a metaphyfical nature. The fubject only furnifhes us with indefinite problems, which may be folved in a thoufand different ways, not one of which, in all probability, is confiftent with nature. I fhall therefore only add upon this fubject, that if, by the term *elements*, we mean to exprefs thofe fimple and indivifible atoms of which matter is compofed, it is extremely probable we know nothing at all about them ; but, if we apply the term *elements*, or *principles of bodies*, to exprefs our idea of the laft point which analyfis is capable of reaching, we muft admit, as elements, all the fubftances into which we are capable, by any means, to reduce bodies by decompofition. Not that we are entitled to affirm, that thefe fubftances we confider as fimple may not be compounded of two, or even of a greater number of principles ; but, fince thefe principles cannot be feparated, or rather fince we have not hitherto difcovered the means of feparating them, they act with regard to us as fimple fubftances, and we ought never to fuppofe them compounded until experiment and obfervation has proved them to be fo.

The

The foregoing reflections upon the progress of chemical ideas naturally apply to the words by which thefe ideas are to be expreffed. Guided by the work which, in the year 1787, Meffrs de Morveau, Berthollet, de Fourcroy, and I compofed upon the Nomenclature of Chemiftry, I have endeavoured, as much as poffible, to denominate fimple bodies by fimple terms, and I was naturally led to name thefe firft. It will be recollected, that we were obliged to retain that name of any fubftance by which it had been long known in the world, and that in two cafes only we took the liberty of making alterations; firft, in the cafe of thofe which were but newly difcovered, and had not yet obtained names, or at leaft which had been known but for a fhort time, and the names of which had not yet received the fanction of the public; and, fecondly, when the names which had been adopted, whether by the ancients or the moderns, appeared to us to exprefs evidently falfe ideas, when they confounded the fubftances, to which they were applied, with others poffeffed of different, or perhaps oppofite qualities. We made no fcruple, in this cafe, of fubftituting other names in their room, and the greateft number of thefe were borrowed from the Greek language. We

d   endeavoured

endeavoured to frame them in fuch a manner
as to exprefs the moft general and the moft
characteriftic quality of the fubftances; and this
was attended with the additional advantage both
of affifting the memory of beginners, who find
it difficult to remember a new word which has
no meaning, and of accuftoming them early to
admit no word without connecting with it fome
determinate idea.

To thofe bodies which are formed by the
union of feveral fimple fubftances we gave new
names, compounded in fuch a manner as the
nature of the fubftances directed; but, as the
number of double combinations is already very
confiderable, the only method by which we
could avoid confufion, was to divide them into
claffes. In the natural order of ideas, the name
of the clafs or genus is that which · expreffes
a quality common to a great number of indi-
viduals: The name of the fpecies, on the contrary, expreffes a quality peculiar to certain in-
dividuals only.

Thefe diftinctions are not, as fome may imagine, merely metaphyfical, but are eftablifhed
by Nature. " A child," fays the Abbé de Condillac,

dillac, " is taught to give the name *tree* to the
" firft one which is pointed out to him. The
" next one he fees prefents the fame idea, and
" he gives it the fame name. This he does like-
" wife to a third and a fourth, till at laft the
" word *tree*, which he firft applied to an indi-
" vidual, comes to be employed by him as the
" name of a clafs or a genus, an abftract idea,
" which comprehends all trees in general. But,
" when he learns that all trees ferve not the
" fame purpofe, that they do not all produce
" the fame kind of fruit, he will foon learn to
" diftinguifh them by fpecific and particular
" names." This is the logic of all the fciences,
and is naturally applied to chemiftry.

The acids, for example, are compounded of
two fubftances, of the order of thofe which we
confider as fimple ; the one conftitutes acidity,
and is common to all acids, and, from this fubftance, the name of the clafs or the genus ought
to be taken ; the other is peculiar to each acid,
and diftinguifhes it from the reft, and from this
fubftance is to be taken the name of the fpecies.
But, in the greateft number of acids, the two
conftituent elements, the acidifying principle,

and

and that which it acidifies, may exift in different proportions, conftituting all the poffible points of equilibrium or of faturation. This is the cafe in the fulphuric and the fulphurous acids ; and thefe two ftates of the fame acid we have marked by varying the termination of the fpecific name.

Metallic fubftances which have been expofed to the joint action of the air and of fire, lofe their metallic luftre, increafe in weight, and affume an earthy appearance. In this ftate, like the acids, they are compounded of a principle which is common to all, and one which is peculiar to each. In the fame way, therefore, we have thought proper to clafs them under a generic name, derived from the common principle ; for which purpofe, we adopted the term *oxyd ;* and we diftinguifh them from each other by the particular name of the metal to which each belongs.

Combuftible fubftances, which in acids and metallic oxyds are a fpecific and particular principle, are capable of becoming, in their turn, common principles of a great number of fubftances. The fulphurous combinations have

been

been long the only known ones in this kind.
Now, however, we know, from the experiments
of Meffrs Vandermonde, Monge, and Berthollet, that charcoal may be combined with iron,
and perhaps with feveral other metals ; and that,
from this combination, according to the proportions, may be produced fteel, plumbago, &c.
We know likewife, from the experiments of M.
Pelletier, that phofphorus may be combined with
a great number of metallic fubftances. Thefe
different combinations we have claffed under
generic names taken from the common fubftance, with a termination which marks this
analogy, fpecifying them by another name taken
from that fubftance which is proper to each.

The nomenclature of bodies compounded of
three fimple fubftances was attended with ftill
greater difficulty, not only on account of their
number, but, particularly, becaufe we cannot
exprefs the nature of their conftituent principles
without employing more compound names. In
the bodies which form this clafs, fuch as the
neutral falts, for inftance, we had to confider,
1ft, The acidifying principle, which is common
to them all ; 2d, The acidifiable principle which
conftitutes their peculiar acid ; 3d, The faline,
earthy,

earthy, or metallic bafis, which determines the
particular fpecies of falt. Here we derived the
name of each clafs of falts from the name of the
acidifiable principle common to all the indivi-
duals of that clafs; and diftinguifhed each fpecies by the name of the faline, earthy, or metallic bafis, which is peculiar to it.

A falt, though compounded of the fame three
principles, may, neverthelefs, by the mere difference of their proportion, be in three different
ftates. The nomenclature we have adopted
would have been defective, had it not expreffed
thefe different ftates; and this we attained chiefly by changes of termination uniformly applied
to the fame ftate of the different falts.

In fhort, we have advanced fo far, that from
the name alone may be inftantly found what
the combuftible fubftance is which enters into
any combination; whether that combuftible fubftance be combined with the acidifying principle,
and in what proportion; what is the ftate of the
acid; with what bafis it is united; whether the
faturation be exact, or whether the acid or the
bafis be in excefs.

It may be eafily fuppofed that it was not poffible to attain all thefe different objects without
departing, in fome inftances, from eftablifhed
cuftom, and adopting terms which at firft fight
will appear uncouth and barbarous. But we
confidered that the ear is foon habituated to
new words, efpecially when they are connected
with a general and rational fyftem. The names,
befides, which were formerly employed, fuch as
powder of algaroth, *falt of alembroth*, *pompholix*,
phagadenic water, *turbith mineral*, *colcothar*, and
many others, were neither lefs barbarous nor
lefs uncommon. It required a great deal of
practice, and no fmall degree of memory, to recollect the fubftances to which they were applied,
much more to recollect the genus of combination to which they belonged. The names of
oil of tartar per deliquium, *oil of vitriol*, *butter of*
arfenic and of antimony, *flowers of zinc*, &c. were
ftill more improper, becaufe they fuggefted falfe
ideas: For, in the whole mineral kingdom, and
particularly in the metallic clafs, there exifts no
fuch thing as butters, oils, or flowers; and, in
fhort, the fubftances to which they give thefe fallacious names, are nothing lefs than rank poifons.

When

When we publifhed our effay on the nomenclature of chemiftry, we were reproached for
having changed the language which was fpoken
by our mafters, which they diftinguifhed by
their authority, and handed down to us. But
thofe who reproach us on this account, have forgotten that it was Bergman and Macquer themfelves who urged us to make this reformation.
In a letter which the learned Profeffor of Upfal,
M. Bergman, wrote, a fhort time before he died,
to M. de Morveau, he bids him *fpare no improper names ; thofe who are learned, will always be
learned, and thofe who are ignorant will thus learn
fooner.*

There is an objection to the work which I am
going to prefent to the public, which is perhaps
better founded, that I have given no account of
the opinion of thofe who have gone before me ;
that I have ftated only my own opinion, without examining that of others. By this I have
been prevented from doing that juftice to my
affociates, and more efpecially to foreign chemifts, which I wifhed to render them. But I
befeech the reader to confider, that, if I had filled an elementary work with a multitude of quo-
tations ; if I had allowed myfelf to enter into
long

long differtations on the hiftory of the fcience,
and the works of thofe who have ftudied it, I
muft have loft fight of the true object I had in
view, and produced a work, the reading of
which muft have been extremely tirefome to
beginners. It is not to the hiftory of the fcience,
or of the human mind, that we are to attend in
an elementary treatife: Our only aim ought to
be eafe and perfpicuity, and with the utmoft care
to keep every thing out of view which might draw
afide the attention of the ftudent; it is a road which
we fhould be continually rendering more fmooth,
and from which we fhould endeavour to remove
every obftacle which can occafion delay. The
fciences, from their own nature, prefent a fufficient number of difficulties, though we add not
thofe which are foreign to them. But, befides
this, chemifts will eafily perceive, that, in the
firft part of my work, I make very little ufe of
any experiments but thofe which were made by
myfelf: If at any time I have adopted, without
acknowledgment, the experiments or the opinions of M. Berthollet, M. Fourcroy, M. de la
Place, M. Monge, or, in general, of any of thofe
whofe principles are the fame with my own, it
is owing to this circumftance, that frequent in-
tercourfe, and the habit of communicating our

ideas.

ideas, our obfervations, and our way of thinking to each other, has eftablifhed between us a
fort of community of opinions, in which it is
often difficult for every one to know his own.

The remarks I have made on the order which
I thought myfelf obliged to follow in the arrangement of proofs and ideas, are to be applied only to the firft part of this work. It is
the only one which contains the general fum of
the doctrine I have adopted, and to which I
wifhed to give a form completely elementary.

The fecond part is compofed chiefly of tables
of the nomenclature of the neutral falts. To
thefe I have only added general explanations, the
object of which was to point out the moft fimple
proceffes for obtaining the different kinds of
known acids. This part contains nothing which
I can call my own, and prefents only a very
fhort abridgment of the refults of thefe proceffes, extracted from the works of different authors.

In the third part, I have given a defcription,
in detail, of all the operations connected with
modern chemiftry. I have long thought that a
work

work of this kind was much wanted, and I am
convinced it will not be without ufe. The method
of performing experiments, and particularly thofe
of modern chemiftry, is not fo generally known
as it ought to be; and had I, in the different
memoirs which I have prefented to the Academy,
been more particular in the detail of the mani-
pulations of my experiments, it is probable I
fhould have made myfelf better underftood, and
the fcience might have made a more rapid progrefs. The order of the different matters contained in this third part appeared to me to be
almoft arbitrary; and the only one I have obferved was to clafs together, in each of the
chapters of which it is compofed, thofe operations which are moft connected with one another. I need hardly mention that this part
could not be borrowed from any other work,
and that, in the principal articles it contains, I
could not derive affiftance from any thing but
the experiments which I have made myfelf.

I fhall conclude this preface by tranferibing,
literally, fome obfervations of the Abbé de Condillac, which I think defcribe, with a good deal
of truth, the ftate of chemiftry at a period not
far diftant from our own. Thefe obfervations

were

were made on a different fubject; but they will
not, on this account, have lefs force, if the application of them be thought juft.

    ' Inftead of applying obfervation to the things
' we wifhed to know, we have chofen rather to
' imagine them. Advancing from one ill found-
' ed fuppofition to another, we have at laft be-
' wildered ourfelves amidft a multitude of errors.
' Thefe errors becoming prejudices, are, of
' courfe, adopted as principles, and we thus be-
' wilder ourfelves more and more. The method,
' too, by which we conduct our reafonings is
' as abfurd; we abufe words which we do not
' underftand, and call this the art of reafoning.
' When matters have been brought this length,
' when errors have been thus accumulated, there
' is but one remedy by which order can be
' reftored to the faculty of thinking; this is,
' to forget all that we have learned, to trace
' back our ideas to their fource, to follow the
' train in which they rife, and, as my Lord Ba-
' con fays, to frame the human underftanding
' anew.

    ' This remedy becomes the more difficult in
' proportion as we think ourfelves more learn-
                   ' ed,

' ed. Might it not be thought that works which
' treated of the fciences with the utmoft perfpi-
' cuity, with great precifion and order, muft be
' underftood by every body? The fact is, thofe
' who have never ftudied any thing will under-
' ftand them better than thofe who have ftudied
' a great deal, and efpecially than thofe who
' have written a great deal.

At the end of the fifth chapter, the Abbé de
Condillac adds : ' But, after all, the fciences
' have made progrefs, becaufe philofophers have
' applied themfelves with more attention to ob-
' ferve, and have communicated to their lan-
' guage that precifion and accuracy which they
' have employed in their obfervations : In cor-
' recting their language they reafon better.'

C O N.

CONTENTS.

PART FIRST.

CHAP.

SECT.

P A R T II.

Of the Combinations of Acids with Salifiable Bafes, and of the Formation
of Neutral Salts, 175

f SECT

TABLE

TABLE

SECT.

SECT.

P A R T   III.

Deſcription of the Inſtruments and Operations of Chemiſtry,   291

INTRO-

SECT.

SECT.

A P P E N D I X.

E L E

ELEMENTS

OF

CHEMISTRY.

PART I.

Of the Formation and Decompofition of Aeriform Fluids—of the
Combuftion of Simple Bodies—
and the Formation of Acids.

CHAP. I.

*Of the Combinations of Caloric, and the Formation
of Elaftic Aëriform Fluids.*

THAT every body, whether folid or fluid,
is augmented in all its dimenfions by any
increafe of its fenfible heat, was long ago fully
eftablifhed as a phyfical axiom, or univerfal pro-
pofition, by the celebrated Boerhaave. Such
facts as have been adduced for controverting the

generality of this principle offer only fallacious
refults, or, at leaft, fuch as are fo complicated
with foreign circumftances as to miflead the
judgment: But, when we feparately confider the
effects, fo as to deduce each from the caufe to
which they feparately belong, it is eafy to perceive that the feparation of particles by heat is
a conftant and general law of nature.

When we have heated a folid body to a certain degree, and have thereby caufed its particles
to feparate from each other, if we allow the
body to cool, its particles again approach each
other in the fame proportion in which they were
feparated by the increafed temperature; the body returns through the fame degrees of expanfion which it before extended through; and, if
it be brought back to the fame temperature from
which we fet out at the commencement of the
experiment, it recovers exactly the fame dimenfions which it formerly occupied. But, as we
are ftill very far from being able to arrive at
the degree of abfolute cold, or deprivation of
all heat, being unacquainted with any degree
of coldnefs which we cannot fuppofe capable
of ftill farther augmentation, it follows, that
we are ftill incapable of caufing the ultimate
particles of bodies to approach each other as
near as is poffible; and, confequently, that
the particles of all bodies do not touch each
other in any ftate hitherto known, which, tho'

a

a very fingular conclufion, is yet impoffible to be denied.

It is fuppofed, that, fince the particles of bodies are thus continually impelled by heat to feparate from each other, they would have no connection between themfelves ; and, of confequence, that there could be no folidity in nature, unlefs they were held together by fome other power which tends to unite them, and, fo to fpeak, to chain them together ; which power, whatever be its caufe, or manner of operation, we name Attraction.

Thus the particles of all bodies may be confidered as fubjected to the action of two oppofite powers, the one repulfive, the other attractive, between which they remain in equilibrio. So long as the attractive force remains ftronger, the body muft continue in a ftate of folidity ; but if, on the contrary, heat has fo far removed thefe particles from each other, as to place them beyond the fphere of attraction, they lofe the adhefion they before had with each other, and the body ceafes to be folid.

Water gives us a regular and conftant example of thefe facts ; whilft below Zero * of the French thermometer, or 32° of Fahrenheit,

it

* Whenever the degree of heat occurs in this work, it is ftated by the author according to Reaumur's fcale. The degrees within brackets are the correfpondent degrees of Fahrenheit's fcale, added by the tranflator. E.

it remains folid, and is called ice. Above that
degree of temperature, its particles being no
longer held together by reciprocal attraction, it
becomes liquid ; and, when we raife its temperature above 80°, (212°) its particles, giving
way to the repulfion caufed by the heat, affume
the ftate of vapour or gas, and the water is
changed into an aëriform fluid.

The fame may be affirmed of all bodies in
nature : They are either folid or liquid, or in
the ftate of elaftic aëriform vapour, according
to the proportion which takes place between
the attractive force inherent in their particles,
and the repulfive power of the heat acting upon
thefe ; or, what amounts to the fame thing, in
proportion to the degree of heat to which they
are expofed.

It is difficult to comprehend thefe phenomena, without admitting them as the effects of
a real and material fubftance, or very fubtile
fluid, which, infinuating itfelf between the particles of bodies, feparates them from each o-
ther ; and, even allowing the exiftence of this
fluid to be hypothetical, we fhall fee in the fequel, that it explains the phenomena of nature
in a very fatisfactory manner.

This fubftance, whatever it is, being the caufe
of heat, or, in other words, the fenfation which
we call *warmth* being caufed by the accumulation of this fubftance, we cannot, in ftrict language,

guage, diftinguifh it by the term *heat ;* becaufe the fame name would then very improperly ex- prefs both caufe and effect. For this reafon, in the memoir which I publifhed in 1777 *, I gave it the names of *igneous fluid* and *matter of heat :* And, fince that time, in the work † publifhed by Mr de Morveau, Mr Berthollet, Mr de Four- croy, and myfelf, upon the reformation of che- mical nomenclature, we thought it neceffary to banifh all periphraftic expreffions, which both lengthen phyfical language, and render it more tedious and lefs diftinct, and which even frequently does not convey fufficiently juft ideas of the fubject intended. Wherefore, we have diftinguifhed the caufe of heat, or that exqui- fitely elaftic fluid which produces it, by the term of *caloric.* Befides, that this expreffion fulfils our object in the fyftem which we have adopted, it poffeffes this farther advantage, that it accords with every fpecies of opinion, fince, ftrictly fpeaking, we are not obliged to fuppofe this to be a real fubftance ; it being fufficient, as will more clearly appear in the fequel of this work, that it be confidered as the repulfive caufe, whatever that may be, which feparates the particles of matter from each other ; fo that

we

---

* Collections of the French Academy of Sciences for that year, p. 420.

† Chemical Nomenclature.

we are ftill at liberty to inveftigate its effects in an abftract and mathematical manner.

In the prefent ftate of our knowledge, we are unable to determine whether light be a modification of caloric, or if caloric be, on the contrary, a modification of light. This, however, is indifputable, that, in a fyftem where only decided facts are admiffible, and where we avoid, as far as poffible, to fuppofe any thing to be that is not really known to exift, we ought provifionally to diftinguifh, by diftinct terms, fuch things as are known to produce different effects. We therefore diftinguifh light from caloric; though we do not therefore deny that thefe have certain qualities in common, and that, in certain circumftances, they combine with other bodies almoft in the fame manner, and produce, in part, the fame effects.

What I have already faid may fuffice to determine the idea affixed to the word *caloric*; but there remains a more difficult attempt, which is, to give a juft conception of the manner in which caloric acts upon other bodies. Since this fubtile matter penetrates through the pores of all known fubftances; fince there are no veffels through which it cannot efcape, and, confequently, as there are none which are capable of retaining it, we can only come at the knowledge of its properties by effects which are fleeting, and difficultly afcertainable. It is in
thefe

thefe things which we neither fee nor feel, that
it is efpecially neceffary to guard againft the
extravagancy of our imagination, which for-
ever inclines to ftep beyond the bounds of
truth, and is very difficultly reftrained within
the narrow line of facts.

We have already feen, that the fame body
becomes folid, or fluid, or aëriform, according
to the quantity of caloric by which it is penetrated ; or, to fpeak more ftrictly, according as
the repulfive force exerted by the caloric is
equal to, ftronger, or weaker, than the attraction of the particles of the body it acts upon.

But, if thefe two powers only exifted, bodies
would become liquid at an indivifible degree of
the thermometer, and would almoft inftantaneoufly pafs from the folid ftate of aggregation to
that of aëriform elafticity. Thus water, for inftance, at the very moment when it ceafes to be
ice, would begin to boil, and would be transformed into an aëriform fluid, having its particles fcattered indefinitely through the furrounding fpace. That this does not happen, muft depend upon the action of fome third power. The
preffure of the atmofphere prevents this feparation, and caufes the water to remain in the liquid ftate till it be raifed to 80° of temperature (212°) above zero of the French thermometer, the quantity of caloric which it receives in the loweft temperature being infufficient

cient to overcome the preſſure of the atmoſphere.

Whence it appears that, without this atmoſpheric preſſure, we ſhould not have any permanent liquid, and ſhould only be able to ſee bodies in that ſtate of exiſtence in the very inſtant
of melting, as the ſmalleſt additional caloric
would inſtantly ſeparate their particles, and diſſi-
pate them through the ſurrounding medium.
Beſides, without this atmoſpheric preſſure, we
ſhould not even have any aëriform fluids, ſtrictly
ſpeaking, becauſe the moment the force of attraction is overcome by the repulſive power of
the caloric, the particles would ſeparate themſelves indefinitely, having nothing to give limits
to their expanſion, unleſs their own gravity
might collect them together, ſo as to form an
atmoſphere.

Simple reflection upon the moſt common experiments is ſufficient to evince the truth of
theſe poſitions. They are more particularly
proved by the following experiment, which I
publiſhed in the Memoirs of the French Academy for 1777, p. 426.

Having filled with ſulphuric ether * a ſmall narrow glaſs veſſel, A, (Plate VII. Fig. 17.), ſtand-

ing

---

* As I ſhall afterwards give a definition, and explain the properties of the liquor called *ether*, I ſhall
only premiſe here, that it is a very volatile inflammable

ing upon its ftalk P, the veffel, which is from twelve to fifteen lines diameter, is to be covered by a wet bladder, tied round its neck with feveral turns of ftrong thread ; for greater fecurity, fix a fecond bladder over the firft. The veffel fhould be filled in fuch a manner with the ether, as not to leave the fmalleft portion of air between the liquor and the bladder. It is now to be placed under the recipient BCD of an air-pump, of which the upper part B ought to be fitted with a leathern lid, through which paffes a wire EF, having its point F very fharp ; and in the fame receiver there ought to be placed the barometer GH. The whole being thus difpofed, let the recipient be exhaufted, and then, by pufhing down the wire EF, we make a hole in the bladder. Immediately the ether begins to boil with great violence, and is changed into an elaftic aëriform fluid, which fills the receiver. If the quantity of ether be fufficient to leave a few drops in the phial after the evaporation is finifhed, the elaftic fluid produced will fuftain the mercury in the barometer attached to the airpump, at eight or ten inches in winter, and from

B   twenty

mable liquor, having a confiderably fmaller fpecific gravity than water, or even fpirit of wine.—A.

twenty to twenty-five in fummer *. To render
this experiment more complete, we may introduce a fmall thermometer into the phial A, containing the ether, which will defcend confiderably during the evaporation.

The only effect produced in this experiment
is, the taking away the weight of the atmofphere,
which, in its ordinary ftate, preffes on the furface of the ether; and the effects refulting from
this removal evidently prove, that, in the ordinary temperature of the earth, ether would always exift in an aëriform ftate, but for the preffure of the atmofphere, and that the paffing of
the ether from the liquid to the aëriform ftate
is accompanied by a confiderable leffening of
heat; becaufe, during the evaporation, a part of
the caloric, which was before in a free ftate, or
at leaft in equilibrio in the furrounding bodies, combines with the ether, and caufes it to
affume the aëriform ftate.

The fame experiment fucceeds with all evaporable fluids, fuch as alkohol, water, and even
mercury ; with this difference, that the atmofphere formed in the receiver by alkohol only
fupports

It would have been more fatisfactory if the Author
had fpecified the degrees of the thermometer at which
thefe heights of the mercury in the barometer are produced.

supports the attached barometer about one inch in winter, and about four or five inches in summer; that formed by water, in the same situation, raises the mercury only a few lines, and that by quickfilver but a few fractions of a line. There is therefore lefs fluid evaporated from alkohol than from ether, lefs from water than from alkohol, and ftill lefs from mercury than from either; confequently there is lefs caloric employed, and lefs cold produced, which quadrates exactly with the refults of thefe experiments.

Another fpecies of experiment proves very evidently that the aëriform ftate is a modification of bodies dependent on the degree of temperature, and on the preffure which thefe bodies undergo. In a Memoir read by Mr de la Place and me to the Academy in 1777, which has not been printed, we have fhown, that, when ether is fubjected to a preffure equal to twenty-eight inches of the barometer, or about the medium preffure of the atmofphere, it boils at the temperature of about 32° (104), or 33° (106.25°), of the thermometer. Mr de Luc, who has made fimilar experiments with fpirit of wine, finds it boils at 67° (182.75°). And all the world knows that water boils at 80° (212°). Now, boiling being only the evaporation of a liquid, or the moment of its paffing from the fluid to the aëriform ftate, it is evident that, if we keep

ether

ether continually at the temperature of 33^q (106.25°), and under the common preſſure of the atmoſphere, we ſhall have it always in an elaſtic aëriform ſtate; and that the ſame thing will happen with alkohol when above 67° (182.75°), and with water when above 80° (212°); all which are perfectly conformable to the following experiment *.

I filled a large veſſel ABCD (Plate VII. Fig. 16.) with water, at 35° (110.75°), or 36° (113°); I ſuppoſe the veſſel tranſparent, that we may ſee what takes place in the experiment; and we can eaſily hold the hands in water at that temperature without inconvenience. Into it I plunged ſome narrow necked bottles F, G, which were filled with the water, after which they were turned up, ſo as to reſt on their mouths on the bottom of the veſſel. Having next put ſome ether into a very ſmall matraſs, with its neck a b c, twice bent as in the Plate, I plunged this matraſs into the water, ſo as to have its neck inſerted into the mouth of one of the bottles F. Immediately upon feeling the effects of the heat communicated to it by the water in the veſſel ABCD it began to boil; and the caloric entering into combination with it, changed it into elaſtic aëriform fluid, with which I filled ſeveral bottles ſucceſſively, F, G, &c.

This

* Vide Memoirs of the French Academy, anno 1780, p. 335.—A.

This is not the place to enter upon the examination of the nature and properties of this aëriform fluid, which is extremely inflammable; but, confining myfelf to the object at prefent in view, without anticipating circumftances, which I am not to fuppofe the reader to know, I fhall only obferve, that the ether, from this experiment, is almoft only capable of exifting in the aëriform ftate in our world; for, if the weight of our atmofphere was only equal to between 20 and 24 inches of the barometer, inftead of 28 inches, we fhould never be able to obtain ether in the liquid ftate, at leaft in fummer; and the formation of ether would confequently be impoffible upon mountains of a moderate degree of elevation, as it would be converted into gas immediately upon being produced, unlefs we employed recipients of extraordinary ftrength, together with refrigeration and compreffion. And, laftly, the temperature of the blood being nearly that at which ether paffes from the liquid to the aëriform ftate, it muft evaporate in the primae viae, and confequently it is very probable the medical properties of this fluid depend chiefly upon its mechanical effect.

Thefe experiments fucceed better with nitrous ether, becaufe it evaporates in a lower temperature than fulphuric ether. It is more difficult to obtain alkohol in the aëriform ftate; becaufe, as it requires 67° (182.75°) to reduce it to vapour,

pour, the water of the bath muſt be almoſt boiling, and conſequently it is impoſſible to plunge the hands into it at that temperature.

It is evident that, if water were uſed in the foregoing experiment, it would be changed into gas, when expoſed to a temperature ſuperior to that at which it boils. Although thoroughly convinced of this, Mr de la Place and myſelf judged it neceſſary to confirm it by the following direct experiment. We filled a glaſs jar A, (Plate VII. Fig. 5.) with mercury, and placed it with its mouth downwards in a diſh B, like-wiſe filled with mercury, and having intro-duced about two groſs of water into the jar, which roſe to the top of the mercury at CD; we then plunged the whole apparatus into an iron boiler EFGH, full of boiling ſea-water of the temperature of 85° (123.25°), placed upon the furnace GHIK. Immediately upon the wa-ter over the mercury attaining the temperature of 80° (212°), it began to boil; and, inſtead of only filling the ſmall ſpace ACD, it was con-verted into an aëriform fluid, which filled the whole jar; the mercury even deſcended below the ſurface of that in the diſh B; and the jar muſt have been overturned, if it had not been very thick and heavy, and fixed to the diſh by means of iron-wire. Immediately after with-drawing the apparatus from the boiler, the va-pour in the jar began to condenſe, and the

mercury

mercury rofe to its former ftation ; but it returned again to the aëriform ftate a few feconds
after replacing the apparatus in the boiler.

We have thus a certain number of fubftances, which are convertible into elaftic aëriform fluids by degrees of temperature, not
much fuperior to that of our atmofphere. We
fhall afterwards find that there are feveral others
which undergo the fame change in fimilar circumftances, fuch as muriatic or marine acid, ammoniac or volatile alkali, the carbonic acid or fixed
air, the fulphurous acid, &c. All of thefe are
permanently elaftic in or about the mean temperature of the atmofphere, and under its common
preffure.

All thefe facts, which could be eafily multiplied if neceffary, give me full right to affume,
as a general principle, that almoft every body
in nature is fufceptible of three feveral ftates of
exiftence, folid, liquid, and aëriform, and
that thefe three ftates of exiftence depend upon
the quantity of caloric combined with the
body. Henceforwards I fhall exprefs thefe
elaftic aëriform fluids by the generic term
gas ; and in each fpecies of gas I fhall diftinguifh between the caloric, which in fome meafure ferves the purpofe of a folvent, and the fubftance, which in combination with the caloric,
forms the bafe of the gas.

To

To thefe bafes of the different gaffes, which are hitherto but little known, we have been o-bliged to affign names ; thefe I fhall point out in Chap. IV. of this work, when I have pre-vioufly given an account of the phenomena at-tendant upon the heating and cooling of bodies, and when I have eftablifhed precife ideas con-cerning the compofition of our atmofphere.

We have already fhown, that the particles of every fubftance in nature exift in a certain ftate of equilibrium, between that attraction which tends to unite and keep the particles to-gether, and the effects of the caloric which tends to feparate them. Hence the caloric not only furrounds the particles of all bo-dies on every fide, but fills up every inter-val which the particles of bodies leave be-tween each other. We may form an idea of this, by fuppofing a veffel filled with fmall fphe-rical leaden bullets, into which a quantity of fine fand is poured, which, infinuating into the intervals between the bullets, will fill up every void. The balls, in this comparifon, are to the fand which furrounds them exactly in the fame fituation as the particles of bodies are with refpect to the caloric; with this difference only, that the balls are fuppofed to touch each other, whereas the particles of bodies are not in con-tact, being retained at a fmall diftance from each other, by the caloric.

If, inſtead of ſpherical balls, we ſubſtitute ſolid
bodies of a hexahedral, octohedral, or any o-
ther regular figure, the capacity of the intervals between them will be leſſened, and conſequently will no longer contain the ſame quantity of ſand. The ſame thing takes place, with
reſpect to natural bodies ; the intervals left between their particles are not of equal capacity,
but vary in conſequence of the different figures
and magnitude of their particles, and of the
diſtance at which theſe particles are maintained, according to the exiſting proportion between their inherent attraction, and the repulſive force exerted upon them by the caloric.

In this manner we muſt underſtand the following expreſſion, introduced by the Engliſh
philoſophers, who have given us the firſt pre-
ciſe ideas upon this ſubject ; *the capacity of bodies
for containing the matter of heat.* As compari-
ſons with ſenſible objects are of great uſe in
aſſiſting us to form diſtinct notions of abſtract
ideas, we ſhall endeavour to illuſtrate this, by
inſtancing the phenomena which take place
between water and bodies which are wetted
and penetrated by it, with a few reflections.

If we immerge equal pieces of different kinds
of wood, ſuppoſe cubes of one foot each, into
water, the fluid gradually inſinuates itſelf into
their pores, and the pieces of wood are augmented both in weight and magnitude : But

C   each

each fpecies of wood will imbibe a different
quantity of water; the lighter and more porous
woods will admit a larger, the compact and clofer
grained will admit of a leffer quantity ; for the
proportional quantities of water imbibed by the
pieces will depend upon the nature of the conftituent particles of the wood, and upon the
greater or leffer affinity fubfifting between them
and water. Very refinous wood, for inftance,
though it may be at the fame time very porous,
will admit but little water. We may therefore
fay, that the different kinds of wood poffefs
different capacities for receiving water ; we
may even determine, by means of the augmentation of their weights, what quantity of water
they have actually abforbed ; but, as we are ignorant how much water they contained, previous to immerfion, we cannot determine the
abfolute quantity they contain, after being taken out of the water.

The fame circumftances undoubtedly take
place, with bodies that are immerfed in caloric;
taking into confideration, however, that water
is an incompreffible fluid, whereas caloric is, on
the contrary, endowed with very great elafticity ; or, in other words, the particles of caloric
have a great tendency to feparate from each
other, when forced by any other power to approach ; this difference muft of neceffity occafion

fion very confiderable diverfities in the refults of experiments made upon thefe two fub-ftances.

Having eftablifhed thefe clear and fimple propofitions, it will be very eafy to explain the ideas which ought to be affixed to the follow-ing expreffions, which are by no means fynoni-mous, but poffefs each a ftrict and determinate meaning, as in the following definitions :

Free caloric, is that which is not combined in any manner with any other body. But, as we live in a fyftem to which caloric has a very ftrong adhefion, it follows that we are never able to obtain it in the ftate of abfolute free-dom.

Combined caloric, is that which is fixed in bodies by affinity or elective attraction, fo as to form part of the fubftance of the body, even part of its folidity.

By the expreffion *fpecific caloric* of bodies, we underftand the refpective quantities of caloric requifite for raifing a number of bodies of the fame weight to an equal degree of tempera-ture. This proportional quantity of caloric de-pends upon the diftance between the conftitu-ent particles of bodies, and their greater or leffer degrees of cohefion ; and this diftance, or rather the fpace or void refulting from it, is, as I have already obferved, called the *capacity of bodies for containing caloric.*

Heat,

Heat, confidered as a fenfation, or, in other words, fenfible heat, is only the effect produced upon our fentient organs, by the motion or paffage of caloric, difengaged from the furrounding bodies. In general, we receive impreffions only in confequence of motion, and we might eftablifh it as an axiom, *That*, WITHOUT MOTION, THERE IS NO SENSATION. This general principle applies very accurately to the fenfations of heat and cold : When we touch a cold body, the caloric which always tends to become in equilibrio in all bodies, paffes from our hand into the body we touch, which gives us the feeling or fenfation of cold. The direct contrary happens, when we touch a warm body, the caloric then paffing from the body into our hand, produces the fenfation of heat. If the hand and the body touched be of the fame temperature, or very nearly fo, we receive no impreffion, either of heat or cold, becaufe there is no motion or paffage of caloric ; and thus no fenfation can take place, without fome correfpondent motion to occafion it.

When the thermometer rifes, it fhows, that free caloric is entering into the furrounding bodies: The thermometer, which is one of thefe, receives its fhare in proportion to its mafs, and to the capacity which it poffeffes for containing caloric. The change therefore which takes place upon the thermometer, only announces a
change

change of place of the caloric in thofe bodies,
of which the thermometer forms one part ; it
only indicates the portion of caloric received,
without being a meafure of the whole quantity
difengaged, difplaced, or abforbed.

The moſt ſimple and moſt exact method for
determining this latter point, is that defcribed
by Mr de la Place, in the Memoirs of the Academy, No. 1780, p. 364 ; a fummary explanation
of which will be found towards the conclufion
of this work. This method confifts in placing
a body, or a combination of bodies, from which
caloric is difengaging, in the midſt of a hollow
fphere of ice ; and the quantity of ice melted
becomes an exact meafure of the quantity of
caloric difengaged. It is poſſible, by means of
the apparatus which we have caufed to be conſtructed upon this plan, to determine, not as has
been pretended, the capacity of bodies for containing heat, but the ratio of the increafe or diminution of capacity produced by determinate
degrees of temperature. It is eafy with the
fame apparatus, by means of divers combinations of experiments, to determine the quantity
of caloric requifite for converting folid fubſtances into liquids, and liquids into elaſtic aëriform fluids ; and, *vice verfa*, what quantity of
caloric efcapes from elaſtic vapours in changing
to liquids, and what quantity efcapes from liquids during their converfion into folids. Perhaps,

haps, when experiments-have been made with
fufficient accuracy, we may one day be able to
determine the proportional quantity of caloric,
neceffary for producing the feveral fpecies of
gaffes. I fhall hereafter, in a feparate chapter,
give an account of the principal refults of fuch
experiments as have been made upon this head.

It remains, before finifhing this article, to
fay a few words relative to the caufe of the
elafticity of gaffes, and of fluids in the ftate of
vapour. It is by no means difficult to perceive
that this elafticity depends upon that of caloric,
which feems to be the moft eminently elaftic
body in nature. Nothing is more readily conceived, than that one body fhould become elaftic by entering into combination with another
body poffeffed of that quality. We muft allow
that this is only an explanation of elafticity, by
an affumption of elafticity, and that we thus
only remove the difficulty one ftep farther, and
that the nature of elafticity, and the reafon for
caloric being elaftic, remains ftill unexplained.
Elafticity in the abftract is nothing more than
that quality of the particles of bodies by which
they recede from each other when forced together. This tendency in the particles of caloric
to feparate, takes place even at confiderable dif-
tances. We fhall be fatisfied of this, when we
confider that air is fufceptible of undergoing
great compreffion, which fuppofes that its particles

ricles were previouſly very diſtant from each
other ; for the power of approaching together
certainly ſuppoſes a previous diſtance, at leaſt
equal to the degree of approach. Conſequently, thoſe particles of the air, which are already
conſiderably diſtant from each other, tend to
ſeparate ſtill farther. In fact, if we produce
Boyle's vacuum in a large receiver, the very
laſt portion of air which remains ſpreads itſelf
uniformly through the whole capacity of the
veſſel, however large, fills it completely throughout, and preſſes every where againſt its ſides :
We cannot, however, explain this effect, without ſuppoſing that the particles make an effort
to ſeparate themſelves on every ſide, and we
are quite ignorant at what diſtance, or what
degree of rarefaction, this effort ceaſes to act.

Here, therefore, exiſts a true repulſion between the particles of elaſtic fluids ; at leaſt,
circumſtances take place exactly as if ſuch a
repulſion actually exiſted ; and we have very
good right to conclude, that the particles of
caloric mutually repel each other. When we
are once permitted to ſuppoſe this repelling
force, the *rationale* of the formation of gaſſes, or
aëriform fluids, becomes perfectly ſimple ; tho'
we muſt, at the ſame time, allow, that it is extremely difficult to form an accurate conception
of this repulſive force acting upon very minute
particles

particles placed at great diſtances from each other.

It is, perhaps, more natural to ſuppoſe, that the particles of caloric have a ſtronger mutual attraction than thoſe of any other ſubſtance, and that theſe latter particles are forced aſunder in conſequence of this ſuperior attraction between the particles of the caloric, which forces them between the particles of other bodies, that they may be able to reunite with each other. We have ſomewhat analogous to this idea in the phenomena which occur when a dry ſponge is dipt into water: The ſponge ſwells; its particles ſeparate from each other; and all its intervals are filled up by the water. It is evident, that the ſponge, in the act of ſwelling, has acquired a greater capacity for containing water than it had when dry. But we cannot certainly maintain, that the introduction of water between the particles of the ſponge has endowed them with a repulſive power, which tends to ſeparate them from each other; on the contrary, the whole phenomena are produced by means of attractive powers; and theſe are, *firſt*, The gravity of the water, and the power which it exerts on every ſide, in common with all other fluids; 2*dly*, The force of attraction which takes place between the particles of the water, cauſing them to unite together; 3*dly*, The mutual attraction of the particles of the ſponge with each other;

and,

and, *laſtly*, The reciprocal attraction which exiſts between the particles of the ſponge and thoſe
of the water. It is eaſy to underſtand, that the
explanation of this fact depends upon properly
appreciating the intenſity of, and connection between, theſe ſeveral powers. It is probable,
that the ſeparation of the particles of bodies, occaſioned by caloric, depends in a ſimilar manner upon a certain combination of different attractive powers, which, in conformity with the
imperfection of our knowledge, we endeavour
to expreſs by ſaying, that caloric communicates
a power of repulſion to the particles of bodies.

D   C H A P.

C H A P.   II.

General Views relative to the Formation and Com-
position of our Atmosphere.

THESE views which I have taken of the
formation of elaſtic aëriform fluids or
gaffes, throw great light upon the original formation of the atmoſpheres of the planets, and
particularly that of our earth. We readily conceive, that it muſt neceſſarily conſiſt of a mixture of the following ſubſtances : *Firſt,* Of all
bodies that are ſuſceptible of evaporation, or,
more ſtrictly ſpeaking, which are capable of retaining the ſtate of aëriform elaſticity in the
temperature of our atmoſphere, and under a
preſſure equal to that of a column of twenty-
eight inches of quickſilver in the barometer ;
and, *ſecondly,* Of all ſubſtances, whether liquid
or ſolid, which are capable of being diſſolved
by this mixture of different gaffes.

The better to determine our ideas relating to
this ſubject, which has not hitherto been ſufficiently conſidered, let us, for a moment, conceive what change would take place in the various

rious fubftances which compofe our earth, if its
temperature were fuddenly altered. If, for inftance, we were fuddenly tranfported into the
region of the planet Mercury, where probably
the common temperature is much fuperior to
that of boiling water, the water of the earth,
and all the other fluids which are fufceptible of
the gaffeous ftate, at a temperature near to that
of boiling water, even quickfilver itfelf, would
become rarified; and all thefe fubftances would
be changed into permanent aëriform fluids or
gaffes, which would become part of the new atmofphere. Thefe new fpecies of airs or gaffes
would mix with thofe already exifting, and certain reciprocal decompofitions and new combinations would take place, until fuch time as all
the elective attractions or affinities fubfifting
amongft all thefe new and old gaffeous fubftances had operated fully; after which, the elementary principles compofing thefe gaffes, being
faturated, would remain at reft. We muft attend
to this, however, that, even in the above hypothetical fituation, certain bounds would occur to
the evaporation of thefe fubftances, produced by
that very evaporation itfelf; for as, in proportion
to the increafe of elaftic fluids, the preffure of the
atmofphere would be augmented, as every degree of preffure tends, in fome meafure, to prevent evaporation, and as even the moft evaporable

rable fluids can refift the operation of a very high temperature without evaporating, if prevented by a proportionally ftronger compreffion, water and all other liquids being able to fuftain a red heat in Papin's digefter ; we muft admit, that the new atmofphere would at laft arrive at fuch a degree of weight, that the water which had not hitherto evaporated would ceafe to boil, and, of confequence, would remain liquid ; fo that, even upon this fuppofition, as in all others of the fame nature, the increafing gravity of the atmofphere would find certain limits which it could not exceed. We might even extend thefe reflections greatly farther, and examine what change might be produced in fuch fituations upon ftones, falts, and the greater part of the fufible fubftances which compofe the mafs of our earth. Thefe would be foftened, fufed, and changed into fluids, &c. : But thefe fpeculations carry me from my object, to which I haften to return.

By a contrary fuppofition to the one we have been forming, if the earth were fuddenly tranfported into a very cold region, the water which at prefent compofes our feas, rivers, and fprings, and probably the greater number of the fluids we are acquainted with, would be converted into folid mountains and hard rocks, at firft diaphanous

aphanous and homogeneous, like rock cryſtal,
but which, in time, becoming mixed with foreign and heterogeneous ſubſtances, would become opake ſtones of various colours.   In
this caſe, the air, or at leaſt ſome part of the
aëriform fluids which now compoſe the maſs of
our atmoſphere, would doubtleſs loſe its elaſticity for want of a ſufficient temperature to retain them in that ſtate: They would return to
the liquid ſtate of exiſtence, and new liquids
would be formed, of whoſe properties we cannot, at preſent, form the moſt diſtant idea.

Theſe two oppoſite ſuppoſitions give a diſtinct proof of the following corollaries: _Firſt_,
That _ſolidity_, _liquidity_, and _aëriform elaſticity_,
are only three different ſtates of exiſtence of the
ſame matter, or three particular modifications
which almoſt all ſubſtances are ſuſceptible of
aſſuming ſucceſſively, and which ſolely depend
upon the degree of temperature to which they
are expoſed; or, in other words, upon the
quantity of caloric with which they are penetrated *.   _2dly_, That it is extremely probable that
air is a fluid naturally exiſting in a ſtate of vapour; or, as we may better expreſs it, that our
atmoſphere is a compound of all the fluids
which

---

* The degree of preſſure which they undergo muſt
be taken into account.   E.

which are fufceptible of the vaporous or permanently elaftic ftate, in the ufual temperature, and under the common preffure. 3*dly,*
That it is not impoffible we may difcover,
in our atmofphere, certain fubftances naturally very compact, even metals themfelves; as
a metallic fubftance, for inftance, only a little
more volatile than mercury, might exift in that
fituation.

Amongft the fluids with which we are acquainted, fome, as water and alkohol, are fufceptible of mixing with each other in all proportions; whereas others, on the contrary, as
quickfilver, water, and oil, can only form a
momentary union; and, after being mixed together, feparate and arrange themfelves according to their fpecific gravities. The fame thing
ought to, or at leaft may, take place in the atmofphere. It is poffible, and even extremely
probable, that, both at the firft creation, and
every day, gaffes are formed, which are difficultly mifcible with atmofpheric air, and are
continually feparating from it. If thefe gaffes
be fpecifically lighter than the general atmofpheric mafs, they muft, of courfe, gather in the
higher regions, and form ftrata that float upon the common air. The phenomena which
accompany igneous meteors induce me to
believe, that there exifts in the upper parts
of

of our atmofphere a ftratum of inflammable fluid in contact with thofe ftrata of air which produce the phenomena of the aurora borealis and other fiery meteors.——I mean hereafter to purfue this fubject in a feparate treatife.

CHAP.

C H A P. III.

Analysis of Atmospheric Air, and its Division into
two Elastic Fluids ; the one fit for Respiration,
the other incapable of being respired.

FROM what has been premised, it follows, that our atmosphere is composed
of a mixture of every substance capable of retaining the gasseous or aëriform state in the
common temperature, and under the usual
pressure which it experiences. These fluids
constitute a mass, in some measure homogeneous, extending from the surface of the earth to
the greatest height hitherto attained, of which
the density continually decreases in the inverse
ratio of the superincumbent weight. But, as I
have before observed, it is possible that this first
stratum is surmounted by several others consisting of very different fluids.

Our business, in this place, is to endeavour
to determine, by experiments, the nature of the
elastic fluids which compose the inferior stratum of air which we inhabit. Modern chemistry has made great advances in this research ;
and it will appear by the following details that
the analysis of atmospherical air has been more

rigo-

rigoroufly determined than that of any other fubftance of the clafs. Chemiftry affords two general methods of determining the conftituent principles of bodies, the method of analyfis, and that of fynthefis. When, for inftance, by combining water with alkohol, we form the fpecies of liquor called, in commercial language, brandy or fpirit of wine, we certainly have a right to conclude, that brandy, or fpirit of wine, is compofed of alkohol combined with water. We can produce the fame refult by the analytical method; and in general it ought to be confidered as a principle in chemical fcience, never to reft fatisfied without both thefe fpecies of proofs.

We have this advantage in the analyfis of atmofpherical air, being able both to decompound it, and to form it a new in the moft fatisfactory manner. I fhall, however, at prefent confine myfelf to recount fuch experiments as are moft conclufive upon this head; and I may confider moft of thefe as my own, having either firft invented them, or having repeated thofe of others, with the intention of analyfing atmofpherical air, in perfectly new points of view.

I took a matrafs (A, fig. 14. plate II.) of about 36 cubical inches capacity, having a long neck B C D E, of fix or feven lines internal diameter, and having bent the neck as in Plate IV. Fig. 2. fo as to allow of its being placed in

E   the

the furnace M M N N, in fuch a manner that
the extremity of its neck E might be inferted
under a bell-glafs F G, placed in a trough of
quickfilver R R S S; I introduced four ounces
of pure mercury into the matrafs, and, by means
of a fyphon, exhaufted the air in the receiver
F G, fo as to raife the quickfilver to L L, and
I carefully marked the height at which it ftood
by pafting on a flip of paper. Having accurately noted the height of the thermometer and barometer, I lighted a fire in the furnace M M N N,
which I kept up almoft continually during
twelve days, fo as to keep the quickfilver always
almoft at its boiling point. Nothing remarkable took place during the firft day : The Mercury, though not boiling, was continually evaporating, and covered the interior furface of the
veffels with fmall drops, at firft very minute,
which gradually augmenting to a fufficient fize,
fell back into the mafs at the bottom of the
veffel. On the fecond day, fmall red particles
began to appear on the furface of the mercury,
which, during the four or five following days,
gradually increafed in fize and number ; after
which they ceafed to increafe in either refpeft.
At the end of twelve days, feeing that the calcination of the mercury did not at all increafe,
I extinguifhed the fire, and allowed the veffels
to cool. The bulk of air in the body and neck
of the matrafs, and in the bell-glafs, reduced to

a

a medium of 28 inches of the barometer and
10° (54.5°) of the thermometer, at the commencement of the experiment was about 50
cubical inches. At the end of the experiment
the remaining air, reduced to the fame medium
preffure and temperature, was only between 42
and 43 cubical inches; confequently it had loft
about $\frac{1}{6}$ of its bulk. Afterwards, having collected all the red particles, formed during the
experiment, from the running mercury in which
they floated, I found thefe to amount to 45
grains.

I was obliged to repeat this experiment feveral times, as it is difficult in one experiment both
to preferve the whole air upon which we operate, and to collect the whole of the red particles, or calx of mercury, which is formed during the calcination. It will often happen in
the fequel, that I fhall, in this manner, give in
one detail the refults of two or three experiments of the fame nature.

The air which remained after the calcination
of the mercury in this experiment, and which
was reduced to $\frac{5}{6}$ of its former bulk, was no
longer fit either for refpiration or for combuftion; animals being introduced into it were
fuffocated in a few feconds, and when a taper
was plunged into it, it was extinguifhed as if it
had been immerfed into water.

In

In the next place, I took the 45 grains of red matter formed during this experiment, which I put into a fmall glafs retort, having a proper apparatus for receiving fuch liquid, or gaffeous product, as might be extracted : Having applied a fire to the retort in a furnace, I obferved that, in proportion as the red matter became heated, the intenfity of its colour augmented. When the retort was almoft red hot, the red matter began gradually to decreafe in bulk, and in a few minutes after it difappeared altogether ; at the fame time $41\frac{1}{2}$ grains of running mercury were collected in the recipient, and 7 or 8 cubical inches of elaftic fluid, greatly more capable of fupporting both refpiration and combuftion than atmofperical air, were collected in the bell-glafs.

A part of this air being put into a glafs tube of about an inch diameter, fhowed the following properties : A taper burned in it with a dazzling fplendour, and charcoal, inftead of confuming quietly as it does in common air, burnt with a flame, attended with a decrepitating noife, like phofphorus, and threw out fuch a brilliant light that the eyes could hardly endure it. This fpecies of air was difcovered almoft at the fame time by Mr Prieftley, Mr Scheele, and myfelf. Mr Prieftley gave it the name of *dephlogifticated air*, Mr Scheele called it *empyreal air*. At firft I named it *highly refpirable air*, to which

which has fince been fubftituted the term of
vital air. We fhall prefently fee what we ought
to think of thefe denominations.

In reflecting upon the circumftances of this
experiment, we readily perceive, that the mercury, during its calcination, abforbs the falu-
brious and refpirable part of the air, or, to
fpeak more ftrictly, the bafe of this refpirable
part ; that the remaining air is a fpecies of mephitis, incapable of fupporting combuftion or
refpiration ; and confequently that atmofpheric
air is compofed of two elaftic fluids of different
and oppofite qualities. As a proof of this important truth, if we recombine thefe two elaftic
fluids, which we have feparately obtained in the
above experiment, viz. the 42 cubical inches of
mephitis, with the 8 cubical inches of refpirable
air, we reproduce an air precifely fimilar to that
of the atmofphere, and poffeffing nearly the fame
power of fupporting combuftion and refpiration,
and of contributing to the calcination of metals.

Although this experiment furnifhes us with
a very fimple means of obtaining the two principal elaftic fluids which compofe our atmofphere, feparate from each other, yet it does not
give us an exact idea of the proportion in which
thefe two enter into its compofition : For the
attraction of mercury to the refpirable part of
the air, or rather to its bafe, is not fufficiently
ftrong to overcome all the circumftances which

oppofe

oppofe this union. Thefe obftacles are the mu‑
tual adhefion of the two conftituent parts of
the atmofphere for each other, and the elective
attraction which unites the bafe of vital air with
caloric; in confequence of thefe, when the cal‑
cination ends, or is at leaft carried as far as is
poffible, in a determinate quantity of atmofphe‑
ric air, there ftill remains a portion of refpi‑
rable air united to the mephitis, which the mer‑
cury cannot feparate. I fhall afterwards fhow,
that, at leaft in our climate, the atmofpheric air
is compofed of refpirable and mephitic airs, in
the proportion of 27 and 73; and I fhall then
difcufs the caufes of the uncertainty which ftill
exifts with refpect to the exactnefs of that pro‑
portion.

Since, during the calcination of mercury, air
is decompofed, and the bafe of its refpirable
part is fixed and combined with the mercury, it
follows, from the principles already eftablifhed,
that caloric and light muft be difengaged du‑
ring the procefs : But the two following caufes
prevent us from being fenfible of this taking
place : As the calcination lafts during feveral
days, the difengagement of caloric and light,
fpread out in a confiderable fpace of time, be‑
comes extremely fmall for each particular mo‑
ment of that time, fo as not to be perceptible ;
and, in the next place, the operation being car‑
ried on by means of fire in a furnace, the heat
produced

produced by the calcination itſelf becomes confounded with that proceeding from the furnace.
I might add the reſpirable part of the air, or
rather its baſe, in entering into combination
with the mercury, does not part with all the
caloric which it contained, but ſtill retains a part
of it after forming the new compound; but the
diſcuſſion of this point, and its proofs from experiment, do not belong to this part of our
ſubject.

It is, however, eaſy to render this diſengagement of caloric and light evident to the ſenſes,
by cauſing the decompoſition of air to take
place in a more rapid manner. And for this
purpoſe, iron is excellently adapted, as it poſſeſſes a much ſtronger affinity for the baſe of
reſpirable air than mercury. The elegant experiment of Mr Ingenhouz, upon the combuſtion of iron, is well known. Take a piece of
fine iron wire, twiſted into a ſpiral, (BC, Plate
IV. Fig. 17.) fix one of its extremities B into
the cork A, adapted to the neck of the bottle
DEFG, and fix to the other extremity of the
wire C, a ſmall morſel of tinder. Matters being thus prepared, fill the bottle DEFG with
air deprived of its mephitic part; then light
the tinder, and introduce it quickly with the
wire upon which it is fixed, into the bottle
which you ſtop up with the cork A, as is ſhown
in the figure (17 Plate IV.) The inſtant the
tinder

tinder comes into contact with the vital air
it begins to burn with great intenfity ; and,
communicating the inflammation to the iron-
wire, it too takes fire, and burns rapidly, throwing out brilliant fparks, which fall to the bottom of the veffel in rounded globules, which
become black in cooling, but retain a degree of
metallic fplendour. The iron thus burnt is
more brittle even than glafs, and is eafily reduced into powder, and is ftill attractable by the
magnet, though not fo powerfully as it was before combuftion. As Mr Ingenhouz has neither examined the change produced on iron, nor
upon the air by this operation, I have repeated
the experiment under different circumftances,
in an apparatus adapted to anfwer my particular views, as follows.

Having filled a bell-glafs (A, Plate IV. Fig. 3.)
of about fix pints meafure, with pure air, or the
highly refpirable part of air, I tranfported this jar
by means of a very flat veffel, into a quickfilver
bath in the bafon BC, and I took care to render the furface of the mercury perfectly dry both
within and without the jar with blotting paper.
I then provided a fmall capfule of china-ware D,
very flat and open, in which I placed fome fmall
pieces of iron, turned fpirally, and arranged in
fuch a way as feemed moft favourable for the
combuftion being communicated to every part.
To the end of one of thefe pieces of iron was
fixed

fixed a small morsel of tinder, to which was added about the sixteenth part of a grain of phosphorus, and, by raising the bell-glass a little, the china capsule, with its contents, were introduced into the pure air. I know that, by this means, some common air must mix with the pure air in the glass; but this, when it is done dexterously, is so very trifling, as not to injure the success of the experiment. This being done, a part of the air is sucked out from the bell-glass, by means of a syphon GHI, so as to raise the mercury within the glass to EF; and, to prevent the mercury from getting into the syphon, a small piece of paper is twisted round its extremity. In sucking out the air, if the motion of the lungs only be used, we cannot make the mercury rise above an inch or an inch and a half; but, by properly using the muscles of the mouth, we can, without difficulty, cause it to rise six or seven inches.

I next took an iron wire, (MN, Plate IV. Fig. 16.) properly bent for the purpose, and making it red hot in the fire, passed it through the mercury into the receiver, and brought it in contact with the small piece of phosphorus attached to the tinder. The phosphorus instantly takes fire, which communicates to the tinder, and from that to the iron. When the pieces have been properly arranged, the whole iron burns, even to the last particle,

F

throwing

throwing out a white brilliant light fimilar to that of Chinefe fireworks. The great heat produced by this combuftion melts the iron into round globules of different fizes, moft of which fall into the China cup ; but fome are thrown out of it, and fwim upon the furface of the mercury. At the beginning of the combuftion, there is a flight augmentation in the volume of the air in the bell-glafs, from the dilatation caufed by the heat ; but, prefently afterwards, a rapid diminution of the air takes place, and the mercury rifes in the glafs ; infomuch that, when the quantity of iron is fufficient, and the air o-perated upon is very pure, almoft the whole air employed is abforbed.

It is proper to remark in this place, that, un-lefs in making experiments for the purpofe of difcovery, it is better to be contented with burn-ing a moderate quantity of iron ; for, when this experiment is pufhed too far, fo as to abforb much of the air, the cup D, which floats upon the quickfilver, approaches too near the bottom of the bell-glafs ; and the great heat produced, which is followed by a very fudden cooling, oc-cafioned by the contact of the cold mercury, is apt to break the glafs. In which cafe, the fud-den fall of the column of mercury, which hap-pens the moment the leaft flaw is produced in the glafs, caufes fuch a wave, as throws a great part of the quickfilver from the bafon. To a-void

void this inconvenience, and to enfure fuccefs to the experiment, one grofs and a half of iron is fufficient to burn in a bell-glafs, which holds about eight pints of air. The glafs ought like-wife to be ftrong, that it may be able to bear the weight of the column of mercury which it has to fupport.

By this experiment, it is not poffible to deter-mine, at one time, both the additional weight acquired by the iron, and the changes which have taken place in the air. If it is wifhed to af-certain what additional weight has been gained by the iron, and the proportion between that and the air abforbed, we muft carefully mark upon the bell-glafs, with a diamond, the height of the mercury, both before and after the expe-riment *. After this, the fyphon (GH, Pl. IV. fig. 3.) guarded, as before, with a bit of paper, to prevent its filling with mercury, is to be in-troduced under the bell-glafs, having the thumb placed upon the extremity, G, of the fyphon, to regulate the paffage of the air ; and by this means the air is gradually admitted, fo as to let the mercury fall to its level. This being done, the bell-glafs is to be carefully removed, the globules

* It will likewife be neceffary to take care that the air contained in the glafs, both before and after the experiment, be reduced to a common temperature and preffure, otherwife the refults of the following calcula-tions will be fallacious.—E.

globules of melted iron contained in the cup,
and thofe which have been fcattered about, and
fwim upon the mercury, are to be accurately
collected, and the whole is to be weighed. The
iron will be found in that ftate called *martial
ethiops* by the old chemifts, poffeffing a degree
of metallic brilliancy, very friable, and readily
reducible into powder, under the hammer, or
with a peftle and mortar. If the experiment
has fucceeded well, from 100 grains of iron will
be obtained 135 or 136 grains of ethiops, which
is an augmentation of 35 per cent.

If all the attention has been paid to this experiment which it deferves, the air will be found
diminifhed in weight exactly equal to what the
iron has gained. Having therefore burnt 100
grains of iron, which has acquired an additional weight of 35 grains, the diminution of air
will be found exactly 70 cubical inches ; and
it will be found, in the fequel, that the weight
of vital air is pretty nearly half a grain for each
cubical inch ; fo that, in effect, the augmentation of weight in the one exactly coincides with
the lofs of it in the other.

I fhall obferve here, once for all, that, in every experiment of this kind, the preffure and
temperature of the air, both before and after
the experiment, muft be reduced, by calculation, to a common ftandard of 10° (54.5°) of
the thermometer, and 28 inches of the barometer.

ter. Towards the end of this work, the manner of performing this very neceffary reduction
will be found accurately detailed.

If it be required to examine the nature of the
air which remains after this experiment, we
muft operate in a fomewhat different manner.
After the combuftion is finifhed, and the veffels
have cooled, we firft take out the cup, and the
burnt iron, by introducing the hand through
the quickfilver, under the bell-glafs ; we next
introduce fome folution of potafh, or cauftic alkali, or of the fulphuret of potafh, or fuch other
fubftance as is judged proper for examining
their action upon the refiduum of air. I fhall,
in the fequel, give an account of thefe methods
of analyfing air, when I have explained the nature of thefe different fubftances, which are only
here in a manner accidentally mentioned. After this examination, fo much water muft be
let into the glafs as will difplace the quickfilver, and then, by means of a fhallow difh placed
below the bell-glafs, it is to be removed into
the common water pneumato-chemical apparatus, where the air remaining may be examined
at large, and with great facility.

When very foft and very pure iron has been
employed in this experiment, and, if the combuftion has been performed in the pureft refpirable or vital air, free from all admixture of the
noxious or mephitic part, the air which remains
after

after the combuſtion will be found as pure as it
was before ; but it is difficult to find iron entirely free from a ſmall portion of charry matter, which is chiefly abundant in ſteel. It is
likewiſe exceedingly difficult to procure the pure
air perfectly free from ſome admixture of mephitis, with which it is almoſt always contaminated ; but this ſpecies of noxious air does not,
in the ſmalleſt degree, diſturb the reſult of the
experiment, as it is always found at the end
exactly in the ſame proportion as at the beginning.

I mentioned before, that we have two ways
of determining the conſtituent parts of atmoſpheric air, the method of analyſis, and that by
ſyntheſis. The calcination of mercury has furniſhed us with an example of each of theſe methods, ſince, after having robbed the reſpirable
part of its baſe, by means of the mercury, we
have reſtored it, ſo as to recompoſe an air preciſely ſimilar to that of the atmoſphere. But
we can equally accompliſh this ſynthetic compoſition of atmoſpheric air, by borrowing the materials of which it is compoſed from different
kingdoms of nature. We ſhall ſee hereafter
that, when animal ſubſtances are diſſolved in
the nitric acid, a great quantity of gas is diſengaged, which extinguiſhes light, and is unfit
for animal reſpiration, being exactly ſimilar to
the noxious or mephitic part of atmoſpheric air.
And, if we take 73 parts, by weight, of this e-
laſtic

laftic fluid, and mix it with 27 parts of highly refpirable air, procured from calcined mercury, we will form an elaftic fluid precifely fimilar to atmofpheric air in all its properties.

There are many other methods of feparating the refpirable from the noxious part of the atmofpheric air, which cannot be taken notice of in this part, without anticipating information, which properly belongs to the fubfequent chapters. The experiments already adduced may fuffice for an elementary treatife ; and, in matters of this nature, the choice of our evidences is of far greater confequence than their number.

I fhall clofe this article, by pointing out the property which atmofpheric air, and all the known gaffes, poffefs of diffolving water, which is of great confequence to be attended to in all experiments of this nature. Mr Sauffure found, by experiment, that a cubical foot of atmofpheric air is capable of holding 12 grains of water in folution : Other gaffes, as the carbonic acid, appear capable of diffolving a greater quantity ; but experiments are ftill wanting by which to determine their feveral proportions. This water, held in folution by gaffes, gives rife to particular phenomena in many experiments, which require great attention, and which has frequently proved the fource of great errors to chemifts in determining the refults of their experiments.

C H A P.

C H A P.   IV.

Nomenclature of the several Constituent Parts of
Atmospheric Air.

HITHERTO I have been obliged to make
use of circumlocution, to express the nature of the several substances which constitute our
atmosphere, having provisionally used the terms
of *respirable* and *noxious*, or *non-respirable parts*
of the air. But the investigations I mean to
undertake require a more direct mode of ex-
pression; and, having now endeavoured to give
simple and distinct ideas of the different substances which enter into the composition of the
atmosphere, I shall henceforth express these ideas
by words equally simple.

The temperature of our earth being very
near to that at which water becomes solid, and
reciprocally changes from solid to fluid, and
as this phenomenon takes place frequently under our observation, it has very naturally followed, that, in the languages of at least every
climate subjected to any degree of winter, a
term has been used for signifying water in the
state of solidity, when deprived of its caloric.
The same, however, has not been found neces-

sary

fary with refpect to water reduced to the ftate
of vapour by an additional dofe of caloric ; fince
thofe perfons who do not make a particular
ftudy of objects of this kind, are ftill ignorant
that water, when in a temperature only a little
above the boiling heat, is changed into an elaftic aëriform fluid, fufceptible, like all other gaffes, of being received and contained in veffels,
and preferving its gaffeous form fo long as it
remains at the temperature of 80° (212°), and
under a preffure not exceeding 28 inches of the
mercurial barometer. As this phenomenon has
not been generally obferved, no language has
ufed a particular term for expreffing water in
this ftate * ; and the fame thing occurs with all
fluids, and all fubftances, which do not evaporate in the common temperature, and under the
ufual preffure of our atmofphere.

For fimilar reafons, names have not been given
to the liquid or concrete ftates of moft of the
aëriform fluids : Thefe were not known to arife
from the combination of caloric with certain
bafes ; and, as they had not been feen either in
the liquid or folid ftates, their exiftence, under
thefe forms, was even unknown to natural philofophers.

G

We

---

* In Englifh, the word *fteam* is exclufively appropriated to water in the ftate of vapour. E.

We have not pretended to make any alteration upon fuch terms as are fanctified by ancient
cuftom; and, therefore, continue to ufe the words
water and *ice* in their common acceptation: We
likewife retain the word *air*, to exprefs that collection of elaftic fluids which compofes our atmofphere; but we have not thought it neceffary to preferve the fame refpect for modern
terms, adopted by latter philofophers, having
confidered ourfelves as at liberty to reject fuch
as appeared liable to occafion erroneous ideas
of the fubftances they are meant to exprefs, and
either to fubftitute new terms, or to employ
the old ones, after modifying them in fuch a
manner as to convey more determinate ideas.
New words have been drawn, chiefly from the
Greek language, in fuch a manner as to make
their etymology convey fome idea of what was
meant to be reprefented; and thefe we have
always endeavoured to make fhort, and of fuch
a nature as to be changeable into adjectives and
verbs.

Following thefe principles, we have, after Mr
Macquer's example, retained the term *gas*, employed by Vanhelmont, having arranged the numerous clafs of elaftic aëriform fluids under that
name, excepting only atmofpheric air. *Gas*,
therefore, in our nomenclature, becomes a generic term, expreffing the fulleft degree of faturation in any body with caloric; being, in
fact,

fact, a term expreffive of a mode of exiftence.
To diftinguifh each fpecies of gas, we employ a
fecond term from the name of the bafe, which,
faturated with caloric, forms each particular gas.
Thus, we name water combined to faturation
with caloric, fo as to form an elaftic fluid, *aqueous gas ;* ether, combined in the fame manner,
etherial gas ; the combination of alkohol with
caloric, becomes *alkoholic gas ;* and, following
the fame principles, we have *muriatic acid gas,
ammoniacal gas,* and fo on of every fubftance
fufceptible of being combined with caloric, in
fuch a manner as to affume the gaffeous or elaftic aëriform ftate.

We have already feen, that the atmofpheric
air is compofed of two gaffes, or aëriform fluids,
one of which is capable, by refpiration, of contributing to animal life, and in which metals
are calcinable, and combuftible bodies may
burn; the other, on the contrary, is endowed
with directly oppofite qualities; it cannot be
breathed by animals, neither will it admit of
the combuftion of inflammable bodies, nor of
the calcination of metals. We have given to
the bafe of the former, or refpirable portion of
the air, the name of *oxygen,* from οξυς, *acidum,*
and γινομαι, *gignor;* becaufe, in reality, one of
the moft general properties of this bafe is to
form acids, by combining with many different
fubftances. The union of this bafe with caloric

ric we term *oxygen gas,* which is the same
with what was formerly called *pure,* or *vital air.*
The weight of this gas, at the temperature of
10° (54.50), and under a preffure equal to 28
inches of the barometer, is half a grain for each
cubical inch, or one ounce and a half to each
cubical foot.

The chemical properties of the noxious portion of atmofpheric air being hitherto but little
known, we have been fatisfied to derive the
name of its bafe from its known quality of kil-
ling fuch animals as are forced to breathe it,
giving it the name of *azote,* from the Greek
privitive particle α and ζωη, *vita;* hence the name
of the noxious part of atmofpheric air is *azotic
gas;* the weight of which, in the fame temperature, and under the fame preffure, is 1 *oz.*
2 *gros.* and 48 *grs.* to the cubical foot, or
0.4444 of a grain to the cubical inch. We
cannot deny that this name appears fomewhat
extraordinary; but this muft be the cafe with
all new terms, which cannot be expected to become familiar until they have been fome time
in ufe. We long endeavoured to find a more
proper defignation without fuccefs; it was at
firft propofed to call it *alkaligen gas,* as, from
the experiments of Mr Berthollet, it appears to
enter into the compofition of ammoniac, or volatile alkali; but then, we have as yet no proof
of its making one of the conftituent elements of
the

the other alkalies ; befide, it is proved to compofe a part of the nitric acid, which gives as
good reafon to have called it *nitrigen.* For thefe
reafons, finding it neceffary to reject any name
upon fyftematic principles, we have confidered
that we run no rifk of miftake in adopting the
terms of *azote,* and *azotic gas,* which only exprefs a matter of fact, or that property which it
poffeffes, of depriving fuch animals as breathe it
of their lives.

I fhould anticipate fubjects more properly re-
ferved for the fubfequent chapters, were I in
this place to enter upon the nomenclature of the
feveral fpecies of gaffes : It is fufficient, in this
part of the work, to eftablifh the principles upon which their denominations are founded. The
principal merit of the nomenclature we have
adopted is, that, when once the fimple elementary fubftance is diftinguifhed by an appropriate term, the names of all its compounds derive
readily, and neceffarily, from this firft denomination.

CHAP.

C H A P. V.

Of the Decompofition of Oxygen Gas by Sulphur, Phofphorus, and Charcoal—and of the Forma-tion of Acids in general.

IN performing experiments, it is a neceffary principle, which ought never to be deviated from, that they be fimplified as much as poffible, and that every circumftance capable of render-ing their refults complicated be carefully remo-ved. Wherefore, in the experiments which form the objeſt of this chapter, we have never employed atmofpheric air, which is not a fimple fubftance. It is true, that the azotic gas, which forms a part of its mixture, appears to be mere-ly paffive during combuftion and calcination; but, befides that it retards thefe operations very confiderably, we are not certain but it may even alter their refults in fome circumftances; for which reafon, I have thought it neceffary to remove even this poffible caufe of doubt, by on-ly making ufe of pure oxygen gas in the follow-ing experiments, which fhow the effeſts produ-ced by combuftion in that gas; and I fhall ad-vert to fuch differences as take place in the re-fults of thefe, when the oxygen gas, or pure

vital

vital air, is mixed, in different proportions, with azotic gas.

Having filled a bell-glafs (A. Pl. iv. fig. 3), of between five and fix pints meafure, with oxygen gas, I removed it from the water trough, where it was filled, into the quickfilver bath, by means of a fhallow glafs difh flipped underneath, and having dried the mercury, I introduced 61¼ grains of Kunkel's phofphorus in two little China cups, like that reprefented at D, fig. 3. under the glafs A; and that I might fet fire to each of the portions of phofphorus feparately, and to prevent the one from catching fire from the other, one of the difhes was covered with a piece of flat glafs. I next raifed the quickfilver in the bell-glafs up to E F, by fucking out a fufficient portion of the gas by means of the fyphon G H I. After this, by means of the crooked iron wire (fig. 16.), made red hot, I fet fire to the two portions of phofphorus fucceffively, firft burning that portion which was not covered with the piece of glafs. The combuftion was extremely rapid, attended with a very brilliant flame, and confiderable difengagement of light and heat. In confequence of the great heat induced, the gas was at firft much dilated, but foon after the mercury returned to its level, and a confiderable abforption of gas took place; at the fame time, the whole

whole infide of the glafs became covered with
white light flakes of concrete phofphoric acid.

At the beginning of the experiment, the
quantity of oxygen gas, reduced, as above directed, to a common ftandard, amounted to 162
cubical inches; and, after the combuftion was
finifhed, only $23\frac{1}{4}$ cubical inches, likewife reduced to the ftandard, remained; fo that the quantity of oxygen gas abforbed during the combuftion was $138\frac{3}{4}$ cubical inches, equal to 69.375
grains.

A part of the phofphorus remained uncon-
fumed in the bottom of the cups, which being
wafhed on purpofe to feparate the acid, weighed about $16\frac{1}{4}$ grains; fo that about 45 grains
of phofphorus had been burned: But, as it is
hardly poffible to avoid an error of one or two
grains, I leave the quantity fo far qualified.
Hence, as nearly 45 grains of phofphorus had,
in this experiment, united with 69.375 grains
of oxygen, and as no gravitating matter could
have efcaped through the glafs, we have a right
to conclude, that the weight of the fubftance refulting from the combuftion in form of white
flakes, muft equal that of the phofphorus and
oxygen employed, which amounts to 114.375
grains. And we fhall prefently find, that thefe
flakes confifted entirely of a folid or concrete
acid. When we reduce thefe weights to hundredth parts, it will be found, that 100 parts of
phofphorus

phofphorus require 154 parts of oxygen for faturation, and that this combination will produce
254 parts of concrete phofphoric acid, in form
of white fleecy flakes.

This experiment proves, in the moft convincing manner, that, at a certain degree of temperature, oxygen poffeffes a ftronger elective attraction, or affinity, for phofphorus than for caloric; that, in confequence of this, the phofphorus attracts the bafe of oxygen gas from the
caloric, which, being fet free, fpreads itfelf over
the furrounding bodies. But, though this experiment be fo far perfectly conclufive, it is not
fufficiently rigorous, as, in the apparatus defcribed, it is impoffible to afcertain the weight
of the flakes of concrete acid which are formed;
we can therefore only determine this by calculating the weights of oxygen and phofphorus
employed; but as, in phyfics, and in chemiftry,
it is not allowable to fuppofe what is capable of
being afcertainea by direct experiment, I thought
it neceffary to rep at this experiment, as follows,
upon a larger fcale, and by means of a different
apparatus.

I took a large glafs baloon (A. Pl. iv. fig. 4.)
with an opening three inches diameter, to which
was fitted a cryftal ftopper ground with emery,
and pierced with two holes for the tubes yyy,
xxx. Before fhutting the baloon with its ftopper, I introduced the fupport BC, furmounted

H   by

by the china cup D, containing 150 *grs.* of
phofphorus; the ftopper was then fitted to the
opening of the baloon, luted with fat lute, and
covered with flips of linen fpread with quick-
lime and white of eggs: When the lute was
perfectly dry, the weight of the whole apparatus was determined to within a grain, or a grain
and a half. I next exhaufted the baloon, by
means of an air pump applied to the tube xxx,
and then introduced oxygen gas by means of
the tube yyy, having a ftop cock adapted to it.
This kind of experiment is moft readily and
moft exactly performed by means of the hydro-
pneumatic machine defcribed by Mr Meufnier
and me in the Memoirs of the Academy for
1782, pag. 466. and explained in the latter
part of this work, with feveral important additions and corrections fince made to it by Mr
Meufnier. With this inftrument we can readily afcertain, in the moft exact manner, both the
quantity of oxygen gas introduced into the baloon, and the quantity confumed during the
courfe of the experiment.

When all things were properly difpofed, I
fet fire to the phofphorus with a burning glafs.
The combuftion was extremely rapid, accompanied with a bright flame, and much heat; as
the operation went on, large quantities of white
flakes attached themfelves to the inner furface
of the baloon, fo that at laft it was rendered
quite

quite opake. The quantity of thefe flakes at
laft became fo abundant, that, although frefh
oxygen gas was continually fupplied, which
ought to have fupported the combuftion, yet the
phofphorus was foon extinguifhed. Having allowed the apparatus to cool completely, I firft
afcertained the quantity of oxygen gas employed, and weighed the baloon accurately, before
it was opened. I next wafhed, dried, and
weighed the fmall quantity of phofphorus remaining in the cup, on purpofe to determine the
whole quantity of phofphorus confumed in the
experiment; this refiduum of the phofphorus
was of a yellow ochrey colour. It is evident,
that by thefe feveral precautions, I could eafily
determine, 1ft, the weight of the phofphorus
confumed; 2d, the weight of the flakes produced by the combuftion; and, 3d, the weight
of the oxygen which had combined with the
phofphorus. This experiment gave very nearly
the fame refults with the former, as it proved
that the phofphorus, during its combuftion, had
abforbed a little more than one and a half its
weight of oxygen; and I learned with more
certainty, that the weight of the new fubftance,
produced in the experiment, exactly equalled
the fum of the weights of the phofphorus confumed, and oxygen abforbed, which indeed was
eafily determinable *a priori*. If the oxygen gas
employed be pure, the refiduum after combuftion

tion is as pure as the gas employed; this proves that nothing efcapes from the phofphorus, capable of altering the purity of the oxygen gas, and that the only action of the phofphorus is to feparate the oxygen from the caloric, with which it was before united.

I mentioned above, that when any combuftible body is burnt in a hollow fphere of ice, or in an apparatus properly conftructed upon that principle, the quantity of ice melted during the combuftion is an exact meafure of the quantity of caloric difengaged. Upon this head, the memoir given by M. de la Place and me, A°. 1780, p. 355, may be confulted. Having fubmitted the combuftion of phofphorus to this trial, we found that one pound of phofphorus melted a little more than 100 pounds of ice during its combuftion.

The combuftion of phofphorus fucceeds equally well in atmofpheric air as in oxygen gas, with this difference, that the combuftion is vaftly flower, being retarded by the large proportion of azotic gas mixed with the oxygen gas, and that only about one-fifth part of the air employed is abforbed, becaufe as the oxygen gas only is abforbed, the proportion of the azotic gas becomes fo great toward the clofe of the experiment, as to put an end to the combuftion.

I

I have already fhown, that phofphorus is
changed by combuftion into an extremely light,
white, flakey matter; and its properties are
entirely altered by this transformation : From
being infoluble in water, it becomes not only
foluble, but fo greedy of moifture, as to attract
the humidity of the air with aftonifhing rapidity ; by this means it is converted into a liquid,
confiderably more denfe, and of more fpecific
gravity than water. In the ftate of phofphorus
before combuftion, it had fcarcely any fenfible
tafte, by its union with oxygen it acquires an
extremely fharp and four tafte : in a word,
from one of the clafs of combuftible bodies, it
is changed into an incombuftible fubftance,
and becomes one of thofe bodies called acids.

This property of a combuftible fubftance to
be converted into an acid, by the addition of
oxygen, we fhall prefently find belongs to a
great number of bodies : Wherefore, ftrict logic requires that we fhould adopt a common
term for indicating all thefe operations which
produce analogous refults ; this is the true way
to fimplify the ftudy of fcience, as it would be
quite impoffible to bear all its fpecifical details
in the memory, if they were not claffically
arranged. For this reafon, we fhall diftinguifh
this converfion of phofphorus into an acid, by
its union with oxygen, and in general every
combination of oxygen with a combuftible fubftance,

ftance, by the term of *oxygenation :* from which I fhall adopt the verb to *oxygenate,* and of confequence fhall fay, that in *oxygenating* phofphorus we convert it into an acid.

Sulphur is likewife a combuftible body, or, in other words, it is a body which poffeffes the power of decompofing oxygen gas, by attracting the oxygen from the caloric with which it was combined. This can very eafily be proved, by means of experiments quite fimilar to thofe we have given with phofphorus ; but it is neceffary to premife, that in thefe operations with fulphur, the fame accuracy of refult is not to be expected as with phofphorus ; becaufe the acid which is formed by the combuftion of fulphur is difficultly condenfible, and becaufe fulphur burns with more difficulty, and is foluble in the different gaffes. But I can fafely affert, from my own experiments, that fulphur in burning abforbs oxygen gas ; that the refulting acid is confiderably heavier than the fulphur burnt ; that its weight is equal to the fum of the weights of the fulphur which has been burnt, and of the oxygen abforbed ; and, laftly, that this acid is weighty, incombuftible, and mifcible with water in all proportions : The only uncertainty remaining upon this head, is with regard to the proportions of fulphur and of oxygen which enter into the compofition of the acid.

Charcoal,

Charcoal, which, from all our prefent knowledge regarding it, muft be confidered as a fimple combuftible body, has likewife the property
of decompofing oxygen gas, by abforbing its
bafe from the caloric : But the acid refulting
from this combuftion does not condenfe in the
common temperature; under the preffure of
our atmofphere, it remains in the ftate of gas,
and requires a large proportion of water to
combine with or be diffolved in. This acid
has, however, all the known properties of other
acids, though in a weaker degree, and combines, like them, with all the bafes which are fufceptible of forming neutral falts.

The combuftion of charcoal in oxygen gas,
may be effected like that of phofphorus in the
bell-glafs, (A. Pl. IV. fig. 3.) placed over mercury : but, as the heat of red hot iron is not
fufficient to fet fire to the charcoal, we muft
add a fmall morfel of tinder, with a minute particle of phofphorus, in the fame manner
as directed in the experiment for the combuftion of iron. A detailed account of this experiment will be found in the memoirs of the
academy for 1781, p. 448. By that experiment it appears, that 28 parts by weight of
charcoal require 72 parts of oxygen for faturation, and that the aëriform acid produced is
precifely equal in weight to the fum of the
weights of the charcoal and oxygen gas employed.

ployed. This aëriform acid was called fixed
or fixable air by the chemifts who firft difcovered it; they did not then know whether it was
air refembling that of the atmofphere, or fome
other elaftic fluid, vitiated and corrupted by combuftion; but fince it is now afcertained to be
an acid, formed like all others by the oxygenation of its peculiar bafe, it is obvious that the
name of fixed air is quite ineligible*.

By burning charcoal in the apparatus mentioned p. 60, Mr de la Place and I found that
one *lib.* of charcoal melted 96 *libs.* 6 *oz.* of
ice; that, during the combuftion, 2 *libs.* 9 *oz.*
1 *gros.* 10 *grs.* of oxygen were abforbed, and
that 3 *libs.* 9 *oz.* 1 *gros.* 10 *grs.* of acid gas
were formed. This gas weighs 0.695 parts
of a grain for each cubical inch, in the common ftandard temperature and preffure mentioned above, fo that 34,242 cubical inches of
acid gas are produced by the combuftion of one
pound of charcoal.

I might multiply thefe experiments, and fhow
by a numerous fucceffion of facts, that all acids
are formed by the combuftion of certain fubftances; but I am prevented from doing fo in
this

* It may be proper to remark, though here omitted by the author, that, in conformity with the general
principles of the new nomenclature, this acid is by Mr
Lavoifier and his coleagues called the carbonic acid,
and when in the aëriform ftate carbonic acid gas. E.

place, by the plan which I have laid down, of proceeding only from facts already afcertained, to fuch as are unknown, and of drawing my examples only from circumftances already explained. In the mean time, however, the three examples above cited may fuffice for giving a clear and accurate conception of the manner in which acids are formed. By thefe it may be clearly feen, that oxygen is an element common to them all, which conftitutes their acidity; and that they differ from each other, according to the nature of the oxygenated or acidified fubftance. We muft therefore, in every acid, carefully diftinguifh between the acidifiable, bafe, which Mr de Morveau calls the radical, and the acidifiing principle or oxygen.

CHAP.

C H A P. VI.

Of the Nomenclature of Acids in general, and parti-
cularly of thofe drawn from Nitre and Sea-Salt.

IT becomes extremely eafy, from the principles laid down in the preceding chapter,
to eftablifh a fyftematic nomenclature for the a-
cids: The word *acid*, being ufed as a generic term,
each acid falls to be diftinguifhed in language, as
in nature, by the name of its bafe or radical.
Thus, we give the generic name of acids to the
products of the combuftion or oxygenation of
phofphorus, of fulphur, and of charcoal ; and
thefe products are refpectively named, the *phof-*
phoric acid, the *fulphuric acid*, and the *carbonic*
acid.

There is however, a remarkable circumftance
in the oxygenation of combuftible bodies, and
of a part of fuch bodies as are convertible into acids, that they are fufceptible of different
degrees of faturation with oxygen, and that the
refulting acids, though formed by the union of
the fame elements, are poffeffed of different properties, depending upon that difference of proportion. Of this, the phofphoric acid, and more
efpecially the fulphuric, furnifhes us with examples.

amples. When fulphur is combined with a fmall proportion of oxygen, it forms, in this firft or lower degree of oxygenation, a volatile acid, having a penetrating odour, and poffeffed of very particular qualities. By a larger proportion of oxygen, it is changed into a fixed, heavy acid, without any odour, and which, by combination with other bodies, gives products quite different from thofe furnifhed by the former. In this inftance, the principles of our nomenclature feem to fail; and it feems difficult to derive fuch terms from the name of the acidifiable bafe, as fhall diftinctly exprefs thefe two degrees of faturation, or oxygenation, without circumlocution. By reflection, however, upon the fubject, or perhaps rather from the neceffity of the cafe, we have thought it allowable to exprefs thefe varieties in the oxygenation of the acids, by fimply varying the termination of their fpecific names. The volatile acid produced from fulphur was anciently known to Stahl under the name of *fulphurous* acid *. We have

pre-

---

* The term formerly ufed by the Englifh chemifts for this acid was written *fulphureous*; but we have thought proper to fpell it as above, that it may better conform with the fimilar terminations of nitrous, carbonous, &c. to be ufed hereafter. In general, we have ufed the Englifh terminations *ic* and *ous* to tranflate the terms of the Author which end with *ique* and *eux*, with hardly any other alterations.—E.

preferved that term for this acid from fulphur un-
der-faturated with oxygen ; and diftinguifh the
other, or completely faturated or oxygenated a-
cid, by the name of *fulphuric* acid. We fhall
therefore fay, in this new chemical language,
that fulphur, in combining with oxygen, is fufceptible of two degrees of faturation ; that the
firft, or lefTer degree, conftitutes fulphurous a-
cid, which is volatile and penetrating ; whilft
the fecond, or higher degree of faturation, produces fulphuric acid, which is fixed and inodo-
rous. We fhall adopt this difference of termination for all the acids which affume feveral degrees of faturation. Hence we have a phofphorous and a phofphoric acid, an acetous and
an acetic acid ; and fo on, for others in fimilar
circumftances.

This part of chemical fcience would have
been extremely fimple, and the nomenclature
of the acids would not have been at all per-
plexed, as it is now in the old nomenclature, if
the bafe or radical of each acid had been known
when the acid itfelf was difcovered. Thus, for
inftance, phofphorus being a known fubftance
before the difcovery of its acid, this latter was
rightly diftinguifhed by a term drawn from the
name of its acidifiable bafe. But when, on the
contrary, an acid happened to be difcovered before its bafe, or rather, when the acidifiable bafe
from which it was formed remained unknown,

names

names were adopted for the two, which have not the fmalleft connection; and thus, not only the memory became burthened with ufelefs appellations, but even the minds of ftudents, nay even of experienced chemifts, became filled with falfe ideas, which time and reflection alone is capable of eradicating. We may give an inftance of this confufion with refpect to the acid fulphur: The former chemifts having procured this acid from the vitriol of iron, gave it the name of the vitriolic acid from the name of the fubftance which produced it ; and they were then ignorant that the acid procured from fulphur by combuftion was exactly the fame.

The fame thing happened with the aëriform acid formerly called *fixed air;* it not being known that this acid was the refult of combining charcoal with oxygen, a variety of denominations have been given to it, not one of which conveys juft ideas of its nature or origin. We have found it extremely eafy to correct and modify the ancient language with refpect to thefe acids proceeding from known bafes, having converted the name of *vitriolic acid* into that of *fulphuric,* and the name of *fixed air* into that of *carbonic acid;* but it is impof- fible to follow this plan with the acids whofe bafes are ftill unknown ; with thefe we have been obliged to ufe a contrary plan, and, inftead of forming the name of the acid from that of its bafe,

bafe, have been forced to denominate the unknown bafe from the name of the known acid,
as happens in the cafe of the acid which is procured from fea falt.

To difengage this acid from the alkaline bafe
with which it is combined, we have only to
pour fulphuric acid upon fea-falt, immediately
a brifk effervefcence takes place, white vapours
arife, of a very penetrating odour, and, by only gently heating the mixture, all the acid is
driven off. As, in the common temperature
and preffure of our atmofphere, this acid is naturally in the ftate of gas, we muft ufe particular precautions for retaining it in proper veffels.
For fmall experiments, the moft fimple and
moft commodious apparatus confifts of a fmall
retort G, (Pl. V. Fig. 5.), into which the fea-
falt is introduced, well dried *, we then pour
on fome concentrated fulphuric acid, and immediately introduce the beak of the retort under
little jars or bell-glaffes A, (fame Plate and Fig.).
previoufly filled with quickfilver. In proportion as the acid gas is difengaged, it paffes into
the jar, and gets to the top of the quickfilver,
which it difplaces. When the difengagement
of

---

* For this purpofe, the operation called *decrepitation*
is ufed, which confifts in fubjecting it to nearly a red
heat, in a proper veffel, fo as to evaporate all its water
of cryftallization.—E.

of the gas flackens, a gentle heat is applied
to the retort, and gradually increafed till nothing more paffes over. This acid gas has a
very ftrong affinity with water, which abforbs
an enormous quantity of it, as is proved by introducing a very thin layer of water into the
glafs which contains the gas ; for, in an inftant,
the whole acid gas difappears, and combines
with the water.

This latter circumftance is taken advantage
of in laboratories and manufactures, on purpofe
to obtain the acid of fea-falt in a liquid form ;
and for this purpofe the apparatus (Pl. IV.
Fig. 1.) is employed. It confifts, 1ft, of a tubulated retort A, into which the fea-falt, and after it the fulphuric acid, are introduced through
the opening H; 2d, of the baloon or recipient
c, b, intended for containing the fmall quantity
of liquid which paffes over during the procefs ;
and, 3d, of a fet of bottles, with two mouths,
L, L, L, L, half filled with water, intended for
abforbing the gas difengaged by the diftillation. This apparatus will be more amply defcribed in the latter part of this work.

Although we have not yet been able, either
to compofe or to decompound this acid of fea-
falt, we cannot have the fmalleft doubt that it,
like all other acids, is compofed by the union
of oxygen with an acidifiable bafe. We have
therefore called this unknown fubftance the
muriatic

muriatic bafe, or *muriatic radical,* deriving this name, after the example of Mr Bergman and Mr de Morveau, from the Latin word *muria,* which was anciently ufed to fignify fea-falt. Thus, without being able exactly to determine the component parts of *muriatic acid,* we defign, by that term, a volatile acid, which retains the form of gas in the common temperature and preffure of our atmofphere, which combines with great facility, and in great quantity, with water, and whofe acidifiable bafe adheres fo very intimately with oxygen, that no method has hitherto been devifed for feparating them. If ever this acidifiable bafe of the muriatic acid is difcovered to be a known fubftance, though now unknown in that capacity, it will be requifite to change its prefent denomination for one analogous with that of its bafe.

In common with fulphuric acid, and feveral other acids, the muriatic is capable of different degrees of oxygenation ; but the excefs of oxygen produces quite contrary effects upon it from what the fame circumftance produces upon the acid of fulphur. The lower degree of oxygenation converts fulphur into a volatile gaffeous acid, which only mixes in fmall proportions with water, whilft a higher oxygenation forms an acid poffeffing much ftronger acid properties, which is very fixed and cannot remain in the ftate of gas but in a very high temperature, which has

no

no fmell, and which mixes in large proportion with water. With muriatic acid, the direct reverfe takes place ; an additional faturation with oxygen renders it more volatile, of
a more penetrating odour, lefs mifcible with
water, and diminifhes its acid properties. We
were at firft inclined to have denominated thefe
two degrees of faturation in the fame manner
as we had done with the acid of fulphur, calling
the lefs oxygenated *muriatous acid*, and that
whieh is more faturated with oxygen *muriatic
acid :* But, as this latter gives very particular
refults in its combinations, and as nothing analogous to it is yet known in chemiftry, we have
left the name of muriatic acid to the lefs faturated, and give the latter the more compounded
appellation of *oxygenated muriatic acid.*

Although the bafe or radical of the acid
which is extracted from nitre or faltpetre be
better known, we have judged proper only
to modify its name in the fame manner
with that of the muriatic acid. It is drawn
from nitre, by the intervention of fulphuric
acid, by a procefs fimilar to that defcribed
for extracting the muriatic acid, and by means
of the fame apparatus (Pl. IV. Fig. 1.). In
proportion as the acid paffes over, it is in part
condenfed in the baloon or recipient, and the
reft is abforbed by the water contained in the
bottles L, L, L, L ; the water becomes firft green,

then blue, and at laſt yellow, in proportion to
the concentration of the acid. During this operation, a large quantity of oxygen gas, mixed
with a ſmall proportion of azotic gas, is diſengaged.

This acid, like all others, is compoſed of oxygen, united to an acidifiable baſe, and is even
the firſt acid in which the exiſtence of oxygen
was well aſcertained. Its two conſtituent elements are but weakly united, and are eaſily ſeparated, by preſenting any ſubſtance with which
oxygen has a ſtronger affinity than with the acidifiable baſe peculiar to this acid. By ſome experiments of this kind, it was firſt diſcovered
that azote, or the baſe of mephitis or azotic gas,
conſtituted its acidifiable baſe or radical ; and
conſequently that the acid of nitre was really an
azotic acid, having azote for its baſe, combined
with oxygen. For theſe reaſons, that we might
be conſiſtent with our principles, it appeared
neceſſary, either to call the acid by the name of
azotic, or to name the baſe *nitric radical ;* but
from either of theſe we were diſſuaded, by the
following conſiderations. In the *firſt* place, it
ſeemed difficult to change the name of nitre or
ſaltpetre, which has been univerſally adopted
in ſociety, in manufactures, and in chemiſtry ;
and, on the other hand, azote having been diſcovered by Mr Berthollet to be the baſe of volatile alkali, or ammoniac, as well as of this a-
cid,

cid, we thought it improper to call it nitric radical. We have therefore continued the term of azote to the bafe of that part of atmofpheric air which is likewife the nitric and ammoniacal radical ; and we have named the acid of nitre, in its lower and higher degrees of oxygenation, *nitrous acid* in the former, and *nitric acid* in the latter ftate ; thus preferving its former appellation properly modified.

Several very refpectable chemifts have difapproved of this deference for the old terms, and wifhed us to have perfevered in perfecting a new chemical language, without paying any refpect for ancient ufage ; fo that, by thus fteering a kind of middle courfe, we have expofed ourfelves to the cenfures of one fect of chemifts, and to the expoftulations of the oppofite party.

The acid of nitre is fufceptible of affuming a great number of feparate ftates, depending upon its degree of oxygenation, or upon the proportions in which azote and oxygen enter into its compofition. By a firft or loweft degree of oxygenation, it forms a particular fpecies of gas, which we fhall continue to name *nitrous gas* ; this is compofed nearly of two parts, by weight, of oxygen combined with one part of azote ; and in this ftate it is not mifcible with water. In this gas, the azote is by no means faturated with oxygen, but, on the contrary, has
ftill

ftill a very great affinity for that element, and
even attracts it from atmofpheric air, immediately upon getting into contact with it. This
combination of nitrous gas with atmofpheric air
has even become one of the methods for determining the quantity of oxygen contained in air,
and confequently for afcertaining its degree of
falubrity.

This addition of oxygen converts the nitrous
gas into a powerful acid, which has a ftrong affinity with water, and which is itfelf fufceptible
of various additional degrees of oxygenation.
When the proportions of oxygen and azote is
below three parts, by weight, of the former, to
one of the latter, the acid is red coloured, and
emits copious fumes. In this ftate, by the application of a gentle heat, it gives out nitrous
gas; and we term it, in this degree of oxygenation, *nitrous acid.* When four parts, by
weight, of oxygen, are combined with one part
of azote, the acid is clear and colourlefs, more
fixed in the fire than the nitrous acid, has lefs
odour, and its conftituent elements are more
firmly united. This fpecies of acid, in conformity with our principles of nomenclature, is
called *nitric acid.*

Thus, nitric acid is the acid of nitre, furcharged with oxygen; nitrous acid is the acid
of nitre furcharged with azote; or, what is the
fame thing, with nitrous gas; and this latter is
azote

azote not fufficiently faturated with oxygen to poffefs the properties of an acid. To this degree of oxygenation, we have afterwards, in the courfe of this work, given the generical name of *oxyd* *.

CHAP.

---

* In ftrict conformity with the principles of the new nomenclature, but which the Author has given his reafons for deviating from in this inftance, the following ought to have been the terms for azote, in its feveral degrees of oxygenation : Azote, azotic gas, (azote combined with caloric), azotic oxyd gas, nitrous acid, and nitric acid.—E.

C H A P.   VII.

Of the Decompofition of Oxygen Gas by means of Metals, and the Formation of Metallic Oxyds.

OXYGEN has a ftronger affinity with metals heated to a certain degree than with
caloric ; in confequence of which, all metallic
bodies, excepting gold, filver, and platina, have
the property of decompofing oxygen gas, by attracting its bafe from the caloric with which it
was combined. We have already fhown in
what manner this decompofition takes place, by
means of mercury and iron ; having obferved,
that, in the cafe of the firft, it muft be confidered as a kind of gradual combuftion, whilft,
in the latter, the combuftion is extremely rapid,
and attended with a brilliant flame. The ufe
of the heat employed in thefe operations is to
feparate the particles of the metal from each
other, and to diminifh their attraction of cohefion or aggregation, or, what is the fame thing,
their mutual attraction for each other.

The abfolute weight of metallic fubftances is
augmented in proportion to the quantity of
oxygen they abforb ; they, at the fame time, lofe
their metallic fplendour, and are reduced into
an

an earthy pulverulent matter. In this ftate metals muft not be confidered as entirely faturated
with oxygen, becaufe their action upon this element is counterbalanced by the power of affinity
between it and caloric. During the calcination
of metals, the oxygen is therefore acted upon
by two feparate and oppofite powers, that of its
attraction for caloric, and that exerted by the
metal, and only tends to unite with the latter in
confequence of the excefs of the latter over the
former, which is, in general, very inconfiderable. Wherefore, when metallic fubftances are
oxygenated in atmofpheric air, or in oxygen
gas, they are not converted into acids like fulphur, phofphorus, and charcoal, but are only
changed into intermediate fubftances, which,
though approaching to the nature of falts, have
not acquired all the faline properties. The old
chemifts have affixed the name of *calx* not only
to metals in this ftate, but to every body which
has been long expofed to the action of fire without being melted. They have converted this
word *calx* into a generical term, under which
they confound calcareous earth, which, from a
neutral falt, which it really was before calcination, has been changed by fire into an earthy
alkali, by *lofing* half of its weight, with metals
which, by the fame means, have joined themfelves to a new fubftance, whofe quantity often
exceeds half their weight, and by which they
have

have been changed almoſt into the nature of acids. This mode of claſſifying ſubſtances of ſo very oppoſite natures, under the ſame generic name, would have been quite contrary to our principles of nomenclature, eſpecially as, by retaining the above term for this ſtate of metallic ſubſtances, we muſt have conveyed very falſe ideas of its nature. We have, therefore, laid aſide the expreſſion *metallic calx* altogether, and have ſubſtituted in its place the term *oxyd*, from the Greek word οξυς.

By this may be ſeen, that the language we have adopted is both copious and expreſſive. The firſt or loweſt degree of oxygenation in bodies, converts them into *oxyds ;* a ſecond degree of additional oxygenation conſtitutes the claſs of acids, of which the ſpecific names, drawn from their particular baſes, terminate in *ous*, as the *nitrous* and *ſulphurous* acids ; the third degree of oxygenation changes theſe into the ſpecies of acids diſtinguiſhed by the termination in *ic*, as the *nitric* and *ſulphuric* acids ; and, laſtly, we can expreſs a fourth, or higheſt degree of oxygenation, by adding the word *oxygenated* to the name of the acid, as has been already done with the *oxygenated muriatic* acid.

We have not confined the term *oxyd* to expreſſing the combinations of metals with oxygen, but have extended it to ſignify that firſt degree of oxygenation in all bodies, which, without

without converting them into acids, caufes them
to approach to the nature of falts. Thus, we give
the name of *oxyd of fulphur* to that foft fubftance
into which fulphur is converted by incipient
combuftion ; and we call the yellow matter left
by phofphorus, after combuftion, by the name
of *oxyd of phofphorus*. In the fame manner, nitrous gas, which is azote in its firft degree of
oxygenation, is the *oxyd of azote*. We have likewife oxyds in great numbers from the vegetable
and animal kingdoms ; and I fhall fhow, in the
fequel, that this new language throws great
light upon all the operations of art and nature.

We have already obferved, that almoft all
the metallic oxyds have peculiar and permanent
colours. Thefe vary not only in the different
fpecies of metals, but even according to the various degrees of oxygenation in the fame metal.
Hence we are under the neceffity of adding two
epithets to each oxyd, one of which indicates
the metal *oxydated* *, while the other indicates

L   the

---

* Here we fee the word *oxyd* converted into the
verb *to oxydate, oxydated, oxydating*, after the fame manner with the derivation of the verb *to oxygenate, oxygenated, oxygenating*, from the word *oxygen*. I am not clear
of the abfolute neceffity of this fecond verb here firft
introduced, but think, in a work of this nature, that it
is the duty of the tranflator to neglect every other confideration for the fake of ftrict fidelity to the ideas of
his author.—E.

the peculiar colour of the oxyd. Thus, we have the black oxyd of iron, the red oxyd of iron, and the yellow oxyd of iron; which expreſſions reſpectively anſwer to the old unmeaning terms of martial ethiops, colcothar, and ruſt of iron, or ochre. We have likewiſe the gray, yellow, and red oxyds of lead, which anſwer to the equally falſe or inſignificant terms, aſhes of lead, maſſicot, and minium.

Theſe denominations ſometimes become rather long, eſpecially when we mean to indicate whether the metal has been oxydated in the air, by detonation with nitre, or by means of acids; but then they always convey juſt and accurate ideas of the correſponding object which we wiſh to expreſs by their uſe. All this will be rendered perfectly clear and diſtinct by means of the tables which are added to this work.

CHAP.

C H A P.   VIII.

Of the Radical Principle of Water, and of its De-
compofition by Charcoal and Iron.

UNTIL very lately, water has always been
thought a fimple fubftance, infomuch
that the older chemifts confidered it as an element. Such it undoubtedly was to them, as
they were unable to decompofe it ; or, at leaft,
fince the decompofition which took place daily
before their eyes was entirely unnoticed. But
we mean to prove, that water is by no means a
fimple or elementary fubftance. I fhall not here
pretend to give the hiftory of this recent, and
hitherto contefted difcovery, which is detailed
in the Memoirs of the Academy for 1781, but
fhall only bring forwards the principal proofs of
the decompofition and compofition of water ;
and, I may venture to fay, that thefe will be
convincing to fuch as confider them impartially.

Experiment Firft.

Having fixed the glafs tube EF, (Pl. vii. fig.
11.) of from 8 to 12 lines diameter, acrofs a
furnace, with a fmall inclination from E to F,

lute

lute the fuperior extremity E to the glafs retort A, containing a determinate quantity of diftilled water, and to the inferior extremity F, the worm SS fixed into the neck of the doubly tubulated bottle H, which has the bent tube KK adapted to one of its openings, in fuch a manner as to convey fuch aëriform fluids or gaffes as may be difengaged, during the experiment, into a proper apparatus for determining their quantity and nature.

To render the fuccefs of this experiment certain, it is neceffary that the tube EF be made of well annealed and difficultly fufible glafs, and that it be coated with a lute compofed of clay mixed with powdered ftone-ware; befides which, it muft be fupported about its middle by means of an iron bar paffed through the furnace, left it fhould foften and bend during the experiment. A tube of China-ware, or porcellain, would anfwer better than one of glafs for this experiment, were it not difficult to procure one fo entirely free from pores as to prevent the paffage of air or of vapours.

When things are thus arranged, a fire is lighted in the furnace EFCD, which is fupported of fuch a ftrength as to keep the tube EF red hot, but not to make it melt; and, at the fame time, fuch a fire is kept up in the furnace VVXX, as to keep the water in the retort A continually boiling.

In

In proportion as the water in the retort A is evaporated, it fills the tube EF, and drives out the air it contained by the tube KK; the aqueous gas formed by evaporation is condenfed by cooling in the worm SS, and falls, drop by drop, into the tubulated bottle H. Having continued this operation until all the water be evaporated from the retort, and having carefully emptied all the veffels employed, we find that a quantity of water has paffed over into the bottle H, exactly equal to what was before contained in the retort A, without any difengagement of gas whatfoever: So that this experiment turns out to be a fimple diftillation; and the refult would have been exactly the fame, if the water had been run from one veffel into the other, through the tube EF, without having undergone the intermediate incandefcence.

Experiment Second.

The apparatus being difpofed, as in the former experiment, 28 *grs.* of charcoal, broken into moderately fmall parts, and which has previoufly been expofed for a long time to a red heat in clofe veffels, are introduced into the tube EF. Every thing elfe is managed as in the preceding experiment.

The water contained in the retort A is diftilled, as in the former experiment, and, being
condenfed

condenfed in the worm, falls into the bottle H;
but, at the fame time, a confiderable quantity
of gas is difengaged, which, efcaping by the tube
KK, is received in a convenient apparatus for
that purpofe. After the operation is finifhed,
we find nothing but a few atoms of afhes remaining in the tube EF; the 28 *grs.* of charcoal
having entirely difappeared.

When the difengaged gaffes are carefully examined, they are found to weigh 113.7 *grs.* *;
thefe are of two kinds, viz. 144 cubical inches
of carbonic acid gas, weighing 100 *grs.* and 380
cubical inches of a very light gas, weighing only 13.7 *grs.* which takes fire when in contact
with air, by the approach of a lighted body;
and, when the water which has paffed over into
the bottle H is carefully examined, it is found
to have loft 85.7 *grs.* of its weight. Thus, in
this experiment, 85.7 *grs.* of water, joined to 28
grs. of charcoal, have combined in fuch a way
as to form 100 *grs.* of carbonic acid, and 13.7
grs. of a particular gas capable of being burnt.

I have already fhown, that 100 *grs.* of carbonic acid gas confifts of 72 *grs.* of oxygen, combined with 28 *grs.* of charcoal; hence the 28

grs.

---

* In the latter part of this work will be found a
particular account of the proceffes neceffary for feparating the different kinds of gaffes, and for determining their quantities.—A.

grs. of charcoal placed in the glaſs tube have acquired 72 *grs.* of oxygen from the water; and it follows, that 85.7 *grs.* of water are compoſed of 72 *grs.* of oxygen, combined with 13.7 *grs.* of a gas fuſceptible of combuſtion. We ſhall ſee preſently that this gas cannot poſſibly have been diſengaged from the charcoal, and muſt, conſequently, have been produced from the water.

I have ſuppreſſed ſome circumſtances in the above account of this experiment, which would only have complicated and obſcured its reſults in the minds of the reader. For inſtance, the inflammable gas diſſolves a very ſmall part of the charcoal, by which means its weight is ſomewhat augmented, and that of the carbonic gas proportionally diminiſhed. Altho' the alteration produced by this circumſtance is very inconſiderable; yet I have thought it neceſſary to determine its effects by rigid calculation, and to report, as above, the reſults of the experiment in its ſimplified ſtate, as if this circumſtance had not happened. At any rate, ſhould any doubts remain reſpecting the conſequences I have drawn from this experiment, they will be fully diſſipated by the following experiments, which I am going to adduce in ſupport of my opinion.

Experiment

Experiment Third.

The apparatus being difpofed exactly as in the former experiment, with this difference, that inftead of the 28 *grs.* of charcoal, the tube EF is filled with 274 *grs.* of foft iron in thin plates, rolled up fpirally. The tube is made red hot by means of its furnace, and the water in the retort A is kept conftantly boiling till it be all evaporated, and has paffed through the tube EF, fo as to be condenfed in the bottle H.

No carbonic acid gas is difengaged in this experiment, inftead of which we obtain 416 cubical inches, or 15 *grs.* of inflammable gas, thirteen times lighter than atmofpheric air. By examining the water which has been diftilled, it is found to have loft 100 *grs.* and the 274 *grs.* of iron confined in the tube are found to have acquired 85 *grs.* additional weight, and its magnitude is confiderably augmented. The iron is now hardly at all attractable by the magnet; it diffolves in acids without effervefcence; and, in fhort, it is converted into a black oxyd, precifely fimilar to that which has been burnt in oxygen gas.

In this experiment we have a true *oxydation* of iron, by means of water, exactly fimilar to that produced in air by the affiftance of heat. One hundred grains of water having been decompofed,

compofed, 85 *grs.* of oxygen have combined
with the iron, fo as to convert it into the ftate
of black oxyd, and 15 *grs.* of a peculiar inflammable gas are difengaged : From all this it clearly follows, that water is compofed of oxygen
combined with the bafe of an inflammable gas,
in the refpective proportions of 85 parts, by
weight of the former, to 15 parts of the latter.

Thus water, befides the oxygen, which is one
of its elements in common with many other
fubftances, contains another element as its conftituent bafe or radical, and for which we muft
find an appropriate term. None that we could
think of feemed better adapted than the word
hydrogen, which fignifies the *generative principle
of water,* from υδωρ *aqua,* and γεινομαι *gignor* *
We call the combination of this element with
caloric *hydrogen gas ;* and the term hydrogen
expreffes the bafe of that gas, or the radical of
water.

M   This

---

* This expreffion Hydrogen has been very feverely
criticifed by fome, who pretend that it fignifies engen-
dered by water, and not that which engenders water.
The experiments related in this chapter prove, that,
when water is decompofed, hydrogen is produced,
and that, when hydrogen is combined with oxygen,
water is produced : So that we may fay, with equal
truth, that water is produced from hydrogen, or hydrogen is produced from water.—A.

This experiment furnifhes us with a new com-buftible body, or, in other words, a body which has fo much affinity with oxygen as to draw it from its connection with caloric, and to decom-pofe air or oxygen gas. This combuftible bo-dy has itfelf fo great affinity with caloric, that, unlefs when engaged in a combination with fome other body, it always fubfifts in the aëri-form or gaffeous ftate, in the ufual temperature and preffure of our atmofphere. In this ftate of gas it is about $\frac{1}{13}$ of the weight of an equal bulk of atmofpheric air; it is not abforbed by water, though it is capable of holding a fmall quantity of that fluid in folution, and it is in-capable of being ufed for refpiration.

As the property this gas poffeffes, in com-mon with all other combuftible bodies, is no-thing more than the power of decompofing air, and carrying off its oxygen from the caloric with which it was combined, it is eafily under-ftood that it cannot burn, unlefs in contact with air or oxygen gas. Hence, when we fet fire to a bottle full of this gas, it burns gently, firft at the neck of the bottle, and then in the infide of it, in proportion as the external air gets in: This combuftion is flow and fucceffive, and on-ly takes place at the furface of contact between the two gaffes. It is quite different when the two gaffes are mixed before they are fet on fire: If, for inftance, after having introduced one part of

oxygen

oxygen gas into a narrow mouthed bottle, we fill it up with two parts of hydrogen gas, and bring a lighted taper, or other burning body, to the mouth of the bottle, the combuſtion of the two gaſſes takes place inſtantaneouſly with a violent exploſion. This experiment ought only to be made in a bottle of very ſtrong green glaſs, holding not more than a pint, and wrapped round with twine, otherwiſe the operator will be expoſed to great danger from the rupture of the bottle, of which the fragments will be thrown about with great force.

If all that has been related above, concerning the decompoſition of water, be exactly conformable to truth ;—if, as I have endeavoured to prove, that ſubſtance be really compoſed of hydrogen, as its proper conſtituent element, combined with oxygen, it ought to follow, that, by reuniting theſe two elements together, we ſhould recompoſe water ; and that this actually happens may be judged of by the following experiment.

Experiment Fourth.

I took a large criſtal baloon, A, Pl. iv. fig. 5. holding about 30 pints, having a large opening, to which was cemented the plate of copper BC, pierced with four holes, in which four tubes terminate. The firſt tube, H h, is intended to

be

be adapted to an air pump, by which the baloon is to be exhaufted of its air. The fecond tube gg, communicates, by its extremity MM, with a refervoir of oxygen gas, with which the baloon is to be filled. The third tube d D d', communicates, by its extremity d NN, with a refervoir of hydrogen gas. The extremity d' of this tube terminates in a capillary opening, through which the hydrogen gas contained in the refervoir is forced, with a moderate degree of quicknefs, by the preffure of one or two inches of water. The fourth tube contains a metallic wire GL, having a knob at its extremity L, intended for giving an electrical fpark from L to d', on purpofe to fet fire to the hydrogen gas: This wire is moveable in the tube, that we may be able to feparate the knob L from the extremity d' of the tube D d'. The three tubes d D d', gg, and H h, are all provided with ftop-cocks.

That the hydrogen gas and oxygen gas may be as much as poffible deprived of water, they are made to pafs, in their way to the baloon A, through the tubes MM, NN, of about an inch diameter, and filled with falts, which, from their deliquefcent nature, greedily attract the moifture of the air: Such are the acetite of potafh, and the muriat or nitrat of lime *. Thefe falts

must

---

* See the nature of thefe falts in the fecond part of this book.—A.

muſt only be reduced to a coarſe powder, leſt they run into lumps, and prevent the gaſſes from geting through their interſtices.

We muſt be provided before hand with a ſufficient quantity of oxygen gas, carefully puriſied from all admixture of carbonic acid, by long contact with a ſolution of potaſh *.

We muſt likewiſe have a double quantity of hydrogen gas, carefully purified in the ſame manner by long contact with a ſolution of potaſh in water. The beſt way of obtaining this gas free from mixture is, by decompoſing water with very pure ſoft iron, as directed in Exp. 3. of this chapter.

Having adjuſted every thing properly, as above directed, the tube H h is adapted to an air-pump, and the baloon A is exhauſted of its air. We next admit the oxygen gas ſo as to fill the baloon, and then, by means of preſſure, as is before mentioned, force a ſmall ſtream of hydrogen gas through its tube D d', which we immediately ſet on fire by an electric ſpark. By means of the above deſcribed apparatus, we can

continue

---

* By potaſh is here meant, pure or cauſtic alkali, deprived of carbonic acid by means of quick-lime: In general, we may obſerve here, that all the alkalies and earths muſt invariably be conſidered as in their pure or cauſtic ſtate, unleſs otherwiſe expreſſed.—E. The method of obtaining this pure alkali of potaſh will be given in the ſequel —A.

continue the mutual combuftion of thefe two gaffes for a long time, as we have the power of fupplying them to the baloon from their refervoirs, in proportion as they are confumed. I have in another place * given a defcription of the apparatus ufed in this experiment, and have explained the manner of afcertaining the quantities of the gaffes confumed with the moft fcrupulous exactitude.

In proportion to the advancement of the combuftion, there is a depofition of water upon the inner furface of the baloon or matrafs A : The water gradually increafes in quantity, and, gathering into large drops, runs down to the bottom of the veffel. It is eafy to afcertain the quantity of water collected, by weighing the baloon both before and after the experiment. Thus we have a twofold verification of our experiment, by afcertaining both the quantities of the gaffes employed, and of the water formed by their combuftion : Thefe two quantities muft be equal to each other. By an operation of this kind, Mr Meufnier and I afcertained that it required 85 parts, by weight, of oxygen, united to 15 parts of hydrogen, to compofe 100 parts of water. This experiment, which has not hitherto been publifhed, was made in prefence of a numerous committee from the Royal Academy.

* See the third part of this work.—A.

demy. We exerted the moft fcrupulous attention to its accuracy; and have reafon to believe
that the above propofitions cannot vary a two
hundredth part from abfolute truth.

From thefe experiments, both analytical and
fynthetic, we may now affirm that we have afcertained, with as much certainty as is poffible
in phyfical or chemical fubjects, that water is
not a fimple elementary fubftance, but is compofed of two elements, oxygen and hydrogen;
which elements, when exifting feparately, have
fo ftrong affinity for caloric, as only to fubfift
under the form of gas in the common temperature and preffure of our atmofphere.

This decompofition and recompofition of water is perpetually operating before our eyes, in
the temperature of the atmofphere, by means of
compound elective attraction. We fhall prefently fee that the phenomena attendant upon
vinous fermentation, putrefaction, and even
vegetation, are produced, at leaft in a certain
degree, by decompofition of water. It is very
extraordinary that this fact fhould have hitherto
been overlooked by natural philofophers and
chemifts: Indeed, it ftrongly proves, that, in
chemiftry, as in moral philofophy, it is extremely difficult to overcome prejudices imbibed in
early education, and to fearch for truth in any
other road than the one we have been accuftomed to follow.

I

I shall finish this chapter by an experiment much less demonstrative than those already related, but which has appeared to make more impression than any other upon the minds of many people. When 16 ounces of alkohol are burnt in an apparatus * properly adapted for collecting all the water disengaged during the combustion, we obtain from 17 to 18 ounces of water. As no substance can furnish a product larger than its original bulk, it follows, that something else has united with the alkohol during its combustion; and I have already shown that this must be oxygen, or the base of air. Thus alkohol contains hydrogen, which is one of the elements of water; and the atmospheric air contains oxygen, which is the other element necessary to the composition of water. This experiment is a new proof that water is a compound substance.

C H A P.

* See an account of this apparatus in the third part of this work.—A.

C H A P.   IX.

Of the quantities of Caloric difengaged from different fpecies of Combuftion.

WE have already mentioned, that, when any body is burnt in the center of a hollow fphere of ice and fupplied with air at the temperature of zero (32°), the quantity of ice melted from the infide of the fphere becomes a meafure of the relative quantities of caloric difengaged.   Mr de la Place and I gave a defcription of the apparatus employed for this kind of experiment in the Memoirs of the Academy for 1780, p. 355.; and a defcription and plate of the fame apparatus will be found in the third part of this work.   With this apparatus, phofphorus, charcoal, and hydrogen gas, gave the following refults :

One pound of phofphorus melted 100 *libs.* of ice.

One pound of charcoal melted 96 *libs.* 8 *oz.*

One pound of hydrogen gas melted 295 *libs.* 9 *oz.* 3½ *gros.*

As a concrete acid is formed by the combuftion of phofphorus, it is probable that very little caloric remains in the acid, and, confe-

N   quently,

quently, that the above experiment gives us very nearly the whole quantity of caloric contained in the oxygen gas. Even if we suppose the phosphoric acid to contain a good deal of caloric, yet, as the phosphorus must have contained nearly an equal quantity before combustion, the error must be very small, as it will only consist of the difference between what was contained in the phosphorus before, and in the phosphoric acid after combustion.

I have already shown in Chap. V. that one pound of phosphorus absorbs one pound eight ounces of oxygen during combustion ; and since, by the same operation, 100 *lib.* of ice are melted, it follows, that the quantity of caloric contained in one pound of oxygen gas is capable of melting 66 *libs.* 10 *oz.* 5 *gros* 24 *grs.* of ice.

One pound of charcoal during combustion melts only 96 *libs.* 8 *oz.* of ice, whilst it absorbs 2 *libs.* 9 *oz.* 1 *gros* 10 *grs.* of oxygen. By the experiment with phosphorus, this quantity of oxygen gas ought to disengage a quantity of caloric sufficient to melt 171 *libs.* 6 *oz.* 5 *gros.* of ice ; consequently, during this experiment, a quantity of caloric, sufficient to melt 74 *libs.* 14 *oz.* 5 *gros* of ice disappears. Carbonic acid is not, like phosphoric acid, in a concrete state after combustion but in the state of gas, and requires to be united with caloric to enable it to subsist fubfift in that ftate; the quantity of caloric miffing in the laft experiment is evidently employed for that purpofe. When we divide that quantity by the weight of carbonic acid, formed by the combuftion of one pound of charcoal, we find that the quantity of caloric neceffary for changing one pound of carbonic acid from the concrete to the gaffeous ftate, would be capable of melting 20 *libs.* 15 *oz.* 5 *gros* of ice.

We may make a fimilar calculation with the combuftion of hydrogen gas and the confequent formation of water. During the combuftion of one pound of hydrogen gas, 5 *libs.* 10 *oz.* 5 *gros* 24 *grs.* of oxygen gas are abforbed, and 295 *libs.* 9 *oz.* $3\frac{1}{2}$ *gros* of ice are melted. But 5 *libs.* 10 *oz.* 5 *gros* 24 *grs.* of oxygen gas, in changing from the aeriform to the folid ftate, lofes, according to the experiment with phofphorus, enough of caloric to have melted 377 *libs.* 12 *oz.* 3 *gros* of ice. There is only difengaged, from the fame quantity of oxygen, during its combuftion with hydrogen gas, as much caloric as melts 295 *libs.* 2 *oz.* $3\frac{1}{2}$ *gros*; wherefore there remains in the water at Zero (32°), formed, during this experiment, as much caloric as would melt 82 *libs.* 9 *oz.* $7\frac{1}{2}$ *gros* of ice.

Hence, as 6 *libs.* 10 *oz.* 5 *gros* 24 *grs.* of water are formed from the combuftion of one pound of hydrogen gas with 5 *libs.* 10 *oz.* 5 *gros* 24 *grs.* of oxygen, it follows that, in each pound

pound of water, at the temperature of *Zero*, (32°), there exifts as much caloric as would melt 12 *libs.* 5 *oz.* 2 *gros* 48 *grs.* of ice, without taking into account the quantity originally contained in the hydrogen gas, which we have been obliged to omit, for want of data to calculate its quantity. From this it appears that water, even in the ftate of ice, contains a confiderable quantity of caloric, and that oxygen, in entering into that combination, retains likewife a good proportion.

From thefe experiments, we may affume the following refults as fufficiently eftablifhed.

Combuftion of Phofphorus.

From the combuftion of phofphorus, as related in the foregoing experiments, it appears, that one pound of phofphorus requires 1 *lib.* 8 *oz.* of oxygen gas for its combuftion, and that 2 *libs.* 8 *oz.* of concrete phofphoric acid are produced.

The quantity of caloric difengaged by the combuftion of one pound of phofphorus, expreffed by the number of pounds of ice melted during that operation, is . 100.00000.

The quantity difengaged from each pound of oxygen, during the combuftion of phofphorus, expreffed in the fame manner, is 66.66667.

The quantity difengaged during the formation

tion of one pound of phofphoric acid, 40.00000
The quantity remaining in each pound of phofphoric acid, 0.00000 *.

Combuftion of Charcoal:

In the combuftion of one pound of charcoal,
2 *libs.* 9 *oz.* 1 *gros* 10 *grs.* of oxygen gas are
abforbed, and 3 *libs.* 9 *oz.* 1 *gros* 10 *grs.* of carbonic acid gas are formed.

Caloric, difengaged during the combuftion
of one pound of charcoal, 96.50000 †.

Caloric difengaged during the combuftion of
charcoal, from each pound of oxygen gas abforbed, 37.52823.

Caloric difengaged during the formation of
one pound of carbonic acid gas, 27.02024.

Caloric retained by each pound of oxygen
after the combuftion 29.13844

Caloric neceffary for fupporting one pound
of carbonic acid in the ftate of gas 20.97960.

Com-

---

* We here fuppofe the phofphoric acid not to contain any caloric, which is not ftrictly true; but, as
I have before obferved, the quantity it really contains
is probably very fmall, and we have not given it a value, for want of a fufficient data to go upon.—A.

† All thefe relative quantities of caloric are expreffed
by the number of pounds of ice, and decimal parts,
melted during the feveral operations.—E.

Combuſtion of Hydrogen Gas.

In the combuſtion of one pound of hydrogen gas, 5 *libs.* 10 *oz.* 5 *gros* 24 *grs.* of oxygen gas are abſorbed, and 6 *libs.* 10 *oz.* 5 *gros* 24 *grs.* of water are formed.

Caloric from each *lib.* of hydrogen gas,	295.58950.
Caloric from each *lib.* of oxygen gas,	52.16280.
Caloric diſengaged during the formation of each pound of water,	44.33840.
Caloric retained by each *lib.* of oxygen after combuſtion with hydrogen	14.50386.
Caloric retained by each *lib.* of water at the temperature of Zero (32°)	12.32823.

Of the Formation of Nitric Acid.

When we combine nitrous gas with oxygen gas, ſo as to form nitric or nitrous acid a degree of heat is produced, which is much leſs conſiderable than what is evolved during the other combinations of oxygen ; whence it follows that oxygen, when it becomes fixed in nitric acid, retains a great part of the heat which it poſſeſſed

feffed in the ftate of gas. It is certainly poffible to determine the quantity of caloric which
is difengaged during the combination of thefe
two gaffes, and confequently to determine what
quantity remains after the combination takes
place. The firft of thefe quantities might be
afcertained, by making the combination of the
two gaffes in an apparatus furrounded by ice;
but, as the quantity of caloric difengaged is
very inconfiderable, it would be neceffary to
operate upon a large quantity of the two gaffes
in a very troublefome and complicated apparatus. By this confideration, Mr de la Place and
I have hitherto been prevented from making
the attempt. In the mean time, the place of
fuch an experiment may be fupplied by calculations, the refults of which cannot be very far
from truth.

Mr de la Place and I deflagrated a convenient quantity of nitre and charcoal in an ice apparatus, and found that twelve pounds of ice
were melted by the deflagration of one pound
of nitre. We fhall fee, in the fequel, that one
pound of nitre is compofed, as under, of

Potafh   7 oz. 6 gros 51.84 grs. = 4515.84 grs.
Dry acid 8   1   21.16   = 4700.16.

The above quantity of dry acid is compofed
of

Oxygen

Oxygen 6 *oz.* 3 *gros* 66.34 *grs.* = 3738.34 *grs.*
Azote 1   5   25.82   = 961.82.

By this we find that, during the above deflagration, 2 *gros* 1⅓ *gr,* of charcoal have fuffered
combuftion, alongft with 3738.34 *grs.* or 6 *oz.*
3 *gros* 66.34 *grs.* of oxygen.   Hence, fince
12 *libs.* of ice were melted during the combuftion, it follows, that one pound of oxygen
burnt in the fame manner would have melted
29.58320 *libs.* of ice.   To which the quantity
of caloric, retained by a pound of oxygen after
combining with charcoal to form carbonic acid
gas, being added, which was already afcertained to
be capable of melting 29.13844 *libs.* of ice, we
have for the total quantity of caloric remaining
in a pound of oxygen, when combined with nitrous gas in the nitric acid 58.72164 ; which is
the number of pounds of ice the caloric remaining in the oxygen in that ftate is capable
of melting.

We have before feen that, in the ftate of oxygen gas, it contained at leaft 66.66667 ; wherefore it follows that, in combining with azote to
form nitric acid, it only lofes 7.94502.   Farther experiments upon this fubject are neceffary
to afcertain how far the refults of this calculation may agree with direct fact.   This enormous quantity of caloric retained by oxygen in
its combination into nitric acid, explains the
caufe

caufe of the great difengagement of caloric during the deflagrations of nitre ; or, more ftrictly fpeaking, upon all occafions of the decompofition of nitric acid.

Of the Combuftion of Wax.

Having examined feveral cafes of fimple
combuftion, I mean now to give a few examples
of a more complex nature.   One pound of wax-
taper being allowed to burn flowly in an ice
apparatus, melted 133 *libs.* 2 *oz.* 5⅓ *gros* of ice.
According to my experiments in the Memoirs
of the Academy for 1784, p. 606, one pound
of wax-taper confifts of 13 *oz.* 1 *gros* 23 *grs.*
of charcoal, and 2 *oz.* 6 *gros* 49 *grs.* of hydrogen.

By the foregoing ex-
periments,   the   above
quantity   of   charcoal
ought to melt   79.39390 *libs.* of ice ;
and the hydrogen fhould
melt   52.37605

In all 131.76995 *libs.*

Thus, we fee the quantity of caloric difengaged from a burning taper, is pretty exactly
conformable to what was obtained by burning
feparately a quantity of charcoal and hydrogen

equal to what enters into its compofition. Thefe experiments with the taper were feveral times repeated, fo that I have reafon to believe them accurate.

Combuftion of Olive Oil.

We included a burning lamp, containing a determinate quantity of olive-oil, in the ordinary apparatus, and, when the experiment was finifhed, we afcertained exactly the quantities of oil confumed, and of ice melted; the refult was, that, during the combuftion of one pound of olive-oil, 148 *libs.* 14 *oz.* 1 *gros* of ice were melted. By my experiments in the Memoirs of the Academy for 1784, and of which the following Chapter contains an abftract, it appears that one pound of olive-oil confifts of 12 *oz.* 5 *gros* 5 *grs.* of charcoal, and 3 *oz.* 2 *gros* 67 *grs.* of hydrogen. By the foregoing experiments, that quantity of charcoal fhould melt 76.18723 *libs.* of ice, and the quantity of hydrogen in a pound of the oil fhould melt 62.15053 *libs.* The fum of thefe two gives 138.33776 *libs.* of ice, which the two conftituent elements of the oil would have melted, had they feparately fuffered combuftion, whereas the oil really melted 148.88330 *libs.* which gives an excefs of 10.54554 in the refult of the experiment

ment above the calculated refult, from data furnifhed by former experiments.

This difference, which is by no means very
confiderable, may arife from errors which are
unavoidable in experiments of this nature, or
it may be owing to the compofition of oil
not being as yet exactly afcertained. It proves,
however, that there is a great agreement between the refults of our experiments, refpecting
the combination of caloric, and thofe which regard its difengagement.

The following defiderata ftill remain to be determined, viz. What quantity of caloric is retained by oxygen, after combining with metals,
fo as to convert them into oxyds ; What quantity is contained by hydrogen, in its different
ftates of exiftence ; and to afcertain, with more
precifion than is hitherto attained, how much
caloric is difengaged during the formation of
water, as there ftill remain confiderable doubts
with refpect to our prefent determination of this
point, which can only be removed by farther
experiments. We are at prefent occupied with
this inquiry ; and, when once thefe feveral
points are well afcertained, which we hope they
will foon be, we fhall probably be under the neceffity of making confiderable corrections upon
moft of the refults of the experiments and calculations in this Chapter. I did not, however,
confider this as a fufficient reafon for with-

holding

holding fo much as is already known from fuch as may be inclined to labour upon the fame fubject. It is difficult, in our endeavours to difcover the principles of a new fcience, to a-void beginning by guefs-work; and it is rarely poffible to arrive at perfection from the firft fet-ting out.

C H A P.

C H A P. X.

Of the Combination of Combuſtible Subſtances with each other.

A S combuſtible ſubſtances in general have a great affinity for oxygen, they ought likewiſe to attract, or tend to combine with each other ; *quae ſunt eadem uni tertio, ſunt eadem inter ſe;* and the axiom is found to be true. Almoſt all the metals, for inſtance, are capable of uniting with each other, and forming what are called *alloys* *, in common language. Moſt of theſe, like all combinations, are ſuſceptible of ſeveral degrees of ſaturation ; the greater number of theſe alloys are more brittle than the pure metals of which they are compoſed, eſpecially when the metals alloyed together are conſiderably different in their degrees of fuſibility. To this difference in fuſibility, part of the phenomena attendant upon *alloyage* are owing, particularly the property of iron, called by workmen

* This term *alloy*, which we have from the language of the arts, ſerves exceedingly well for diſtinguiſhing all the combinations or intimate unions of metals with each other, and is adopted in our new nomenclature for that purpoſe.—A.

men *hotſhort*. This kind of iron muſt be conſidered as an alloy, or mixture of pure iron,
which is almoſt infuſible, with a ſmall portion
of ſome other metal which fuſes in a much
lower degree of heat. So long as this alloy remains cold, and both metals are in the ſolid
ſtate, the mixture is malleable ; but, if heated
to a ſufficient degree to liquify the more fuſible
metal, the particles of the liquid metal, which
are interpoſed between the particles of the metal remaining ſolid, muſt deſtroy their continu-
ity, and occaſion the alloy to become brittle.
The alloys of mercury, with the other metals,
have uſually been called *amalgams*, and we ſee
no inconvenience from continuing the uſe of that
term.

Sulphur, phoſphorus, and charcoal, readily
unite with metals. Combinations of ſulphur
with metals are uſually named *pyrites*. Their
combinations with phoſphorus and charcoal are
either not yet named, or have received new
names only of late ; ſo that we have not ſcru-
pled to change them according to our principles. The combinations of metal and ſulphur
we call *ſulphurets*, thoſe with phoſphorus *phoſ-
phurets*, and thoſe formed with charcoal *carbu-
rets*. Theſe denominations are extended to all
the combinations into which the above three
ſubſtances enter, without being previouſly oxygenated.

genated. Thus, the combination of fulphur with potafh, or fixed vegetable alkali, is called *fulphuret of potafh ;* that which it forms with ammoniac, or volatile alkali, is termed *fulphuret of ammoniac.*

Hydrogen is likewife capable of combining with many combuftible fubftances. In the ftate of gas, it diffolves charcoal, fulphur, phofpho-rus, and feveral metals ; we diftinguifh thefe combinations by the terms, *carbonated hydrogen gas, fulphurated hydrogen gas,* and *phofphorated hydrogen gas.* The fulphurated hydrogen gas was called *hepatic air* by former chemifts, or *foetid air from fulphur,* by Mr Scheele. The virtues of feveral mineral waters, and the foetid fmell of animal excrements, chiefly arife from the pre-fence of this gas. The phofphorated hydrogen gas is remarkable for the property, difcovered by Mr Gengembre, of taking fire fpontaneoufly upon getting into contact with atmofpheric air, or, what is better, with oxygen gas. This gas has a ftrong flavour, refembling that of putrid fifh ; and it is very probable that the phofpho-refcent quality of fifh, in the ftate of putrefac-tion, arifes from the efcape of this fpecies of gas. When hydrogen and charcoal are combi-ned together, without the intervention of calo-ric, to bring the hydrogen into the ftate of gas, they form oil, which is either fixed or volatile, according to the proportions of hydrogen and

char.

charcoal in its compofition. The chief difference between fixed or fat oils drawn from vegetables by expreffion, and volatile or effential
oils, is, that the former contains an excefs of
charcoal, which is feparated when the oils are
heated above the degree of boiling water ;
whereas the volatile oils, containing a juft proportion of thefe two conftituent ingredients, are
not liable to be decompofed by that heat, but,
uniting with caloric into the gaffeous ftate, pafs
over in diftillation unchanged.

In the Memoirs of the Academy for 1784,
p. 593. I gave an account of my experiments
upon the compofition of oil and alkohol, by the
union of hydrogen with charcoal, and of their
combination with oxygen. By thefe experiments, it appears that fixed oils combine with
oxygen during combuftion, and are thereby
converted into water and carbonic acid. By
means of calculation applied to the products of
thefe experiments, we find that fixed oil is compofed of 21 parts, by weight, of hydrogen combined with 79 parts of charcoal. Perhaps the
folid fubftances of an oily nature, fuch as wax,
contain a proportion of oxygen, to which they
owe their ftate of folidity. I am at prefent engaged in a feries of experiments, which I hope
will throw great light upon this fubject.

It is worthy of being examined, whether hydrogen in its concrete ftate, uncombined with
calorie,

caloric, be fufceptible of combination with fulphur, phofphorus, and the metals. There is
nothing that we know of, which, *a priori*,
fhould render thefe combinations impoffible;
for combuftible bodies being in general fufceptible of combination with each other, there is
no evident reafon for hydrogen being an exception to the rule : However, no direct experiment as yet eftablifhes either the poffibility or
impoffibility of this union. Iron and zinc are
the moft likely, of all the metals, for entering
into combination with hydrogen ; but, as thefe
have the property of decompofing water, and
as it is very difficult to get entirely free from
moifture in chemical experiments, it is hardly
poffible to determine whether the fmall portions
of hydrogen gas, obtained in certain experiments with thefe metals, were previoufly combined with the metal in the ftate of folid hydrogen, or if they were produced by the decompofition of a minute quantity of water. The more
care we take to prevent the prefence of water
in thefe experiments, the lefs is the quantity of
hydrogen gas procured ; and, when very accurate precautions are employed, even that quantity becomes hardly fenfible.

However this inquiry may turn out refpecting
the power of combuftible bodies, as fulphur,
phofphorus, and metals, to abforb hydrogen,
we are certain that they only abforb a very fmall

P   por-

portion ; and that this combination, inftead of being effential to their conftitution, can only be confidered as a foreign fubftance, which contaminates their purity. It is the province of the advocates * for this fyftem to prove, by decifive experiments, the real exiftence of this combined hydrogen, which they have hitherto only done by conjectures founded upon fuppofitions.

CHAP.

* By thefe are meant the fupporters of the phlogiftic theory, who at prefent confider hydrogen, or the bafe of inflammable air, as the phlogifton of the celebrated Stahl.—E.

CHAP. XI.

Obfervations upon Oxyds and Acids with feveral Bafes—and upon the Compofition of Animal and Vegetable Subftances.

W E have, in Chap. V. and VIII. examined the products refulting from the combuftion of the four fimple combuftible fubftances, fulphur, phofphorus, charcoal, and hydrogen: We have fhown, in Chap. X that the fimple combuftible fubftances are capable of combining with each other into compound combuftible fubftances, and have obferved that oils in general, and particularly the fixed vegetable oils, belong to this clafs, being compofed of hydrogen and charcoal. It remains, in this chapter, to treat of the oxygenation of thefe compound combuftible fubftances, and to fhow that there exift acids and oxyds having double and triple bafes. Nature furnifhes us with numerous examples of this kind of combinations, by means of which, chiefly, fhe is enabled to produce a vaft variety of compounds from a very limited number of elements, or fimple fubftances.

It

It was long ago well known, that, when muriatic and nitric acids were mixed together, a compound acid was formed, having properties quite diftinct from thofe of either of the acids taken feparately. This acid was called *aqua regia*, from its moft celebrated property of diffolving gold, called *king of metals* by the alchymifts. Mr Berthollet has diftinctly proved that the peculiar properties of this acid arife from the combined action of its two acidifiable bafes; and for this reafon we have judged it neceffary to diftinguifh it by an appropriate name : That of *nitro-muriatic* acid appears extremely applicable, from its expreffing the nature of the two fubftances which enter into its compofition.

This phenomenon of a double bafe in one acid, which had formerly been obferved only in the nitro-muriatic acid, occurs continually in the vegetable kingdom, in which a fimple acid, or one poffeffed of a fingle acidifiable bafe, is very rarely found. Almoft all the acids procurable from this kingdom have bafes compofed of charcoal and hydrogen, or of charcoal, hydrogen, and phofphorus, combined with more or lefs oxygen. All thefe bafes, whether double or triple, are likewife formed into oxyds, having lefs oxygen than is neceffary to give them the properties of acids. The acids and oxyds from the animal kingdom are ftill more compound, as their bafes generally confift of a combination

bination of charcoal, phofphorus, hydrogen, and
azote.

As it is but of late that I have acquired any
clear and diftinct notions of thefe fubftances, I
fhall not, in this place, enlarge much upon the
fubject, which I mean to treat of very fully in
fome memoirs I am preparing to lay before the
Academy. Moft of my experiments are already
performed; but, to be able to give exact reports
of the refulting quantities, it is neceffary that
they be carefully repeated, and increafed in
number: Wherefore, I fhall only give a fhort
enumeration of the vegetable and animal acids
and oxyds, and terminate this article by a few
reflections upon the compofition of vegetable
and animal bodies.

Sugar, mucus, under which term we include
the different kinds of gums, and ftarch, are vegetable oxyds, having hydrogen and charcoal
combined, in different proportions, as their radicals or bafes, and united with oxygen, fo as
to bring them to the ftate of oxyds. From the
ftate of oxyds they are capable of being changed into acids by the addition of a frefh quantity of oxygen; and, according to the degrees
of oxygenation, and the proportion of hydrogen
and charcoal in their bafes, they form the feveral kinds of vegetable acids.

It would be eafy to apply the principles of
our nomenclature to give names to thefe vege-

table acids and oxyds, by ufing the names of the two fubftances which compofe their bafes: They would thus become hydro-carbonous acids and oxyds: In this method we might indicate which of their elements exifted in excefs, without circumlocution, after the manner ufed by Mr Rouelle for naming vegetable extracts: He calls thefe extracto-refinous when the extractive matter prevails in their compofition, and refino-extractive when they contain a larger proportion of refinous matter. Upon that plan, and by varying the terminations according to the formerly eftablifhed rules of our nomenclature, we have the following denominations: Hydro-carbonous, hydro-carbonic; carbono-hydrous, and carbono-hydric oxyds. And for the acids: Hydro-carbonous, hydro carbonic, oxygenated hydro-carbonic; carbono-hydrous, carbono-hydric, and oxygenated carbono-hydric. It is probable that the above terms would fuffice for indicating all the varieties in nature, and that, in proportion as the vegetable acids become well underftood, they will naturally arrange themfelves under thefe denominations. But, though we know the elements of which thefe are compofed, we are as yet ignorant of the proportions of thefe ingredients, and are ftill far from being able to clafs them in the above methodical manner; wherefore, we have determined to retain

the

the ancient names provifionally. I am fomewhat farther advanced in this inquiry than at
the time of publifhing our conjunct effay upon
chemical nomenclature; yet it would be improper to draw decided confequences from experiments not yet fufficiently precife : Though I acknowledge that this part of chemiftry ftill remains in fome degree obfcure, I muft exprefs
my expectations of its being very foon elucidated.

I am ftill more forcibly neceffitated to follow
the fame plan in naming the acids, which have
three or four elements combined in their bafes;
of thefe we have a confiderable number from
the animal kingdom, and fome even from vegetable fubftances.   Azote, for inftance, joined
to hydrogen and charcoal, form the bafe or radical of the Pruffic acid; we have reafon to believe that the fame happens with the bafe of the
Gallic acid; and almoft all the animal acids have
their bafes compofed of azote, phofphorus, hydrogen, and charcoal.   Were we to endeavour
to exprefs at once all thefe four component parts
of the bafes, our nomenclature would undoubtedly be methodical; it would have the property
of being clear and determinate; but this affem-
blage of Greek and Latin fubftantives and adjectives, which are not yet univerfally admitted
by chemifts, would have the appearance of a
                                    barbarous

barbarous language, difficult both to pronounce and to be remembered. Befides, this part of chemiftry being ftill far from that accuracy it muft arrive to, the perfection of the fcience ought certainly to precede that of its language; and we muft ftill, for fome time, retain the old names for the animal oxyds and acids. We have only ventured to make a few flight modifications of thefe names, by changing the termination into *ous*, when we have reafon to fuppofe the bafe to be in excefs, and into *ic*, when we fufpect the oxygen predominates.

The following are all the vegetable acids hitherto known:

1. Acetous acid.	8. Pyro-mucous acid.
2. Acetic acid.	9. Pyro-lignous acid.
3. Oxalic acid.	10. Gallic acid.
4. Tartarous acid.	11. Benzoic acid.
5. Pyro-tartarous acid.	12. Camphoric acid.
6. Citric acid.	13. Succinic acid.
7. Malic acid.	

Though all thefe acids, as has been already faid, are chiefly, and almoft entirely, compofed of hydrogen, charcoal, and oxygen, yet, properly fpeaking, they contain neither water carbonic acid nor oil, but only the elements neceffary for forming thefe fubftances. The power of affinity reciprocally exerted by the hydrogen, charcoal, and oxygen, in thefe acids, is in a ftate

of

of equilibrium only capable of exifting in the ordinary temperature of the atmofphere; for, when they are heated but a very little above the temperature of boiling water, this equilibrium is deftroyed, part of the oxygen and hydrogen unite, and form water; part of the charcoal and hydrogen combine into oil; part of the charcoal and oxygen unite to form carbonic acid; and, laftly, there generally remains a fmall portion of charcoal, which, being in excefs with refpect to the other ingredients, is left free. I mean to explain this fubject fomewhat farther in the fucceeding chapter.

The oxyds of the animal kingdom are hitherto lefs known than thofe from the vegetable kingdom, and their number is as yet not at all determined. The red part of the blood, lymph, and moft of the fecretions, are true oxyds, under which point of view it is very important to confider them. We are only acquainted with fix animal acids, feveral of which, it is probable, approach very near each other in their nature, or, at leaft, differ only in a fcarcely fenfible degree. I do not include the phofphoric acid amongft thefe, becaufe it is found in all the kingdoms of nature. They are,

1. Lactic acid.   4. Formic acid.
2. Saccho-lactic acid.   5. Sebacic acid.
3. Bombic acid.   6. Pruffic acid.

Q   The

The connection between the conftituent elements of the animal oxyds and acids is not more permanent than in thofe from the vegetable kingdom, as a fmall increafe of temperature is fufficient to overturn it. I hope to render this fubject more diftinct than has been done hitherto in the following chapter.

CHAP.

C H A P.   XII.

Of the Decompofition of Vegetable and Animal Sub-
ftances by the Action of Fire.

BEFORE we can thoroughly comprehend
what takes place during the decompofition of vegetable fubftances by fire, we muft
take into confideration the nature of the elements which enter into their compofition, and
the different affinities which the particles of thefe
elements exert upon each other, and the affinity which caloric poffeffes with them. The true
conftituent elements of vegetables are hydrogen, oxygen, and charcoal : Thefe are common
to all vegetables, and no vegetable can exift
without them : Such other fubftances as exift
in particular vegetables are only effential to the
compofition of thofe in which they are found,
and do not belong to vegetables in general.

Of thefe elements, hydrogen and oxygen have
a ftrong tendency to unite with caloric, and be
converted into gas, whilft charcoal is a fixed
element, having but little affinity with caloric.
On the other hand, oxygen, which, in the ufual
temperature, tends nearly equally to unite with
hydrogen and with charcoal, has a much ftronger

er affinity with charcoal when at the red heat *, and then unites with it to form carbonic acid.

Although we are far from being able to appreciate all thefe powers of affinity, or to exprefs their proportional energy by numbers, we are certain, that, however variable they may be when confidered in relation to the quantity of caloric with which they are combined, they are all nearly in equilibrium in the ufual temperature of the atmofphere; hence vegetables neither contain oil †, water, nor carbonic acid, tho' they contain all the elements of thefe fubftances. The hydrogen is neither combined with the oxygen nor with the charcoal, and reciprocally; the particles of thefe three fubftances form a triple combination, which remains in equilibrium

* Though this term, red heat, does not indicate any abfolutely determinate degree of temperature, I fhall ufe it fometimes to exprefs a temperature confiderably above that of boiling water.—A.

† I muft be underftood here to fpeak of vegetables reduced to a perfectly dry ftate; and, with refpect to oil, I do not mean that which is procured by expreffion either in the cold, or in a temperature not exceeding that of boiling water; I only allude to the empyreumatic oil procured by diftillation with a naked fire, in a heat fuperior to the temperature of boiling water; which is the only oil declared to be produced by the operation of fire. What I have publifhed upon this fubject in the Memoirs of the Academy for 1786 may be confulted.—A.

brium whilſt undiſturbed by caloric but a very
flight increaſe of temperature is fufficient to
overturn this ſtructure of combination.

If the increaſed temperature to which the vegetable is expoſed does not exceed the heat of
boiling water, one part of the hydrogen combines with the oxygen, and forms water, the
reſt of the hydrogen combines with a part of
the charcoal, and forms volatile oil, whilſt the
remainder of the charcoal, being ſet free from
its combination with the other elements, remains fixed in the bottom of the diſtilling veſſel.

When, on the contrary, we employ a red
heat, no water is formed, or, at leaſt, any that
may have been produced by the firſt application of the heat is decompoſed, the oxygen having a greater affinity with the charcoal at this
degree of heat, combines with it to form carbonic acid, and the hydrogen being left free
from combination with the other elements, u-
nites with caloric, and eſcapes in the ſtate of
hydrogen gas. In this high temperature, either
no oil is formed, or, if any was produced during the lower temperature at the beginning cf
the experiment, it is decompoſed by the action
of the red heat. Thus the decompoſition of vegetable matter, under a high temperature, is
produced by the action of double and triple affinities ; while the charcoal attracts the oxygen,

on

on purpofe to form carbonic acid, the caloric attracts the hydrogen, and converts it into hydrogen gas.

The diftillation of every fpecies of vegetable fubftance confirms the truth of this theory, if we can give that name to a fimple relation of facts. When fugar is fubmitted to diftillation, fo long as we only employ a heat but a little below that of boiling water, it only lofes its water of criftallization, it ftill remains fugar, and retains all its properties; but, immediately upon raifing the heat only a little above that degree, it becomes blackened, a part of the charcoal feparates from the combination, water flightly acidulated paffes over accompanied by a little oil, and the charcoal which remains in the retort is nearly a third part of the original weight of the fugar.

The operation of affinities which take place during the decompofition, by fire, of vegetables which contain azote, fuch as the cruciferous plants, and of thofe containing phofphorus, is more complicated; but, as thefe fubftances only enter into the compofition of vegetables in very fmall quantities, they only, apparently, produce flight changes upon the products of diftillation; the phofphorus feems to combine with the charcoal, and, acquiring fixity from that union, remains behind in the retort, while the

azote,

azote, combining with a part of the hydrogen, forms ammoniac, or volatile alkali.

Animal fubftances, being compofed nearly of the fame elements with cruciferous plants, give the fame products in diftillation, with this difference, that, as they contain a greater quantity of hydrogen and azote, they produce more oil and more ammoniac. I fhall only produce one fact as a proof of the exactnefs with which this theory explains all the phenomena which occur during the diftillation of animal fubftances, which is the rectification and total decompofition of volatile animal oil, commonly known by the name of Dippel's oil. When thefe oils are procured by a firft diftillation in a naked fire they are brown, from containing a little charcoal almoft in a free ftate; but they become quite colourlefs by rectification. Even in this ftate the charcoal in their compofition has fo flight a connection with the other elements as to feparate by mere expofure to the air. If we put a quantity of this animal oil, well rectified, and confequently clear, limpid, and tranfparent, into a bell-glafs filled with oxygen gas over mercury, in a fhort time the gas is much diminifhed, being abforbed by the oil, the oxygen combining with the hydrogen of the oil forms water, which finks to the bottom, at the fame time the charcoal which was combined with the hydrogen being fet free, manifefts itfelf

by

by rendering the oil black. Hence the only way of preferving thefe oils colourlefs and tranfparent, is by keeping them in bottles perfectly full and accurately corked, to hinder the contact of air, which always difcolours them.

Succeffive rectifications of this oil furnifh another phenomenon confirming our theory. In each diftillation a fmall quantity of charcoal remains in the retort, and a little water is formed by the union of the oxygen contained in the air of the diftilling veffels with the hydrogen of the oil. As this takes place in each fucceffive diftillation, if we make ufe of large veffels and a confiderable degree of heat, we at laft decompofe the whole of the oil, and change it entirely into water and charcoal. When we ufe fmall veffels, and efpecially when we employ a flow fire, or degree of heat little above that of boiling water, the total decompofition of thefe oils, by repeated diftillation, is greatly more tedious, and more difficultly accomplifhed. I fhall give a particular detail to the Academy, in a feparate memoir, of all my experiments upon the decompofition of oil; but what I have related above may fuffice to give juft ideas of the compofition of animal and vegetable fubftances, and of their decompofition by the action of fire.

C H A P.

C H A P. XIII.

*Of the Decompofition of Vegetable Oxyds by the Vi-
nous Fermentation.*

THE manner in which wine, cyder, mead,
and all the liquors formed by the fpiritous fermentation, are produced, is well known
to every one. The juice of grapes or of apples
being expreffed, and the latter being diluted
with water, they are put into large vats, which
are kept in a temperature of at leaft 10° (54.5°)
of the thermometer. A rapid inteftine motion,
or fermentation, very foon takes place, numerous globules of gas form in the liquid and
burft at the furface; when the fermentation is
at its height, the quantity of gas difengaged is
fo great as to make the liquor appear as if boiling violently over a fire. When this gas is
carefully gathered, it is found to be carbonic
acid perfectly pure, and free from admixture
with any other fpecies of air or gas whatever.

When the fermentation is completed, the
juice of grapes is changed from being fweet,
and full of fugar, into a vinous liquor which no
longer contains any fugar, and from which we
procure, by diftillation, an inflammable liquor,

R known

known in commerce under the name of Spirit of Wine. As this liquor is produced by the fermentation of any faccharine matter whatever diluted with water, it muft have been contrary to the principles of our nomenclature to call it fpirit of wine rather than fpirit of cyder, or of fermented fugar; wherefore, we have adopted a more general term, and the Arabic word *alkohol* feems extremely proper for the purpofe.

This operation is one of the moft extraordinary in chemiftry: We muft examine whence proceed the difengaged carbonic acid and the inflammable liquor produced, and in what manner a fweet vegetable oxyd becomes thus converted into two fuch oppofite fubftances, whereof one is combuftible, and the other eminently the contrary. To folve thefe two queftions, it is neceffary to be previoufly acquainted with the analyfis of the fermentable fubftance, and of the products of the fermentation. We may lay it down as an inconteftible axiom, that, in all the operations of art and nature, nothing is created; an equal quantity of matter exifts both before and after the experiment; the quality and quantity of the elements remain precifely the fame; and nothing takes place beyond changes and modifications in the combination of thefe elements. Upon this principle the whole art of performing chemical experiments

periments depends : We muſt always ſuppoſe an
exact equality between the elements of the body
examined and thoſe of the products of its analyſis.

Hence, ſince from muſt of grapes we procure
alkohol and carbonic acid, I have an undoubted
right to ſuppoſe that muſt conſiſts of carbonic
acid and alkohol.   From theſe premiſes, we
have two methods of aſcertaining what paſſes
during vinous fermentation, by determining the
nature of, and the elements which compoſe, the
fermentable ſubſtances, or by accurately examining the products reſulting from fermentation ;
and it is evident that the knowledge of either of
theſe muſt lead to accurate concluſions concerning the nature and compoſition of the other.
From theſe conſiderations, it became neceſſary accurately to determine the conſtituent elements of
the fermentable ſubſtances ; and, for this purpoſe, I did not make uſe of the compound juices
of fruits, the rigorous analyſis of which is perhaps impoſſible, but made choice of ſugar,
which is eaſily analyſed, and the nature of which
I have already explained.   This ſubſtance is a
true vegetable oxyd with two baſes, compoſed
of hydrogen and charcoal brought to the ſtate
of an oxyd, by a certain proportion of oxygen ;
and theſe three elements are combined in ſuch
a way, that a very ſlight force is ſufficient to
deſtroy the equilibrium of their connection. By

a

a long train of experiments, made in various ways, and often repeated, I afcertained that the proportion in which thefe ingredients exift in fugar, are nearly eight parts of hydrogen, 64 parts of oxygen, and 28 parts of charcoal, all by weight, forming 100 parts of fugar.

Sugar muft be mixed with about four times its weight of water, to render it fufceptible of fermentation; and even then the equilibrium of its elements would remain undifturbed, without the affiftance of fome fubftance, to give a commencement to the fermentation. This is accomplifhed by means of a little yeaft from beer; and, when the fermentation is once excited, it continues of itfelf until completed. I fhall, in another place, give an account of the effects of yeaft, and other ferments, upon fermentable fubftances. I have ufually employed 10 *libs.* of yeaft, in the ftate of pafte, for each 100 *libs.* of fugar, with as much water as is four times the weight of the fugar. I fhall give the refults of my experiments exactly as they were obtained, preferving even the fractions produced by calculation.

TABLE

TABLE I. *Materials of Fermentation.*

		libs.	oz.	gros	grs.
Water - - -		400	0	0	0
Sugar - - -		100	0	0	0
Yeaſt in paſte, 10 *libs.* ⎰ Water -		7	3	6	44
compoſed of ⎱ Dry yeaſt -		2	12	1	28

Total 510

TABLE II. *Conſtituent Elements of the Materials of Fermentation.*

		libs.	oz.	gros	grs.
407 *libs*, 3 *oz.* 6 *gros* 44 *grs.* ⎰ Hydrogen		61	1	2	71.40
of water, compoſed of ⎱ Oxygen		346	2	3	44.60
	Hydrogen	8	0	0	0
100 *libs.* ſugar, compoſed of	Oxygen	64	0	0	0
	Charcoal	28	0	0	0
	Hydrogen	0	4	5	9.30
2 *libs.* 12 *oz.* 1 *gros* 28 *grs.* of	Oxygen	1	10	2	28.76
dry yeaſt, compoſed of	Charcoal	0	12	4	59
	Azote	0	0	5	2.94

Total weight 510 0 0 0

TABLE

TABLE III. *Recapitulation of thefe Elements.*

		libs.	oz.	gros	grs.				
Oxygen	of the water	340	0	0	0				
	of the water in the yeaft	6	2	3	44.60	*libs. oz. gros grs.* 411 12 6 1.36			
	of the fugar	64	0	0	0				
	of the dry yeaft	1	10	2	28.76				
Hydrogen	of the water	60	0	0	0				
	of the water in the yeaft	1	1	2	71.40	69 6 0 8.70			
	of the fugar	8	0	0	0				
	of the dry yeaft	0	4	5	9.30				
Charcoal	of the fugar	28	0	0	0	28 12 4 59.00			
	of the yeaft	0	12	4	59.00				
Azote of the yeaft	-	-	-	-		0 0 5 2.94			
					In all	510 0 0 0			

Having thus accurately determined the nature and quantity of the conftituent elements of the materials fubmitted to fermentation, we have next to examine the products refulting from that procefs. For this purpofe, I placed the above 510 *libs.* of fermentable liquor in a proper * apparatus, by means of which I could accurately determine the quantity and quality of gas difengaged during the fermentation, and could even weigh every one of the products fepa-

* The above apparatus is defcribed in the Third Part.—A.

feparately, at any period of the procefs I judged proper. An hour or two after the fubftances are mixed together, efpecially if they are kept in a temperature of from 15° (65.75°) to 18° (72.5°) of the thermometer, the firft marks of fermentation commence ; the liquor turns thick and frothy, little globules of air are difengaged, which rife and burft at the furface ; the quantity of thefe globules quickly increafes, and there is a rapid and abundant produ&tion of very pure carbonic acid, accompanied with a fcum, which is the yeaft feparating from the mixture. After fome days, lefs or more according to the degree of heat, the inteftine motion and difengagement of gas diminifh ; but thefe do not ceafe entirely, nor is the fermentation completed for a confiderable time. During the procefs, 35 *libs.* 5 *oz.* 4 *gros* 19 *grs.* of dry carbonic acid are difengaged, which carry alongft with them 13 *libs.* 14 *oz.* 5 *gros* of water. There remains in the veffel 460 *libs.* 11 *oz.* 6 *gros* 53 *grs.* of vinous liquor, flightly acidulous. This is at firft muddy, but clears of itfelf, and depofits a portion of yeaft. When we feparately analife all thefe fubftances, which is effe&ted by very troublefome proceffes, we have the refults as given in the following Tables. This procefs, with all the fubordinate calculations and analyfes, will be detailed at large in the Memoirs of the Academy.

TABLE

TABLE IV. *Products of Fermentation.*

		libs.	oz.	gros	grs.
35 *libs.* 5 *oz.* 4 *gros* 19 *grs.* of carbonic acid, compoſed of	Oxygen - -	25	7	1	34
	Charcoal - -	9	14	2	57
408 *libs.* 15 *oz.* 5 *gros* 14 *grs.* of water, compoſed of	Oxygen -	347	10	0	59
	Hydrogen -	61	5	4	27
57 *libs.* 11 *oz.* 1 *gros* 58 *grs.* of dry alkohol, compoſed of	Oxygen, combined with hydrogen	31	6	1	64
	Hydrogen, combined with oxygen	5	8	5	3
	Hydrogen, combined with charcoal	4	0	5	0
	Charcoal, combined with hydrogen	16	11	5	63
2 *libs.* 8 *oz.* of dry acetous acid, compoſed of	Hydrogen -	0	2	4	0
	Oxygen -	1	11	4	0
	Charcoal - -	0	10	0	0
4 *libs.* 1 *oz.* 4 *gros* 3 *grs.* of reſiduum of ſugar, compoſed of	Hydrogen - -	0	5	1	67
	Oxygen - -	2	9	7	27
	Charcoal - -	1	2	2	53
1 *lib.* 6 *oz.* 0 *gros* 5 *grs.* of dry yeaſt, compoſed of	Hydrogen -	0	2	2	41
	Oxygen - -	0	13	1	14
	Chareoal - -	0	6	2	30
	Azote - -	0	0	2	37

510 *libs.*   Total   510   0   0   0

TABLE

TABLE V. *Recapitulation of the Products.*

		libs.	oz.	gros	grs.
409 *libs.* 10 *oz.* 0 *gros* 54 *grs.* of oxygen contained in the	Water - -	347	10	0	59
	Carbonic acid -	25	7	1	34
	Alkohol - -	31	6	1	64
	Acetous acid -	1	11	4	0
	Refiduum of fugar	2	9	7	27
	Yeaft - -	0	13	1	14
28 *libs.* 12 *oz.* 5 *gros* 59 *grs.* of charcoal contained in the	Carbonic acid -	9	14	2	57
	Alkohol -	16	11	5	63
	Acetous acid -	0	10	0	0
	Refiduum of fugar	1	2	2	53
	Yeaft - -	0	6	2	30
71 *libs.* 8 *oz.* 6 *gros* 66 *grs.* of hydrogen, contained in the	Water - -	61	5	4	27
	Water of the alkohol	5	8	5	3
	Combined with the charcoal of the alko.	4	0	5	0
	Acetous acid -	0	2	4	0
	Refiduum of fugar	0	5	1	67
	Yeaft - -	0	2	2	41
2 *gros* 37 *grs.* of azote in the yeaft		0	0	2	37
510 *libs.*	Total	510	0	0	0

In thefe refults, I have been exact, even to
grains ; not that it is poffible, in experiments
of this nature, to carry our accuracy fo far, but
as the experiments were made only with a few
pounds of fugar, and as, for the fake of comparifon, I reduced the refults of the actual experiments to the quintal or imaginary hundred

S   pounds,

pounds, I thought it neceſſary to leave the fractional parts precifely as produced by calculation.

When we confider the refults prefented by
thefe tables with attention, it is eafy to difcover
exactly what occurs during fermentation. In
the firft place, out of the 100 *libs.* of fugar
employed, 4 *libs.* 1 *oz.* 4 *gros* 3 *grs.* remain, without having fuffered decompofition ;
fo that, in reality, we have only operated upon
95 *libs.* 14 *oz.* 3 *gros* 69 *grs.* of fugar ; that is
to fay, upon 61 *libs.* 6 *oz.* 45 *grs.* of oxygen,
7 *libs.* 10 *oz.* 6 *gros* 6 *grs.* of hydrogen, and 26
libs. 13 *oz.* 5 *gros* 19 *grs.* of charcoal. By comparing thefe quantities, we find that they are
fully fufficient for forming the whole of the alkohol, carbonic acid and acetous acid produced
by the fermentation. It is not, therefore, neceſſary to fuppofe that any water has been decompofed during the experiment, unlefs it be
pretended that the oxygen and hydrogen exift
in the fugar in that ftate. On the contrary, I
have already made it evident that hydrogen,
oxygen and charcoal, the three conftituent elements of vegetables, remain in a ftate of equilibrium or mutual union with each other
which fubfifts fo long as this union remains
undifturbed by increafed temperature, or by
fome new compound attraction ; and that then
only

only thefe elements combine, two and two together, to form water and carbonic acid.

The effects of the vinous fermentation upon
fugar is thus reduced to the mere feparation of
its elements into two portions ; one part is oxygenated at the expence of the other, fo as to
form carbonic acid, whilft the other part, being
difoxyginated in favour of the former, is converted into the combuftible fubftance alkohol ;
therefore, if it were poffible to reunite alkohol
and carbonic acid together, we ought to form
fugar. It is evident that the charcoal and hydrogen in the alkohol do not exift in the ftate
of oil, they are combined with a portion of
oxygen, which renders them mifcible with water ; wherefore thefe three fubftances, oxygen,
hydrogen, and charcoal, exift here likewife in
a fpecies of equilibrium or reciprocal combination ; and in fact, when they are made to pafs
through a red hot tube of glafs or porcelain,
this union or equilibrium is deftroyed, the elements become combined, two and two, and water and carbonic acid are formed.

I had formally advanced, in my firft Memoirs upon the formation of water, that it was
decompofed in a great number of chemical experiments, and particularly during the vinous
fermentation. I then fuppofed that water exifted ready formed in fugar, though I am now
convinced that fugar only contains the elements

proper

proper for compofing it. It may be readily conceived, that it muft have coft me a good deal to
abandon my firft notions, but by feveral years
reflection, and after a great number of experiments and obfervations upon vegetable fubftances, I have fixed my ideas as above.

I fhall finifh what I have to fay upon vinous
fermentation, by obferving, that it furnifhes us
with the means of analyfing fugar and every
vegetable fermentable matter. We may confider the fubftances fubmitted to fermentation,
and the products refulting from that operation,
as forming an algebraic equation ; and, by fucceffively fuppofing each of the elements in this
equation unknown, we can calculate their values in fucceffion, and thus verify our experiments by calculation, and our calculation by experiment reciprocally. I have often fuccefsfully
employed this method for correcting the firft
refults of my experiments, and to direct me in
the proper road for repeating them to advantage. I have explained myfelf at large upon
this fubject, in a Memoir upon vinous fermentation already prefented to the Academy, and
which will fpeedily be publifhed.

CHAP.

C H A P.   XIV.

Of the Putrefactive Fermentation.

THE phenomena of putrefaction are caufed, like thofe of vinous fermentation, by the operation of very complicated affinities. The conftituent elements of the bodies fubmitted to this procefs ceafe to continue in equilibrium in the threefold combination, and form themfelves anew into binary combinations *, or compounds, confifting of two elements only ; but thefe are entirely different from the refults produced by the vinous fermentation. Inftead of one part of the hydrogen remaining united with part of the water and charcoal to form alkohol, as in the vinous fermentation, the whole of the hydrogen is diffipated, during putrefaction, in the form of hydrogen gas, whilft, at the fame time, the oxygen and charcoal, uniting with caloric, efcape in the form of carbonic acid gas ; fo that, when the whole procefs is finifhed, efpeci-

ally

---

* Binary combinations are fuch as confift of two fimple elements combined together.   Ternary, and quaternary, confift of three and four elements.—E.

ally if the materials have been mixed with a sufficient quantity of water, nothing remains but the earth of the vegetable mixed with a small portion of charcoal and iron. Thus putrefaction is nothing more than a complete analysis of vegetable substance, during which the whole of the constituent elements is disengaged in form of gas, except the earth, which remains in the state of mould *.

Such is the result of putrefaction when the substances submitted to it contain only oxygen, hydrogen, charcoal and a little earth. But this case is rare, and these substances putrify imperfectly and with difficulty, and require a considerable time to complete their putrefaction. It is otherwise with substances containing azote, which indeed exists in all animal matters, and even in a considerable number of vegetable substances. This additional element is remarkably favourable to putrefaction ; and for this reason animal matter is mixed with vegetable, when the putrefaction of these is wished to be hastened. The whole art of forming composts and dunghills, for the purposes of agriculture, consists in the proper application of this admixture.

The addition of azote to the materials of putrefaction not only accelerates the process,

that

* In the Third Part will be given the description of an apparatus proper for being used in experiments of this kind.—A.

that element likewife combines with part of the hydrogen, and forms a new fubftance called *volatile alkali* or *ammoniac*. The refults obtained by analyfing animal matters, by different proceffes, leave no room for doubt with regard to the conftituent elements of ammoniac ; whenever the azote has been previoufly feparated from thefe fubftances, no ammoniac is produced ; and in all cafes they furnifh ammoniac only in proportion to the azote they contain. This compofition of ammoniac is likewife fully proved by Mr Berthollet, in the Memoirs of the Academy for 1785, p. 316. where he gives a variety of analytical proceffes by which ammoniac is decompofed, and its two elements, azote and hydrogen, procured feparately.

I already mentioned in Chap. X. that almoft all combuftible bodies were capable of combining with each other ; hydrogen gas poffeffes this quality in an eminent degree, it diffolves charcoal, fulphur, and phofphorus, producing the compounds named *carbonated hydrogen gas*, *fulphurated hydrogen gas*, and *phofphorated hydrogen gas*. The two latter of thefe gaffes have a peculiarly difagreeable flavour ; the fulphurated hydrogen gas has a ftrong refemblance to the fmell of rotten eggs, and the phofphorated fmells exactly like putrid fifh. Ammoniac has likewife a peculiar odour, not lefs penetrating, or lefs difagreeable, than thefe other gaffes. From the

the mixture of thefe different flavours proceeds
the fetor which accompanies the putrefaction of
animal fubftances.   Sometimes ammoniac predominates, which is eafily perceived by its
fharpnefs upon the eyes; fometimes, as in fecu-
lent matters, the fulphurated gas is moft preva-
lent ; and fometimes, as in putrid herrings, the
phofphorated hydrogen gas is moft abundant,

I long fuppofed that nothing could derange
or interrupt the courfe of putrefaction ; but Mr
Fourcroy and Mr Thouret have obferved fome
peculiar phenomena in dead bodies, buried at a
certain depth, and preferved to a certain degree, from contact with air ; having found the
mufcular flefh frequently converted into true a-
nimal fat.   This muft have arifen from the difengagement of the azote, naturally contained in
the animal fubftance, by fome unknown caufe,
leaving only the hydrogen and charcoal remaining, which are the elements proper for producing fat or oil.   This obfervation upon the
poffibility of converting animal fubftances into
fat may fome time or other lead to difcoveries
of great importance to fociety.   The faeces of
animals, and other excrementitious matters, are
chiefly compofed of charcoal and hydrogen, and
approach confiderably to the nature of oil, of
which they furnifh a confiderable quantity by
diftillation with a naked fire ; but the intole-
rable foetor which accompanies all the products
of

of thefe fubftances prevents our expecting that, at leaft for a long time, they can be rendered ufeful in any other way than as manures.

I have only given conjectural approximations in this Chapter upon the compofition of animal fubftances, which is hitherto but imperfectly underftood. We know that they are compofed of hydrogen, charcoal, azote, phofphorus, and fulphur, all of which, in a ftate of quintuple combination, are brought to the ftate of oxyd by a larger or fmaller quantity of oxygen. We are, however, ftill unacquainted with the proportions in which thefe fubftances are combined, and muft leave it to time to complete this part of chemical analyfis, as it has already done with feveral others.

T   C H A P.

C H A P. XV.

Of the Acetous Fermentation.

THE acetous fermentation is nothing more than the acidification or oxygenation of wine *, produced in the open air by means of the abſorption of oxygen. The reſulting acid is the acetous acid, commonly called Vinegar, which is compoſed of hydrogen and charcoal united together in proportions not yet aſcertained, and changed into the acid ſtate by oxygen. As vinegar is an acid, we might conclude from analogy that it contains oxygen, but this is put beyond doubt by direct experiments: In the firſt place, we cannot change wine into vinegar without the contact of air containing oxygen; ſecondly, this proceſs is accompanied by a diminution of the volume of the air in which it is carried on from the abſorption of its oxygen; and, thirdly, wine may be changed into vinegar by any other means of oxygenation.

Independent

---

* The word Wine, in this chapter, is uſed to ſignify the liquor produced by the vinous fermentation, whatever vegetable ſubſtance may have been uſed for obtaining it.—E.

Independent of the proofs which thefe facts
furnifh of the acetous acid being produced by
the oxygenation of wine, an experiment made
by Mr Chaptal, Profeffor of Chemiftry at Mont-
pellier, gives us a diftinct view of what takes
place in this procefs. He impregnated water
with about its own bulk of carbonic acid from
fermenting beer, and placed this water in a cellar in veffels communicating with the air, and
in a fhort time the whole was converted into
acetous acid. The carbonic acid gas procured
from beer vats in fermentation is not perfectly
pure, but contains a fmall quantity of alkohol
in folution, wherefore water impregnated with
it contains all the materials neceffary for forming the acetous acid. The alkohol furnifhes
hydrogen and one portion of charcoal, the carbonic acid furnifhes oxygen and the reft of the
charcoal, and the air of the atmofphere furnifhes the reft of the oxygen neceffary for changing
the mixture into acetous acid. From this obfervation it follows, that nothing but hydrogen
is wanting to convert carbonic acid into acetous
acid; or more generally, that, by means of hydrogen, and according to the degree of oxygenation, carbonic acid may be changed into all
the vegetable acids; and, on the contrary, that,
by depriving any of the vegetable acids of their
hydrogen, they may be converted into carbonic
acid.

Although

Although the principal facts relating to the acetous acid are well known, yet numerical exactitude is ftill wanting, till furnifhed by more exact experiments than any hitherto performed; wherefore I fhall not enlarge any farther upon the fubject. It is fufficiently fhown by what has been faid, that the conftitution of all the vegetable acids and oxyds is exactly conformable to the formation of vinegar; but farther experiments are neceffary to teach us the proportion of the conftituent elements in all thefe acids and oxyds. We may eafily perceive, however, that this part of chemiftry, like all the reft of its divifions, makes rapid progrefs towards perfection, and that it is already rendered greatly more fimple than was formerly believed.

C H A P.

C H A P. XVI.

Of the Formation of Neutral Salts, and of their different Bases.

WE have juft feen that all the oxyds and acids from the animal and vegetable kingdoms are formed by means of a fmall number of fimple elements, or at leaft of fuch as have not hitherto been fufceptible of decompofition, by means of combination with oxygen; thefe are azote, fulphur, phofphorus, charcoal, hydrogen, and the muriatic radical *. We may juftly admire the fimplicity of the means employed by nature to multiply qualities and forms, whether by combining three or four acidifiable bafes in different proportions, or by altering the dofe of oxygen employed for oxydating or acidifying them. We fhall find the means no lefs fimple and diverfified, and as abundantly productive of forms and qualities, in the order of bodies we are now about to treat of.

Acidifiable

---

* I have not ventured to omit this element, as here enumerated with the other principles of animal and vegetable fubftances, though it is not at all taken notice of in the preceding chapters as entering into the compofition of thefe bodies.—E.

Acidifiable fubftances, by combining with oxygen, and their confequent converfion into acids, acquire great fufceptibility of farther combination; they become capable of uniting with earthy and metallic bodies, by which means neutral falts are formed. Acids may therefore be confidered as true *falifying* principles, and the fubftances with which they unite to form neutral falts may be called *falifiable* bafes : The nature of the union which thefe two principles form with each other is meant as the fubject of the prefent chapter.

This view of the acids prevents me from confidering them as falts, though they are poffeffed of many of the principal properties of faline bodies, as folubility in water, &c. I have already obferved that they are the refult of a firft order of combination, being compofed of two fimple elements, or at leaft of elements which act as if they were fimple, and we may therefore rank them, to ufe the language of Stahl, in the order of *mixts*. The neutral falts, on the contrary, are of a fecondary order of combination, being formed by the union of two *mixts* with each other, and may therefore be termed *compounds*. Hence I fhall not arrange the alkalies * or earths in the clafs of falts, to which I
allot

* Perhaps my thus rejecting the alkalies from the clafs of falts may be confidered as a capital defect in
the

allot only fuch as are compofed of an oxygenated fubftance united to a bafe.

I have already enlarged fufficiently upon the
formation of acids in the preceding chapter,
and fhall not add any thing farther upon that
fubject; but having as yet given no account of
the falifiable bafes which are capable of uniting
with them to form neutral falts, I mean, in this
chapter, to give an account of the nature and
origin of each of thefe bafes. Thefe are potafh,
foda, ammoniac, lime, magnefia, barytes, argill *, and all the metallic bodies.

§ 1. Of Potafh.

We have already fhown, that, when a vegetable fubftance is fubmitted to the action of fire
in diftilling veffels, its component elements, oxygen, hydrogen, and charcoal, which formed a
threefold combination in a ftate of equilibrium,
unite, two and two, in obedience to affinities
which act conformable to the degree of heat
employed.

the method I have adopted, and I am ready to admit
the charge; but this inconvenience is compenfated
by fo many advantages, that I could not think it of
fufficient confequence to make me alter my plan.—A.

* Called Alumine by Mr Lavoifier; but as Argill
has been in a manner naturalized to the language for
this fubftance by Mr Kirwan, I have ventured to ufe
it in preference.—E.

employed. Thus, at the firſt application of the fire, whenever the heat produced exceeds the temperature of boiling water, part of the oxygen and hydrogen unite to form water; ſoon after the reſt of the hydrogen, and part of the charcoal, combine into oil; and, laſtly, when the fire is puſhed to the red heat, the oil and water, which had been formed in the early part of the procefs, become again decompofed, the oxygen and charcoal unite to form carbonic acid, a large quantity of hydrogen gas is ſet free, and nothing but charcoal remains in the retort.

A great part of thefe phenomena occur during the combuſtion of vegetables in the open air; but, in this cafe, the prefence of the air introduces three new fubſtances, the oxygen and azote of the air and caloric, of which two at leaſt produce confiderable changes in the refults of the operation. In proportion as the hydrogen of the vegetable, or that which refults from the decompofition of the water, is forced out in the form of hydrogen gas by the progrefs of the fire, it is ſet on fire immediately upon getting in contaƈt with the air, water is again formed, and the greater part of the caloric of the two gaffes becoming free produces flame. When all the hydrogen gas is driven out, burnt, and again reduced to water, the remaining charcoal continues to burn, but without flame; it is
formed

formed into carbonic acid, which carries off a portion of caloric fufficient to give it the gaffeous form; the reft of the caloric, from the oxygen of the air, being fet free, produces the heat and light obferved during the combuftion of charcoal. The whole vegetable is thus reduced into water and carbonic acid, and nothing remains but a fmall portion of gray earthy matter called afhes, being the only really fixed principles which enter into the conftitution of vegetables.

The earth, or rather afhes, which feldom exceeds a twentieth part of the weight of the vegetable, contains a fubftance of a particular nature, known under the name of fixed vegetable alkali, or potafh. To obtain it, water is poured upon the afhes, which diffolves the potafh, and leaves the afhes which are infoluble; by afterwards evaporating the water, we obtain the potafh in a white concrete form : It is very fixed even in a very high degree of heat. I do not mean here to defcribe the art of preparing potafh, or the method of procuring it in a ftate of purity, but have entered upon the above detail that I might not ufe any word not previoufly explained.

The potafh obtained by this procefs is always lefs or more faturated with carbonic acid, which is eafily accounted for: As the potafh does not form, or at leaft is not fet free, but in propor-

U   tion

tion as the charcoal of the vegetable is converted into carbonic acid by the addition of oxygen,
either from the air or the water, it follows, that
each particle of potafh, at the inftant of its formation, or at leaft of its liberation, is in contact
with a particle of carbonic acid, and, as there
is a confiderable affinity between thefe two fubftances, they naturally combine together.   Although the carbonic acid has lefs affinity with
potafh than any other acid, yet it is difficult to
feparate the laft portions from it.   The moft
ufual method of accomplifhing this is to diffolve
the potafh in water; to this folution add two or
three times its weight of quicklime, then filtrate
the liquor and evaporate it in clofe veffels; the
faline fubftance left by the evaporation is potafh almoft entirely deprived of carbonic acid.
In this ftate it is foluble in an equal weight of
water, and even attracts the moifture of the air
with great avidity; by this property it furnifhes
us with an excellent means of rendering air or
gas dry by expofing them to its action.   In this
ftate it is foluble in alkohol, though not when
combined with carbonic acid; and Mr Berthollet
employs this property as a method of procuring
potafh in the ftate of perfect purity.

All vegetables yield lefs or more of potafh in
confequence of combuftion, but it is furnifhed
in various degrees of purity by different vegetables; ufually, indeed, from all of them it is
mixed

mixed with different falts from which it is eafily
feparable. We can hardly entertain a doubt
that the afhes, or earth which is left by vegetables in combuftion, pre-exifted in them before
they were burnt, forming what may be called
the fkeleton, or offeous part of the vegetable.
But it is quite otherwife with potafh ; this fubftance has never yet been procured from vegetables but by means of proceffes or intermedia
capable of furnifhing oxygen and azote, fuch
as combuftion, or by means of nitric acid ; fo
that it is not yet demonftrated that potafh may
not be a produce from thefe operations. I have
begun a feries of experiments upon this object,
and hope foon to be able to give an account of
their refults.

§ 2. *Of Soda.*

Soda, like potafh, is an alkali procured by
lixiviation from the afhes of burnt plants, but
only from thofe which grow upon the fea-fide,
and efpecially from the herb *kali,* whence is derived the name *alkali,* given to this fubftance by
the Arabians. It has fome properties in common with potafh, and others which are entirely
different : In general, thefe two fubftances have
peculiar characters in their faline combinations
which are proper to each, and confequently
diftinguifh them from each other; thus foda,
which,

which, as obtained from marine plants, is ufually entirely faturated with carbonic acid, does
not attract the humidity of the atmofphere like
potafh, but, on the contrary, deficcates, its criftals efflorefce, and are converted into a white
powder having all the properties of foda, which
it really is, having only loft its water of criftallization.

Hitherto we are not better acquainted with
the conftituent elements of foda than with thofe
of potafh, being equally uncertain whether it
previoufly exifted ready formed in the vegetable
or is a combination of elements effected by combuftion. Analogy leads us to fufpect that azote
is a conftituent element of all the alkalies, as is
the cafe with ammoniac; but we have only flight
prefumptions, unconfirmed by any decifive experiments, refpecting the compofition of potafh
and foda.

§ 3. *Of Ammoniac.*

We have, however, very accurate knowledge
of the compofition of ammoniac, or volatile alkali, as it is called by the old chemifts. Mr
Berthollet, in the Memoirs of the Academy for
1784, p. 316. has proved by analyfis, that 1000
parts of this fubftance confift of about 807 parts
of azote combined with 193 parts of hydrogen.

Ammoniac

Ammoniac is chiefly procurable from animal fubftances by diftillation, during which procefs the azote and hydrogen neceffary to its formation unite in proper proportions ; it is not, however, procured pure by this procefs, being mixed with oil and water, and moftly faturated with carbonic acid. To feparate thefe fubftances it is firft combined with an acid, the muriatic for inftance, and then difengaged from that combination by the addition of lime or potafh. When ammoniac is thus produced in its greateft degree of purity it can only exift under the gaffeous form, at leaft in the ufual temperature of the atmofphere, it has an exceffively penetrating fmell, is abforbed in large quantities by water, efpecially if cold and affifted by compreffion. Water thus faturated with ammoniac has ufually been termed volatile alkaline fluor ; we fhall call it either fimply ammoniac, or liquid ammoniac, and ammoniacal gas when it exifts in the aëriform ftate.

§ 4. *Of Lime, Magnefia, Barytes, and Argill.*

The compofition of thefe four earths is totally unknown, and, until by new difcoveries their conftituent elements are afcertained, we are certainly authorifed to confider them as fimple bodies. Art has no fhare in the production of thefe earths, as they are all procured ready form-

ed

ed from nature; but, as they have all, efpecially
the three firft, great tendency to combination,
they are never found pure. Lime is ufually faturated with carbonic acid in the ftate of chalk,
calcarious fpars, moft of the marbles, &c.;
fometimes with fulphuric acid, as in gypfum
and plafter ftones; at other times with fluoric
acid forming vitreous or fluor fpars; and, laftly, it is found in the waters of the fea, and of
faline fprings, combined with muriatic acid. Of
all the falifiable bafes it is the moft univerfally
fpread through nature.

Magnefia is found in mineral waters, for the
moft part combined with fulphuric acid; it is
likewife abundant in fea-water, united with muriatic acid; and it exifts in a great number of
ftones of different kinds.

Barytes is much lefs common than the three
preceding earths; it is found in the mineral
kingdom, combined with fulphuric acid, forming heavy fpars, and fometimes, though rarely,
united to carbonic acid.

Argill, or the bafe of alum, having lefs tendency to combination than the other earths, is
often found in the ftate of argill, uncombined
with any acid. It is chiefly procurable from
clays, of which, properly fpeaking, it is the bafe,
or chief ingredient.

§ 5. *Of*

§ 5.   Of Metallic Bodies.

The metals, except gold, and fometimes filver, are rarely found in the mineral kingdom in
their metallic ftate, being ufually lefs or more
faturated with oxygen, or combined with fulphur, arfenic, fulphuric acid, muriatic acid, carbonic acid, or phofphoric acid. Metallurgy, or
the docimaftic art, teaches the means of feparating them from thefe foreign matters; and
for this purpofe we refer to fuch chemical books
as treat upon thefe operations.

We are probably only acquainted as yet with
a part of the metallic fubftances exifting in nature, as all thofe which have a ftronger affinity
to oxygen, than charcoal poffeffes, are incapable
of being reduced to the metallic ftate, and, confequently, being only prefented to our obfervation under the form of oxyds, are confounded
with earths.   It is extremely probable that barytes, which we have juft now arranged with
earths, is in this fituation; for in many experiments it exhibits properties nearly approaching
to thofe of metallic bodies.   It is even poffible
that all the fubftances we call earths may be
only metallic oxyds, irreducible by any hitherto
known procefs.

Thofe metallic bodies we are at prefent acquainted with, and which we can reduce to the

metallic

metallic or reguline ſtate, are the following ſe-
venteen :

1. Arſenic.	7. Biſmuth.	13. Copper.
2. Molybdena.	8. Antimony.	14. Mercury.
3. Tungſtein.	9. Zinc.	15. Silver.
4. Manganeſe.	10. Iron.	16. Platina.
5. Nickel.	11. Tin.	17. Gold.
6. Cobalt.	12. Lead.	

I only mean to conſider theſe as ſaliſiable
baſes, without entering at all upon the conſideration of their properties in the arts, and for
the uſes of ſociety. In theſe points of view each
metal would require a complete treatiſe, which
would lead me far beyond the bounds I have
preſcribed for this work.

C H A P.

CHAP. XVII.

Continuation of the Obfervations upon Salifiable Bafes, and the Formation of Neutral Salts.

IT is neceffary to remark, that earths and alkalies unite with acids to form neutral falts without the intervention of any medium, whereas metallic fubftances are incapable of forming this combination without being previoufly lefs or more oxygenated; ftrictly fpeaking, therefore, metals are not foluble in acids, but only metallic oxyds. Hence, when we put a metal into an acid for folution, it is neceffary, in the firft place, that it become oxygenated, either by attracting oxygen from the acid or from the water; or, in other words, that a metal cannot be diffolved in an acid unlefs the oxygen, either of the acid, or of the water mixed with it, has a ftronger affinity to the metal than to the hydrogen or the acidifiable bafe; or, what amounts to the fame thing, that no metallic folution can take place without a previous decompofition of the water, or the acid in which it is made. The explanation of the principal phenomena of metallic folution depends entire-

X   ly

ly upon this fimple obfervation, which was overlooked even by the illuftrious Bergman.

The firft and moft ftriking of thefe is the ef-fervefcence, or, to fpeak lefs equivocally, the difengagement of gas which takes place during the folution; in the folutions made in nitric acid this effervefcence is produced by the difen-gagement of nitrous gas; in folutions with ful-phuric acid it is either fulphurous acid gas or hydrogen gas, according as the oxydation of the metal happens to be made at the expence of the fulphuric acid or of the water. As both nitric acid and water are compofed of elements which, when feparate, can only exift in the gaf-feous form, at leaft in the common temperature of the atmofphere, it is evident that, whenever either of thefe is deprived of its oxygen, the remain-ing element muft inftantly expand and affume the ftate of gas; the effervefcence is òccafioned by this fudden converfion from the liquid to the gafieous ftate. The fame decompofition, and confequent formation of gas, takes place when folutions of metals are made in fulphuric acid : In general, efpecially by the humid way, metals do not attract all the oxygen it contains; they therefore reduce it, not into fulphur, but into ful-phurous acid, and as this acid can only exift as gas in the ufual temperature, it is difengaged, and occafions effervefcence.

The

The second phenomenon is, that, when the metals have been previously oxydated, they all diffolve in acids without effervefcence: This is eafily explained; becaufe, not having now any occafion for combining with oxygen, they neither decompofe the acid nor the water by which, in the former cafe, the effervefcence is occafioned.

A third phenomenon, which requires particular confideration, is, that none of the metals produce effervefcence by folution in oxygenated muriatic acid. During this procefs the metal, in the firft place, carries off the excefs of oxygen from the oygenated muriatic acid, by which it becomes oxydated, and reduces the acid to the ftate of ordinary muriatic acid. In this cafe there is no production of gas, not that the muriatic acid does not tend to exift in the gaffeous ftate in the common temperature, which it does equally with the acids formerly mentioned, but becaufe this acid, which otherwife would expand into gas, finds more water combined with the oxygenated muriatic acid than is neceffary to retain it in the liquid form; hence it does not difengage like the fulphurous acid, but remains, and quietly diffolves and combines with the metallic oxyd previoufly formed from its faperabundant oxygen.

The fourth phenomenon is, that metals are abfolutely infoluble in fuch acids as have their bafes joined to oxygen by a ftronger affinity than thefe metals are capable of exerting upon that acidifying principle. Hence filver, mercury, and lead, in their metallic ftates, are infoluble in muriatic acid, but, when previoufly oxydated, they become readily foluble without effervef- cence.

From thefe phenomena it appears that oxygen is the bond of union between metals and acids ; and from this we are led to fuppofe that oxygen is contained in all fubftances which have a ftrong affinity with acids : Hence it is very probable the four eminently falifiable earths contain oxygen, and their capability of uniting with acids is produced by the intermediation of that element. What I have formerly noticed relative to thefe earths is confiderably ftrengthened by the above confiderations, viz. that they may very poffibly be metallic oxyds, with which oxygen has a ftronger affinity than with charcoal, and confequently not reducible by any known means.

All the acids hitherto known are enumerated in the following table, the firft column of which contains the names of the acids according to the new nomenclature, and in the fecond column are placed the bafes or radicals of thefe acids, with obfervations.

Names

Names of the Acids.	*Names of the Bases, with Observations.*

1. Sulphurous
2. Sulphuric } Sulphur.

3. Phosphorous
4. Phosphoric } Phosphorus.

5. Muriatic
6. Oxygenated muriatic } Muriatic radical or base, hitherto unknown.

7. Nitrous
8. Nitric
9. Oxygenated nitric } Azote.

10. Carbonic — Charcoal

11. Acetous
12. Acetic
13. Oxalic
14. Tartarous
15. Pyro-tartarous
16. Citric
17. Malic
18. Pyro-lignous
19. Pyro-mucous

The bases or radicals of all these acids seem to be formed by a combination of charcoal and hydrogen; and the only difference seems to be owing to the different proportions in which these elements combine to form their bases, and to the different doses of oxygen in their acidification. A connected series of accurate experiments is still wanted upon this subject.

20. Gallic
21. Prussic
22. Benzoic
23. Succinic
24. Camphoric
25. Lactic
26. Saccho-lactic

Our knowledge of the bases of these acids is hitherto imperfect; we only know that they contain hydrogen and charcoal as principal elements, and that the prussic acid contains azote.

27. Bombic
28. Formic
29. Sebacic

The base of these and all the acids procured from animal substances seems to consist of charcoal, hydrogen, phosphorus, and azote.

30. Boracic
31. Fluoric } The bases of these two are hitherto entirely unknown.

32. Antimonic — Antimony.
33. Argentic — Silver.
34. Arseniac * — Arsenic.

Names

* This term swerves a little from the rule in making the name of this acid terminate in *ac* instead of *ic*. The base and acid are distinguished in French by *arsenic* and *arsenique*; but, having chosen the English termination *ic* to translate the French *ique*, I was obliged to use this small deviation.—E.

Names of the Acids.	*Names of the Bases.*
35. Bifmuthic	Bifmuth.
36. Cobaltic	Cobalt.
37. Cupric	Copper.
38. Stannic	Tin.
39. Ferric	Iron.
40. Munganic	Manganefe.
41. Mercuric *	Mercury.
42. Molybdic	Molybdena.
43. Nickolic	Nickel,
44. Auric	Gold.
45. Platinic	Platina.
46. Plumbic	Lead.
47. Tungftic	Tungftein.
48. Zincic	Zinc.

In this lift, which contains 48 acids, I have enumerated 17 metallic acids hitherto very imperfectly known, but upon which Mr Berthollet is about to publifh a very important work. It cannot be pretended that all the acids which exift in nature, or rather all the acidifiable bafes, are yet difcovered ; but, on the other hand, there are confiderable grounds for fuppofing that a more accurate inveftigation than has hitherto been attempted will diminifh the number of the vegetable acids, by fhowing that feveral of thefe, at prefent confidered as diftinct acids, are only modi-

* Mr Lavoifier has hydrargirique ; but mercurius being ufed for the bafe or metal, the name of the acid, as above, is equally regular, and lefs harfh.—E.

modifications of others. All that can be done in the prefent ftate of our knowledge is, to give a view of chemiftry as it really is, and to eftablifh fundamental principles, by which fuch bodies as may be difcovered in future may receive names, in conformity with one uniform fyftem.

The known falifiable bafes, or fubftances capable of being converted into neutral falts by union with acids, amount to 24; viz. 3 alkalies, 4 earths, and 17 metallic fubftances; fo that, in the prefent ftate of chemical knowledge, the whole poffible number of neutral falts amounts to 1152 *. This number is upon the fuppofition that the metallic acids are capable of diffolving other metals, which is a new branch of chemiftry not hitherto inveftigated, upon which depends all the metallic combinations named *vitreous*. There is reafon to believe that many of thefe fuppofable faline combinations are not capable of being formed, which muft greatly reduce the real number of neutral falts producible by nature and art. Even if we fuppofe the real number to amount only to five or fix hundred fpecies of poffible neutral falts, it is evident that, were we to diftinguifh them, after the

* This number excludes all triple falts, or fuch as contain more than one falifiable bafe, all the falts whofe bafes are over or under faturated with acid, and thofe formed by the nitro-muriatic acid.—E.

the manner of the ancients, either by the names of their firſt diſcoverers, or by terms derived from the ſubſtances from which they are procured, we ſhould at laſt have ſuch a confuſion of arbitrary deſignations, as no memory could poſſibly retain. This method might be tolerable in the early ages of chemiſtry, or even till within theſe twenty years, when only about thirty ſpecies of ſalts were known ; but, in the preſent times, when the number is augmenting daily, when every new acid gives us 24 or 48 new ſalts, according as it is capable of one or two degrees of oxygenation, a new method is certainly neceſſary. The method we have adopted, drawn from the nomenclature of the acids, is perfectly analogical, and, following nature in the ſimplicity of her operations, gives a natural and eaſy nomenclature applicable to every poſſible neutral ſalt.

In giving names to the different acids, we expreſs the common property by the generical term *acid*, and diſtinguiſh each ſpecies by the name of its peculiar acidifiable baſe. Hence the acids formed by the oxygenation of ſulphur, phoſphorus, charcoal, &c. are called *ſulphuric acid, phoſphoric acid, carbonic acid*, &c. We thought it likewiſe proper to indicate the different degrees of ſaturation with oxygen, by different terminations of the ſame ſpecific names. Hence

Hence we diftinguifh between fulphurous and fulphuric, and between phofphorous and phofphoric acids, &c.

By applying thefe principles to the nomenclature of neutral falts, we give a common term to all the neutral falts arifing from the combination of one acid, and diftinguifh the fpecies by adding the name of the falifiable bafe. Thus, all the neutral falts having fulphuric acid in their compofition are named *fulphats ;* thofe formed by the phofphoric acid, *phofphats,* &c. The fpecies being diftinguifhed by the names of the falifiable bafes gives us *fulphat of potafh, fulphat of foda, fulphat of ammoniac, fulphat of lime, fulphat of iron,* &c. As we are acquainted with 24 falifiable bafes, alkaline, earthy, and metallic, we have confequently 24 fulphats, as many phofphats, and fo on through all the acids. Sulphur is, however, fufceptible of two degrees of oxygenation, the firft of which produces fulphurous, and the fecond, fulphuric acid ; and, as the neutral falts produced by thefe two acids, have different properties, and are in fact different faits, it becomes neceffary to diftinguifh thefe by peculiar terminations ; we have therefore diftinguifhed the neutral falts formed by the acids in the firft or leffer degree of oxygenation, by changing the termination *at* into *ite,* as *fulphites, phofphites* *, &c. Thus, oxy-

Y   genated

---

* As all the fpecific names of the acids in the new
nomen-

genated or acidified fulphur, in its two degrees
of oxygenation is capable of forming 48 neutral falts, 24 of which are fulphites, and as many fulphats ; which is likewife the cafe with all
the acids capable of two degrees of oxygenation *.

It were both tirefome and unneceffary to follow thefe denominations through all the varieties of their poffible application ; it is enough
to have given the method of naming the various
falts, which, when once well underftood, is eafily applied to every poffible combination. The
name of the combuftible and acidifiable body
being once known, the names of the acid it is
capable of forming, and of all the neutral com-
binatious

nomenclature are adjectives, they would have applied
feverally to the various falifiable bafes, without the invention of other terms, with perfect diftinctnefs. Thus,
fulphurous potafh, and *fulphuric potafh,* are equally diftinct
as *fulphite of potafh,* and *fulphat of potafh* ; and have the
advantage of being more eafily retained in the memory, becaufe more naturally arifing from the acids themfelves, than the arbitrary terminations adopted by Mr
Lavoifier.—E.

* There is yet a third degree of oxygenation of a-
cids, as the oxygenated muriatic and oxygenated nitric
acids. The terms applicable to the neutral falts refulting from the union of thefe acids with falifiable bafes
is fupplied by the Author in the Second Part of this
Work. Thefe are formed by prefixing the word *oxygenated* to the name of the falt produced by the fecond
degree of oxygenation. Thus, *oxygenated* muriat of potafh, *oxygenated* nitrat of foda, &c.—E.

binations the acid is fufceptible of entering into, are moft readily remembered. Such as require a more complete illuftration of the methods in which the new nomenclature is applied
will, in the Second Part of this book, find
Tables which contain a full enumeration of all
the neutral falts, and, in general, all the poffible chemical combinations, fo far as is confiftent with the prefent ftate of our knowledge.
To thefe I fhall fubjoin fhort explanations, containing the beft and moft fimple means of procuring the different fpecies of acids, and fome
account of the general properties of the neutral
falts they produce.

I fhall not deny, that, to render this work
more complete, it would have been neceffary to
add particular obfervations upon each fpecies of
falt, its folubility in water and alkohol, the proportions of acid and of falifiable bafe in its
compofition, the quantity of its water of criftallization, the different degrees of faturation
it is fufceptible of, and, finally, the degree
of force or affinity with which the acid adheres to the bafe. This immenfe work has been
already begun by Meffrs Bergman, Morveau,
Kirwan, and other celebrated chemifts, but is
hitherto only in a moderate ftate of advancement, even the principles upon which it is
founded are not perhaps fufficiently accurate.

Thefe

Thefe numerous details would have fwelled this elementary treatife to much too great a fize; befides that, to have gathered the neceffary materials, and to have completed all the feries of experiments requifite, muft have retarded the publication of this book for many years. This is a vaft field for employing the zeal and abilities of young chemifts, whom I would advife to endeavour rather to do well than to do much, and to afcertain, in the firft place, the compofition of the acids, before entering upon that of the neutral falts. Every edifice which is intended to refift the ravages of time fhould be built upon a fure foundation; and, in the prefent ftate of chemiftry, to attempt difcoveries by experiments, either not perfectly exact, or not fufficiently rigorous, will ferve only to interrupt its progrefs, inftead of contributing to its advancement.

PART

PART II.

Of the Combination of Acids with Salifiable Bafes, and of the Formation of Neutral Salts.

INTRODUCTION.

IF I had ftrictly followed the plan I at firft laid down for the conduct of this work, I would have confined myfelf, in the Tables and accompanying obfervations which compofe this Second Part, to fhort definitions of the feveral known acids, and abridged accounts of the proceffes by which they are obtainable, with a mere nomenclature or enumeration of the neutral falts which refult from the combination of thefe acids with the various falifiable bafes. But I afterwards found that the addition of fimilar Tables of all the fimple fubftances which enter

into

into the compofition of the acids and oxyds,
together with the various poffible combinations
of thefe elements, would add greatly to the utility of this work, without being any great increafe to its fize. Thefe additions, which are
all contained in the twelve firft fections of this
Part, and the Tables annexed to thefe, form a
kind of recapitulation of the firft fifteen Chapters of the Firft Part : The reft of the Tables
and Sections contain all the faline combinations.

It muft be very apparent that, in this Part of
the Work, I have borrowed greatly from what
has been already publifhed by Mr de Morveau
in the Firft Volume of the *Encyclopedie par ordre
des Matières*. I could hardly have difcovered
a better fource of information, efpecially when
the difficulty of confulting books in foreign
languages is confidered. I make this general
acknowledgment on purpofe to fave the trouble
of references to Mr de Morveau's work in the
courfe of the following part of mine.

TABLE

TABLE OF SIMPLE SUBSTANCES.

Simple fubftances belonging to all the kingdoms of nature, which may be confidered as the elements of bodies.

New Names.	Correfpondent old Names.
Light - - -	Light.
Caloric - - -	{ Heat. Principle or element of heat. Fire. Igneous fluid. Matter of fire and of heat.
Oxygen - - -	{ Dephlogifticated air. Empyreal air. Vital air, or Bafe of vital air.
Azote - - -	{ Phlogifticated air or gas. Mephitis, or its bafe.
Hydrogen - -	{ Inflammable air or gas, or the bafe of inflammable air.

Oxydable and Acidifiable fimple Subftances not Metallic.

New Names.	Correfpondent old names.
Sulphur - - -	}
Phofphorus - - -	} The fame names.
Charcoal - - -	}
Muriatic radical -	}
Fluoric radical - -	} Still unknown.
Boracic radical - -	}

Oxydable and Acidifiable fimple Metallic Bodies.

New Names.		Correfpondent Old Names.
Antimony -		Antimony.
Arfenic - -		Arfenic.
Bifmuth - -		Bifmuth.
Cobalt - -		Cobalt.
Copper - -		Copper.
Gold - -		Gold.
Iron - - -		Iron.
Lead - - -	Regulus of	Lead.
Manganefe - -		Manganefe.
Mercury - -		Mercury.
Molybdena - -		Molybdena.
Nickel - - -		Nickel.
Platina - -		Platina.
Silver - -		Silver.
Tin - -		Tin.
Tungftein - -		Tungftein.
Zinc - -		Zinc.

Salifiable

Salifiable fimple Earthy Subftances.

New Names.	*Correfpondent old Names.*
Lime	{ Chalk, calcareous earth. { Quicklime.
Magnefia	{ Magnefia, bafe of Epfom falt. { Calcined or cauftic magnefia.
Barytes	Barytes, or heavy earth.
Argill	Clay, earth of alum.
Silex	Siliceous or vitrifiable earth.

SECT. I.—*Obfervations upon the Table of Simple Subftances.*

The principle objeƈt of chemical experiments is to decompofe natural bodies, fo as feparately to examine the different fubftances which enter into their compofition. By confulting chemical fyftems, it will be found that this fcience of chemical analyfis has made rapid progrefs in our own times. Formerly oil and falt were confidered as elements of bodies, whereas later obfervation and experiment have fhown that all falts, inftead of being fimple, are compofed of an acid united to a bafe. The bounds of analyfis have been greatly enlarged by modern difcoveries * ; the acids are fhown to be compofed of oxygen, as an acidifying principle common to all, united in each to a particular bafe. I have proved what Mr Haffenfratz had before

---

* See Memoirs of the Academy for 1776, p. 671. and for 1778, p. 535.—A.

before advanced, that thefe radicals of the acids are not all fimple elements, many of them being, like the oily principle, compofed of hydrogen and charcoal. Even the bafes of neutral falts have been proved by Mr Berthollet to be compounds, as he has fhown that ammoniac is compofed of azote and hydrogen.

Thus, as chemiflry advances towards perfection, by dividing and fubdividing, it is impoffible to fay where it is to end ; and thefe things we at prefent fuppofe fimple may foon be found quite otherwife. All we dare venture to affirm of any fubftance is, that it muft be confidered as fimple in the prefent ftate of our knowledge, and fo far as chemical analyfis has hitherto been able to fhow. We may even prefume that the earths muft foon ceafe to be confidered as fimple bodies; they are the only bodies of the falifiable clafs which have no tendency to unite with oxygen ; and I am much inclined to believe that this proceeds from their being already faturated with that element. If fo, they will fall to be confidered as compounds confifting of fimple fubftances, perhaps metallic, oxydated to a certain degree. This is only hazarded as a conjecture ; and I truft the reader will take care not to confound what I have related as truths, fixed on the firm bafis of obfervation and experiment, with mere hypothetical conjectures.

Z   The

The fixed alkalies, potaſh, and ſoda, are o-
mitted in the foregoing Table, becauſe they
are evidently compound ſubſtances, though we
are ignorant as yet what are the elements they
are compoſed of.

TABLE

TABLE *of compound oxydable and acidifiable bases.*

Names *of the radicals.*

Oxydable or acidifiable base, from the mineral kingdom. ⎰ Nitro-muriatic radical or base of the acid formerly called aqua regia.

Oxydable or acidifiable hydro-carbonous or carbono-hydrous radicals from the vegetable kingdom.

⎧ Tartarous radical or base.
| Malic.
| Citric.
| Pyro-lignous.
| Pyro-mucous.
| Pyro-tartarous.
⎨ Oxalic.
| Acetous.
| Succinic.
| Benzoic.
| Camphoric.
⎩ Gallic.

Oxydable or acidifiable radicals from the animal kingdom, which mostly contain azote, and frequently phosphorus.

⎧ Lactic.
| Saccholactic.
| Formic.
⎨ Bombic.
| Sebacic.
| Lithic.
⎩ Pruffic.

Radicals

SECT.

* *Note.*——The radicals from the vegetable kingdom are converted by a first degree of oxygenation into vegetable oxyds, such as sugar, starch, and gum or mucus : Those of the animal kingdom by the same means form animal oxyds, as lymph, &c.—A.

SECT. II.—*Observations upon the Table of Compound Radicals.*

The older chemifts being unacquainted with the compofition of acids, and not fufpecting them to be formed by a peculiar radical or bafe for each, united to an acidifying principle or element common to all, could not confequently give any name to fubftances of which they had not the moft diftant idea. We had therefore to invent a new nomenclature for this fubjeft, though we were at the fame time fenfible that this nomenclature muft be fufceptible of great modification when the nature of the compound radicals fhall be better underftood *.

The compound oxydable and acidifiable radicals from the vegetable and animal kingdoms, enumerated in the foregoing table, are not hitherto reducible to fyftematic nomenclature, becaufe their exact analyfis is as yet unknown. We only know in general, by fome experiments of my own, and fome made by Mr Haffenfratz, that moft of the vegetable acids, fuch as the tartarous, oxalic, citric, malic, acetous, pyrotartarous, and pyromucous, have radicals compofed of hydrogen and charcoal, combined in fuch

* See Part I. Chap. XI. upon this fubject.—A.

fuch a way as to form fingle bafes, and that
thefe acids only differ from each other by the
proportions in which thefe two fubftances enter
into the compofition of their bafes, and by the
degree of oxygenation which thefe bafes have
received. We know farther, chiefly from the
experiments of Mr Berthollet, that the radicals
from the animal kingdom, and even fome of
thofe from vegetables, are of a more compound
nature, and, befides hydrogen and charcoal,
that they often contain azote, and fometimes
phofphorus ; but we are not hitherto poffeffed
of fufficiently accurate experiments for calculating the proportions of thefe feveral fubftances.
We are therefore forced, in the manner of the
older chemifts, ftill to name thefe acids after the
fubftances from which they are procured. There
can be little doubt that thefe names will be laid
afide when our knowledge of thefe fubftances
becomes more accurate and extenfive ; the
terms *hydro-carbonous*, *hydro-carbonic*, *carbono-
hydrous*, and *carbono hydric* *, will then become
fubftituted for thofe we now employ, which will
then only remain as teftimonies of the imperfect
ftate in which this part of chemiftry was tranf-
mitted to us by our predeceffors.

It

* See Part I. Chap. XI. upon the application of
thefe names according to the proportions of the two
ingredients.--A

It is evident that the oils, being compofed of hydrogen and charcoal combined, are true carbono-hydrous or hydro-carbonous radicals; and, indeed, by adding oxygen, they are convertible into vegetable oxyds and acids, according to their degrees of oxygenation. We cannot, however, affirm that oils enter in their entire ftate into the compofition of vegetable oxyds and acids; it is poffible that they previoufly lofe a part either of their hydrogen or charcoal, and that the remaining ingredients no longer exift in the proportions neceffary to conftitute oils. We ftill require farther experiments to elucidate thefe points.

Properly fpeaking, we are only acquainted with one compound radical from the mineral kingdom, the nitro-muriatic, which is formed by the combination of azote with the muriatic radical. The other compound mineral acids have been much lefs attended to, from their producing lefs ftriking phenomena.

Sect. III.—*Obfervations upon the Combinations of Light and Caloric with different Subftances.*

I have not conftructed any table of the combinations of light and caloric with the various fimple and compound fubftances, becaufe our conceptions of the nature of thefe combinations are not hitherto fufficiently accurate. We know,

know, in general, that all bodies in nature are
imbued, furrounded, and penetrated in every
way with caloric, which fills up every interval
left between their particles; that, in certain
cafes, caloric becomes fixed in bodies, fo as to
conftitute a part even of their folid fubftance,
though it more frequently acts upon them with
a repulfive force, from which, or from its accumulation in bodies to a greater or leffer degree, the transformation of folids into fluids,
and of fluids to aeriform elafticity, is entirely
owing. We have employed the generic name
gas to indicate this aëriform ftate of bodies produced by a fufficient accumulation of caloric;
fo that, when we wifh to exprefs the aëriform
ftate of muriatic acid, carbonic acid, hydrogen,
water, alkohol, &c. we do it by adding the word
gas to their names; thus muriatic acid gas,
carbonic acid gas, hydrogen gas, aqueous gas,
alkoholic gas, &c.

The combinations of light, and its mode of
acting upon different bodies, is ftill lefs known.
By the experiments of Mr Berthollet, it appears
to have great affinity with oxygen, is fufceptible
of combining with it, and contributes alongft
with caloric to change it into the ftate of gas.
Experiments upon vegetation give reafon to believe that light combines with certain parts of
vegetables, and that the green of their leaves,
and the various colours of their flowers, is chief-

ly

ly owing to this combination. This much is certain, that plants which grow in darknefs are perfectly white, languid; and unhealthy, and that to make them recover vigour, and to acquire their natural colours, the direct influence of light is abfolutely neceffary. Somewhat fimilar takes place even upon animals : Mankind degenerate to a certain degree when employed in fedentary manufactures, or from living in crowded houfes, or in the narrow lanes of large cities; whereas they improve in their nature and conftitution in moft of the country labours which are carried on in the open air. Organization, fenfation, fpontaneous motion, and all the operations of life, only exift at the furface of the earth, and in places expofed to the influence of light. Without it nature itfelf would be lifelefs and inanimate. By means of light, the benevolence of the Deity hath filled the furface of the earth with organization, fenfation, and intelligence. The fable of Promotheus might perhaps be confidered as giving a hint of this philofophical truth, which had even prefented itfelf to the knowledge of the ancients. I have intentionally avoided any difquifitions relative to organized bodies in this work, for which reafon the phenomena of refpiration, fanguification, and animal heat, are not confidered; but I hope, at fome future time, to be able to elucidate thefe curious fubjects.

SECT.

TABLE of the binary Combinations of Oxygen with simple Substances.

Names of the simple substances.	First degree of oxygenation. New Names.	First degree. Ancient Names.	Second degree of oxygenation. New Names.	Second degree. Ancient Names.	Third degree of oxygenation. New Names.	Third degree. Ancient Names.	Fourth degree of oxygenation. New Names.	Fourth degree. Ancient Names.
Combinations of oxygen with simple non-metallic substances.								
Caloric	Oxygen gas	Vital or dephlogisticated air						
Hydrogen	Water *							
Azote	Nitrous oxyd, or base of nitrous gas	Nitrous gas or air	Nitrous acid	Smoaking nitrous acid	Nitric acid	Pale, or not smoaking nitrous acid	Oxygenated nitric acid	Unknown
Charcoal	Oxyd of charcoal, or carbonic oxyd	Unknown	Carbonous acid	Unknown	Carbonic acid	Fixed air	Oxygenated carbonic acid	Unknown
Sulphur	Oxyd of sulphur	Soft sulphur	Sulphurous acid	Sulphureous acid	Sulphuric acid	Vitriolic acid	Oxygenated sulphuric acid	Unknown
Phosphorus	Oxyd of phosphorus	Residuum from the combustion of phosphorus	Phosphorous acid	Volatile acid of phosphorus	Phosphoric acid	Phosphoric acid	Oxygenated phosphoric acid	Unknown
Muriatic radical	Muriatic oxyd	Unknown	Muriatic acid	Unknown	Muriatic acid	Marine acid	Oxygenated muriatic acid	Dephlogisticated marine acid
Fluoric radical	Fluoric oxyd	Unknown	Fluoric acid	Unknown	Fluoric acid	Unknown till lately		
Boracic radical	Boracic oxyd	Unknown	Boracic acid	Unknown	Boracic acid	Homberg's sedative salt		
Combinations of oxygen with the simple metallic substances.								
Antimony	Grey oxyd of antimony	Grey calx of antimony	White oxyd of antimony	White calx of antimony, diaphoretic antimony	Antimonic acid			
Silver	Oxyd of silver	Calx of silver			Argentic acid			
Arsenic	Grey oxyd of arsenic	Grey calx of arsenic	White oxyd of arsenic	White calx of arsenic	Arseniac acid	Acid of arsenic	Oxygenated arseniac acid	Unknown
Bismuth	Grey oxyd of bismuth	Grey calx of bismuth	White oxyd of bismuth	White calx of bismuth	Bismuthic acid			
Cobalt	Grey oxyd of cobalt	Grey calx of cobalt			Cobaltic acid			
Copper	Brown oxyd of copper	Brown calx of copper	Blue and green oxyds of copper	Blue and green calces of copper	Cupric acid			
Tin	Grey oxyd of tin	Grey calx of tin	White oxyd of tin	White calx of tin, or putty of tin	Stannic acid			
Iron	Black oxyd of iron	Martial ethiops	Yellow and red oxyds of iron	Ochre and rust of iron	Ferric acid			
Manganese	Black oxyd of manganese	Black calx of manganese	White oxyd of manganese	White calx of manganese	Manganesic acid			
Mercury	Black oxyd of mercury	Ethiops mineral †	Yellow and red oxyds of mercury	Turbith mineral, red precipitate, calcined mercury, precipitate per se	Mercuric acid			
Molybdena	Oxyd of molybdena	Calx of molybdena			Molybdic acid	Acid of molybdena	Oxygenated molybdic acid	Unknown
Nickel	Oxyd of nickel	Calx of nickel			Nickelic acid			
Gold	Yellow oxyd of gold	Yellow calx of gold	Red oxyd of gold	Red calx of gold, purple precipitate of cassius	Auric acid			
Platina	Yellow oxyd of platina	Yellow calx of platina			Platinic acid			
Lead	Grey oxyd of lead	Grey calx of lead	Yellow and red oxyds of lead	Massicot and minium	Plumbic acid			
Tungstein	Oxyd of Tungstein	Calx of Tungstein			Tungstic acid	Acid of Tungstein	Oxygenated Tungstic acid	Unknown
Zinc	Grey oxyd of zinc	Grey calx of zinc	White oxyd of zinc	White calx of zinc, pompholix	Zincic acid			

* Only one degree of oxygenation of hydrogen is hitherto known.—A

† Ethiops mineral is the sulphuret of mercury; this should have been called black precipitate of mercury.—E.

SECT. IV.—*Obfervations upon the Combinations of Oxygen with the fimple Subftances.*

Oxygen forms almoft a third of the mafs of our atmofphere, and is confequently one of the moft plentiful fubftances in nature. All the animals and vegetables live and grow in this immenfe magazine of oxygen gas, and from it we procure the greateft part of what we employ in experiments. So great is the reciprocal affinity between this element and other fubftances, that we cannot procure it difengaged from all combination. In the atmofphere it is united with caloric, in the ftate of oxygen gas, and this again is mixed with about two thirds of its weight of azotic gas.

Several conditions are requifite to enable a body to become oxygenated, or to permit oxygen to enter into combination with it. In the firft place, it is neceffary that the particles of the body to be oxygenated fhall have lefs reciprocal attraction with each other than they have for the oxygen, which otherwife cannot poffibly combine with them. Nature, in this cafe, may be affifted by art, as we have it in our power to diminifh the attraction of the particles of bodies almoft at will by heating them, or, in other words, by introducing caloric into the inter-

A a   ftices

ſtices between their particles ; and, as the attraction of theſe particles for each other is diminiſhed in the inverſe ratio of their diſtance, it
is evident that there muſt be a certain point of
diſtance of particles when the affinity they poſ-
ſeſs with each other becomes leſs than that they
have for oxygen, and at which oxygenation
muſt neceſſarily take place if oxygen be preſent.

We can readily conceive that the degree of
heat at which this phenomenon begins muſt be
different in different bodies. Hence, on purpoſe to oxygenate moſt bodies, eſpecially the
greater part of the ſimple ſubſtances, it is only
neceſſary to expoſe them to the influence of the
air of the atmoſphere in a convenient degree of
temperature. With reſpect to lead, mercury,
and tin, this needs be but little higher than the
medium temperature of the earth ; but it requires a more conſiderable degree of heat to
oxygenate iron, copper, &c. by the dry way, or
when this operation is not aſſiſted by moiſture.
Sometimes oxygenation takes place with great
rapidity, and is accompanied by great ſenſible
heat, light, and flame ; ſuch is the combuſtion
of phoſphorus in atmoſpheric air, and of iron
in oxygen gas. That of ſulphur is leſs rapid ;
and the oxygenation of lead, tin, and moſt of
the metals, takes place vaſtly ſlower, and conſequently the diſengagement of caloric, and
more eſpecially of light, is hardly ſenſible.

Some

Some fubftances have fo ftrong an affinity
with oxygen, and combine with it in fuch low
degrees of temperature, that we cannot procure
them in their unoxygenated ftate ; fuch is the
muriatic acid, which has not hitherto been decompofed by art, perhaps even not by nature,
and which confequently has only been found in
the ftate of acid. It is probable that many o-
ther fubftances of the mineral kingdom are neceffarily oxygenated in the common temperature of the atmofphere, and that being already
faturated with oxygen, prevents their farther action upon that element.

There are other means of oxygenating fimple
fubftances befides expofure to air in a certain
degree of temperature, fuch as by placing them
in contact with metals combined with oxygen,
and which have little affinity with that element.
The red oxyd of mercury is one of the beft fubftances for this purpofe, efpecially with bodies
which do not combine with that metal. In this
oxyd the oxygen is united with very little force
to the metal, and can be driven out by a degree
of heat only fufficient to make glafs red hot ;
wherefore fuch bodies as are capable of uniting
with oxygen are readily oxygenated, by means
of being mixed with red oxyd of mercury, and
moderately heated. The fame effect may be,
to a certain degree, produced by means of the
black oxyd of manganefe, the red oxyd of lead,
the

the oxyds of filver, and by moſt of the metallic
oxyds, if we only take care to chooſe ſuch as
have leſs affinity with oxygen than the bodies
they are meant to oxygenate. All the metallic
reductions and revivifications belong to this
claſs of operations, being nothing more than
oxygenations of charcoal, by means of the ſeveral metallic oxyds. The charcoal combines
with the oxygen and with caloric, and eſcapes
in form of carbonic acid gas, while the metal
remains pure and revivified, or deprived of the
oxygen which before combined with it in the
form of oxyd.

All combuſtible ſubſtances may likewiſe be
oxygenated by means of mixing them with nitrat of potaſh or of ſoda, or with oxygenated
muriat of pot-aſh, and ſubjecting the mixture to
a certain degree of heat ; the oxygen, in this
caſe, quits the nitrat or the muriat, and combines with the combuſtible body. This ſpecies
of oxygenation requires to be performed with
extreme caution, and only with very ſmall quantities ; becauſe, as the oxygen enters into the
compoſition of nitrats, and more eſpecially of
oxygenated muriats, combined with almoſt as
much caloric as is neceſſary for converting it
into oxygen gas, this immenſe quantity of caloric becomes ſuddenly free the inſtant of the
combination of the oxygen with the combuſtible
body,

body, and produces such violent explosions as are perfectly irresistible.

By the humid way we can oxygenate most combustible bodies, and convert most of the oxyds of the three kingdoms of nature into acids. For this purpose we chiefly employ the nitric acid, which has a very slight hold of oxygen, and quits it readily to a great number of bodies by the assistance of a gentle heat. The oxygenated muriatic acid may be used for several operations of this kind, but not in them all.

I give the name of *binary* to the combinations of oxygen with the simple substances, because in these only two elements are combined. When three substances are united in one combination I call it *ternary*, and *quaternary* when the combination consists of four substances united.

TABLE

TABLE *of the combinations of Oxygen with the compound radicals.*

Names of the radi-cals.	Names of the refulting acids.	
	New nomenclature.	*Old nomenclature.*
Nitro muriatic radical }	Nitro muriatic acid	Aqua regia.

*

Tartaric	Tartarous acid	Unknown till lately.
Malic	Malic acid	Ditto.
Citric	Citric acid	Acid of lemons.
Pyro-lignous	Pyro-lignous acid	{ Empyreumatic acid of wood.
Pyro-mucous	Pyro-mucous acid	Empyr. acid of fugar.
Pyro-tartarous	Pyro-tartarous acid	Empyr. acid of tartar.
Oxalic	Oxalic acid	Acid of forel.
Acetic	{ Acetous acid	{ Vinegar, or acid of vinegar.
	{ Acetic acid	Radical vinegar.
Succinic	Succinic acid	Volatile falt of amber.
Benzoic	Benzotic acid	Flowers of benzoin.
Camphoric	Camphoric acid	Unknown till lately.
Gallic	Gallic acid	{ The aftringent princi-ple of vegetables.

* *

Lactic	Lactic acid	Acid of four whey.
Saccholactic	Saccholactic acid	Unknown till lately.
Formic	Formic acid	Acid of ants.
Bombic	Bombic acid	Unknown till lately.
Sebacic	Sebacic acid	Ditto.
Lithic	Lithic acid	Urinary calculus.
Pruffic	Pruffic acid	{ Colouring matter of Pruffian blue.

SECT.

* Thefe radicals by a firft degree of oxygenation form vegetable oxyds, as fugar, ftarch, mucus, &c.—A.

* * Thefe radicals by a firft degree of oxygenation form the
animal oxyds, as lymph, red part of the blood, animal fecretions, &c.—A.

SECT. V.—*Obfervations upon the Combinations of Oxygen with the Compound Radicals.*

I publifhed a new theory of the nature and formation of acids in the Memoirs of the Academy for 1776, p. 671. and 1778, p. 535. in which I concluded, that the number of acids muft be greatly larger than was till then fuppofed. Since that time, a new field of inquiry has been opened to chemifts ; and, inftead of five or fix acids which were then known, near thirty new acids have been difcovered, by which means the number of known neutral falts have been increafed in the fame proportion. The nature of the acidifiable bafes, or radicals of the acids, and the degrees of oxygenation they are fufceptible of, ftill remain to be inquired into. I have already fhown, that almoft all the oxydable and acidifiable radicals from the mineral kingdom are fimple, and that, on the contrary, there hardly exifts any radical in the vegetable, and more efpecially in the animal kingdom, but is compofed of at leaft two fubftances, hydrogen and charcoal, and that azote and phofphorus are frequently united to thefe, by which we have compound radicals of two, three, and four bafes or fimple elements united.

From

From thefe obfervations, it appears that the vegetable and animal oxyds and acids may differ from each other in three feveral ways : 1ft, According to the number of fimple acidifiable elements of which their radicals are compofed : 2dly, According to the proportions in which thefe are combined together: And, 3dly, According to their different degrees of oxygenation : Which circumftances are more than fufficient to explain the great variety which nature produces in thefe fubftances. It is not at all furprifing, after this, that moft of the vegetable acids are convertible into each other, nothing more being requifite than to change the proportions of the hydrogen and charcoal in their compofition, and to oxygenate them in a greater or leffer degree. This has been done by Mr Crell in fome very ingenious experiments, which have been verified and extended by Mr Haffenfratz. From thefe it appears, that charcoal and hydrogen, by a firft oxygenation, produce tartarous acid, oxalic acid by a fecond degree, and acetous or acetic acid by a third, or higher oxygenation ; only, that charcoal feems to exift in a rather fmaller proportion in the acetous and acetic acids. The citric and malic acids differ little from the preceding acids.

Ought we then to conclude that the oils are the radicals of the vegetable and animal acids ? I have already expreffed my doubts upon this

fubject :

fubject: 1ft, Although the oils appear to be formed of nothing but hydrogen and charcoal, we do not know if thefe are in the precife proportion neceffary for conftituting the radicals of the acids: 2dly, Since oxygen enters into the compofition of thefe acids equally with hydrogen and charcoal, there is no more reafon for fuppofing them to be compofed of oil rather than of water or of carbonic acid. It is true that they contain the materials neceffary for all thefe combinations, but then thefe do not take place in the common temperature of the atmofphere; all the three elements remain combined in a ftate of equilibrium, which is readily deftroyed by a temperature only a little above that of boiling water *.

B b   TABLE

* See Part I. Chap. XII. upon this fubject.—A.

TABLE *of the Binary Combinations of Azote with the Simple Substances.*

Simple Substances.	Results of the Combinations.	
	New Nomenclature.	*Old Nomenclature.*

Caloric	Azotic gas	{ Phlogisticated air, or Mephitis.
Hydrogen	Ammoniac	Volatile alkali.
Oxygen	⎧ Nitrous oxyd	Base of Nitrous gas.
	⎨ Nitrous acid	Smoaking nitrous acid.
	⎬ Nitric acid	Pale nitrous acid.
	⎩ Oxygenated nitric acid	Unknown.

Charcoal ⎧ This combination is hitherto unknown; should it ever be discovered, it will be called, according to the principles of our nomenclature, Azuret of Charcoal. Charcoal dissolves in azotic gas, and forms carbonated azotic gas. ⎭

Phosphorus.   Azuret of phosphorus.   Still unknown.

Sulphur ⎰ Azuret of sulphur. Still unknown. We know that sulphur dissolves in azotic gas, forming sulphurated azotic gas. ⎱

Compound radicals ⎰ Azote combines with charcoal and hydrogen, and sometimes with phosphorus, in the compound oxydable and acidifiable bases, and is generally contained in the radicals of the animal acids. ⎱

Metallic substances ⎰ Such combinations are hitherto unknown; if ever discovered, they will form metallic azurets, as azuret of gold, of silver, &c. ⎱

Lime
Magnesia
Barytes ⎱ Entirely unknown. If ever discovered, they will form azuret of lime, azuret of magnesia, &c.
Argill
Potash
Soda

SECT.

SECT. VI.—*Obfervations upon the Combinations of Azote with the Simple Subftances.*

Azote is one of the moft abundant elements; combined with caloric it forms azotic gas, or mephitis, which compofes nearly two thirds of the atmofphere. This element is always in the ftate of gas in the ordinary preffure and temperature, and no degree of compreffion or of cold has been hitherto capable of reducing it either to a folid or liquid form. This is likewife one of the effential conftituent elements of animal bodies, in which it is combined with charcoal and hydrogen, and fometimes with phofphorus; thefe are united together by a certain portion of oxygen, by which they are formed into oxyds or acids according to the degree of oxygenation. Hence the animal fubftances may be varied, in the fame way with vegetables, in three different manners : 1ft, According to the number of elements which enter into the compofition of the bafe or radical : 2dly, According to the proportions of thefe elements : 3dly, According to the degree of oxygenation.

When combined with oxygen, azote forms the nitrous and nitric oxyds and acids ; when with hydrogen, ammoniac is produced. Its combinations with the other fimple elements

are

are very little known ; to thefe we give the name of Azurets, preferving the termination in *uret* for all nonoxygenated compounds. It is extremely probable that all the alkaline fubftances may hereafter be found to belong to this genus of azurets.

The azotic gas may be procured from atmofpheric air, by abforbing the oxygen gas which is mixed with it by means of a folution of fulphuret of potafh, or fulphuret of lime. It requires twelve or fifteen days to complete this procefs, during which time the furface in contact muft be frequently renewed by agitation, and by breaking the pellicle which forms on the top of the folution. It may likewife be procured by diffolving animal fubftances in dilute nitric acid very little heated. In this operation the azote is difengaged in form of gas, which we receive under bell glaffes filled with water in the pneumato-chemical apparatus. We may procure this gas by deflagrating nitre with charcoal, or any other combuftible fubftance; when with charcoal, the azotic gas is mixed with carbonic acid gas, which may be abforbed by a folution of cauftic alkali, or by lime water, after which the azotic gas remains pure. We can procure it in a fourth manner from combinations of ammoniac with metallic oxyds, as pointed out by Mr de Fourcroy: The hydrogen of the ammoniac combines with the oxygen of the oxyd,

oxyd, and forms water, whilft the azote being left free efcapes in form of gas.

The combinations of azote were but lately difcovered : Mr Cavendifh firft obferved it in nitrous gas and acid, and Mr Berthollet in ammoniac and the pruffic acid. As no evidence of its decompofition has hitherto appeared, we are fully entitled to confider azote as a fimple elementary fubftance.

TABLE

TABLE *of the Binary Combinations of Hydrogen with Simple Substances.*

Simple Substances.	Resulting Compounds.	
	New Nomenclature.	Old Names.
Caloric	Hydrogen gas	Inflammable air.
Azote	Ammoniac	Volatile Alkali,
Oxygen	Water	Water.
Sulphur	Hydruret of sulphur, or sulphuret of hydrogen	Hitherto unknown *.
Phosphorus	Hydruret of phosphorus, or phosphuret of hydrogen	
Charcoal	Hydro-carbonous, or carbono-hydrous radicals †	Not known till lately,
Metallic substances, as iron, &c.	Metallic hydrurets ‡, as hydruret of iron, &c.	Hitherto unknown,

* These combinations take place in the state of gas, and form, respectively, sulphurated and phosphorated oxygen gas.— A.

† This combination of hydrogen with charcoal includes the fixed and volatile oils, and forms the radicals of a considerable part of the vegetable and animal oxyds and acids. When it takes place in the state of gas it forms carbonated hydrogen gas.—A.

‡ None of these combinations are known, and it is probable that they cannot exist, at least in the usual temperature of the atmosphere, owing to the great affinity of hydrogen for caloric.—A.

Sect. VII.—*Obſervations upon Hydrogen, and its Combinations with Simple Subſtances.*

Hydrogen, as its name expreſſes, is one of the conſtituent elements of water, of which it forms fifteen hundredth parts by weight, combined with eighty-five hundredth parts of oxygen. This ſubſtance, the properties and even exiſtence of which was unknown till lately, is very plentifully diſtributed in nature, and acts a very confiderable part in the proceſſes of the animal and vegetable kingdoms. As it poſſeſſes ſo great affinity with caloric as only to exiſt in the ſtate of gas, it is conſequently impoſſible to procure it in the concrete or liquid ſtate, independent of combination.

To procure hydrogen, or rather hydrogen gas, we have only to ſubject water to the action of a ſubſtance with which oxygen has greater affinity than it has to hydrogen; by this means the hydrogen is ſet free, and, by uniting with caloric, aſſumes the form of hydrogen gas. Red hot iron is uſually employed for this purpoſe: The iron, during the proceſs, becomes oxydated, and is changed into a ſubſtance reſembling the iron ore from the iſland of Elba. In this ſtate of oxyd it is much leſs attractible by

the

the magnet, and diffolves in acids without effervefcence.

Charcoal, in a red heat, has the fame power
of decompofing water, by attracting the oxygen
from its combination with hydrogen. In this
procefs carbonic acid gas is formed, and mixes
with the hydrogen gas, but is eafily feparated
by means of water or alkalies, which abforb the
carbonic acid, and leave the hydrogen gas pure.
We may likewife obtain hydrogen gas by diffolving iron or zinc in dilute fulphuric acid.
Thefe two metals decompofe water very flowly,
and with great difficulty, when alone, but do it
with great eafe and rapidity when affifted by fulphuric acid; the hydrogen unites with caloric
during the procefs, and is difengaged in form
of hydrogen gas, while the oxygen of the water
unites with the metal in the fôrm of oxyd, which
is immediately diffolved in the acid, forming a
fulphat of iron or of zinc.

Some very diftinguifhed chemifts confider hydrogen as the *phlogifton* of Stahl; and as that
celebrated chemift admitted the exiftence of
phlogifton in fulphur, charcoal, metals, &c. they
are of courfe obliged to fuppofe that hydrogen
exifts in all thefe fubftances, though they cannot prove their fuppofition; even if they could,
it would not avail much, fince this difengagement of hydrogen is quite infufficient to explain
the phenomena of calcination and combuftion.

We

We muſt always recur to the examination of this queſtion, " Are the heat and light, which are diſengaged during the different ſpecies of com-buſtion, furniſhed by the burning body, or by the oxygen which combines in all theſe opera-tions ?" And certainly the ſuppoſition of hydro-gen being diſengaged throws no light whatever upon this queſtion. Béſides, it belongs to thoſe who make ſuppoſitions to prove them ; and, doubtleſs, a doctrine which without any ſuppo-ſition explains the phenomena as well, and as naturally, as theirs does by ſuppoſition, has at leaſt the advantage of greater ſimplicity *.

C c   TABLE

---

* Thoſe who wiſh to ſee what has been ſaid upon this great chemical queſtion by Meſſrs de Morveau, Berthollet, De Fourcroy, and myſelf, may conſult our tranſlation of Mr Kirwan's Eſſay upon Phlogiſton —A.

TABLE *of the Binary Combinations of Sulphur with Simple Substances.*

Simple Substances.	Resulting Compounds. New Nomenclature.	Old Nomenclature.
Caloric	Sulphuric gas	
Oxygen	{ Oxyd of sulphur { Sulphurous acid { Sulphuric acid	Soft sulphur. Sulphureous acid. Vitriolic acid.
Hydrogen	Sulphuret of hydrogen	}
Azote	azote	} Unknown Combina-
Phosphorus	phosphorus	} tions.
Charcoal	charcoal	}
Antimony	antimony	Crude antimony.
Silver	filver	
Arfenic	arfenic	Orpiment, realgar.
Bifmuth	bifmuth	
Cobalt	cobalt	
Copper	copper	Copper pyrites.
Tin	tin	
Iron	iron	Iron pyrites.
Manganefe	manganefe	
Mercury	mercury	{ Ethiops mineral, { cinnabar.
Molybdena	molybdena	
Nickel	nickel	
Gold	gold	
Platina	platina	
Lead	lead	Galena.
Tungftein	tungftein	
Zinc	zinc	Blende.
Potafh	potafh	{ Alkaline liver of ful- { phur with fixed ve- { getable alkali.
Soda	foda	{ Alkaline liver of ful- { phur with fixed mi- { neral alkali.
Ammoniac	ammoniac	{ Volatile liver of ful- { phur, fmoaking li- { quor of Boyle.
Lime	lime	{ Calcareous liver of ful- { phur.
Magnefia	magnefia	{ Magnefian liver of ful- { phur.
Barytes	barytes	{ Barytic liver of ful- { phur.
Argill	argill	Yet unknown.

SECT.

SECT. VIII.—*Obfervations on Sulphur, and its Combinations.*

Sulphur is a combuftible fubftance, having a very great tendency to combination; it is naturally in a folid ftate in the ordinary temperature, and requires a heat fomewhat higher than boiling water to make it liquify. Sulphur is formed by nature in a confiderable degree of purity in the neighbourhood of volcanos; we find it likewife, chiefly in the ftate of fulphuric acid, combined with argill in aluminous fchiftus, with lime in gypfum, &c. From thefe combinations it may be procured in the ftate of fulphur, by carrying off its oxygen by means of charcoal in a red heat; carbonic acid is formed, and efcapes in the ftate of gas; the fulphur remains combined with the clay, lime, &c. in the ftate of fulphuret, which is decompofed by acids; the acid unites with the earth into a neutral falt, and the fulphur is precipitated.

TABLE

Table *of the Binary Combinations of Phofphorus
with the Simple Subftances.*

Simple Subftances.	Refulting Compounds.
Caloric - -	Phofphoric gas.
Oxygen - -	{ Oxyd of phofphorus. Phofphorous acid. Phofphoric acid.
Hydrogen - -	Phofphuret of hydrogen.
Azote - -	Phofphuret of azote.
Sulphur - -	Phofphuret of Sulphur.
Charcoal - -	Phofphuret of charcoal.
Metallic fubftances	Phofphuret of metals *.
Potafh - - Soda - - Ammoniac - - Lime - - Barytes - - Magnefia - - Argill - -	} Phofphuret of Potafh, Soda, &c. †

Sect.

* Of all thefe combinations of phofphorus with
metals, that with iron only is hitherto known, forming
the fubftance formerly called Siderite ; neither is it yet
afccrtained whether, in this combination, the phofphorus be oxygenated or not.—A.

† Thefe combinations of phofphorus with the alkalies and earths are not yet known ; and, from the experiments of Mr Gengembre, they appear to be impoffible.—A.

SECT. IX.—*Obfervations upon Phofphorus, and its Combinations.*

Phofphorus is a fimple combuftible fubftance, which was unknown to chemifts till 1667, when it was difcovered by Brandt, who kept the procefs fecret; foon after Kunkel found out Brandt's method of preparation, and made it public. It has been ever fince known by the name of Kunkel's phofphorus. It was for a long time procured only from urine; and, though Homberg gave an account of the procefs in the Memoirs of the Academy for 1692, all the philofophers of Europe were fupplied with it from England. It was firft made in France in 1737, before a committee of the Academy at the Royal Garden. At prefent it is procured in a more commodious and more oeconomical manner from animal bones, which are real calcareous phofphats, according to the procefs of Meffrs Gahn, Scheele, Rouelle, &c. The bones of adult animals being calcined to whitenefs, are pounded, and paffed through a fine filk fieve; pour upon the fine powder a quantity of dilute fulphuric acid, lefs than is fufficient for diffolving the whole. This acid unites with the calcareous earth of the bones into a fulphat of lime, and the phofphoric acid remains free in the liquor. The liquid

is

is decanted off, and the refiduum wafhed with boiling water; this water which has been ufed to wafh out the adhering acid is joined with what was before decanted off, and the whole is gradually evaporated; the diffolved fulphat of lime criftallizes in form of filky threads, which are removed, and by continuing the evaporation we procure the phofphoric acid under the appearance of a white pellucid glafs. When this is powdered, and mixed with one third its weight of charcoal, we procure very pure phofphorus by fublimation. The phofphoric acid, as procured by the above procefs, is never fo pure as that obtained by oxygenating pure phofphorus either by combuftion or by means of nitric acid; wherefore this latter fhould always be employed in experiments of refearch.

Phofphorus is found in almoft all animal fubftances, and in fome plants which give a kind of animal analyfis. In all thefe it is ufually combined with charcoal, hydrogen, and azote, forming very compound radicals, which are, for the moft patt, in the ftate of oxyds by a firft degree of union with oxygen. The difcovery of Mr Haffenfratz, of phofphorus being contained in charcoal, gives reafon to fufpect that it is more common in the vegetable kingdom than has generally been fuppofed: It is certain, that, by proper proceffes, it may be procured from every individual of fome of the families of plants.

As

As no experiment has hitherto given reafon to
fufpect that phofphorus is a compound body, I
have arranged it with the fimple or elementary
fubftances. It takes fire at the temperature of
32° (104°) of the thermometer.

TABLE *of the Binary Combinations of Charcoal.*

Simple

Subftances.   *Refulting Compounds.*

Oxygen	{ Oxyd of charcoal	Unknown.
	{ Carbonic acid	Fixed air, chalky acid.
Sulphur	Carburet of fulphur	
Phofphorus	Carburet of phofphorus	} Unknown.
Azote	Carburet of azote	
Hydrogen	{ Carbono-hydrous radical	
	{ Fixed and volatile oils	
Metallic fub-ftances	} Carburets of metals	Of thefe only the carburets of iron and zinc are known, and were formerly called Plumbago.
Alkalies and earths	} Carburet of potafh, &c.	Unknown.

SECT. X.—*Obfervations upon Charcoal, and its Combinations with Simple Subftances.*

As charcoal has not been hitherto decompofed, it muft, in the prefent ftate of our knowledge, be confidered as a fimple fubftance. By
modern experiments it appears to exift ready
formed in vegetables; and I have already re-
marked, that, in thefe, it is combined with hydrogen, fometimes with azote and phofphorus,
forming compound radicals, which may be
changed into oxyds or acids according to their
degree of oxygenation.

To obtain the charcoal contained in vegetable
or animal fubftances, we fubject them to the
action of fire, at firft moderate, and afterwards
very ftrong, on purpofe to drive off the laft portions of water, which adhere very obftinately to
the charcoal. For chemical purpofes, this is
ufually done in retorts of ftone-ware or porcellain, into which the wood, or other matter, is
introduced, and then placed in a reverberatory
furnace, raifed gradually to its greateft heat:
The heat volatilizes, or changes into gas, all the
parts of the body fufceptible of combining with
caloric into that form, and the charcoal, being
more fixed in its nature, remains in the retort
combined

combined with a little earth and some fixed salts.

In the businefs of charring wood, this is done by a lefs expenfive procefs. The wood is difpofed in heaps, and covered with earth, fo as to prevent the accefs of any more air than is abfolutely neceffary for fupporting the fire, which is kept up till all the water and oil is driven off, after which the fire is extinguifhed by fhutting up all the air-holes.

We may analyfe charcoal either by combuftion in air, or rather in oxygen gas, or by means of nitric acid. In either cafe we convert it into carbonic acid, and fometimes a little potafh and fome neutral falts remain. This analyfis has hitherto been but little attended to by chemifts; and we are not even certain if potafh exifts in charcoal before combuftion, or whether it be formed by means of fome unknown combination during that procefs.

Sect. XI.—*Obfervations upon the Muriatic, Fluoric, and Boracic Radicals, and their Combinations.*

As the combinations of thefe fubftances, either with each other, or with the other combuftible bodies, are hitherto entirely unknown, we have

D d

not

not attempted to form any table for their nomenclature. We only know that thefe radicals
are fufceptible of oxygenation, and of forming
the muriatic, fluoric, and boracic acids, and
that in the acid ftate they enter into a number
of combinations, to be afterwards detailed.
Chemiftry has hitherto been unable to difoxygenate any of them, fo as to produce them in a
fimple ftate. For this purpofe, fome fubftance
muft be employed to which oxygen has a
ftronger affinity than to their radicals, either by
means of fingle affinity, or by double elective
attraction. All that is known relative to the
origin of the radicals of thefe acids will be mentioned in the fections fet apart for confidering
their combinations with the falifiable bafes.

SECT. XII.—*Obfervations upon the Combinations
of Metals with each other.*

Before clofing our account of the fimple or
elementary fubftances, it might be fuppofed neceffary to give a table of alloys or combinations
of metals with each other; but, as fuch a table
would be both exceedingly voluminous and
very unfatisfactory, without going into a feries
of experiments not yet attempted, I have
thought it advifeable to omit it altogether. All
that

that is neceffary to be mentioned is, that thefe alloys fhould be named according to the metal in largeft proportion in the mixture or combibination ; thus the term *alloy of gold and filver,* or gold alloyed with filver, indicates that gold is the predominating metal.

Metallic alloys, like all other combinations, have a point of faturation. It would even appear, from the experiments of Mr de la Briche, that they have two perfectly diftinct degrees of faturation.

TABLE

TABLE *of the Combinations of Azote in the ſtate of Nitrous Acid with the Salifiable Baſes, arranged according to the affinities of theſe Baſes with the Acid.*

Names of the baſes.	*Names of the neutral ſalts.*	*Notes.*
	New nomenclature.	

Barytes	Nitrite of barytes.	⎧ Theſe ſalts are only
Potaſh	potaſh.	⎪ known of late, and
Soda	ſoda.	⎨ have received no par-
Lime	lime.	⎪ ticular name in the old
Magneſia	magneſia.	⎪ nomenclature.
Ammoniac	ammoniac.	⎩
Argill	argill.	

		⎧ As metals diſſolve
		⎪ both in nitrous and
		⎪ nitric acids, metallic
Oxyd of zinc	zinc.	⎪ ſalts muſt of conſe-
iron	iron.	⎪ quence be formed ha-
manganeſe	manganeſe.	⎪ ving different degrees
cobalt	cobalt.	⎪ of oxygenation. Thoſe
nickel	nickel.	⎪ wherein the metal is
lead	lead.	⎨ leaſt oxygenated muſt
tin	tin.	⎪ be called Nitrites, when
copper	copper.	⎪ more ſo, Nitrats; but
biſmuth	biſmuth.	⎪ the limits of this di-
antimony	antimony.	⎪ ſtinction are difficultly
arſenic	arſenic.	⎪ aſcertainable. The old-
mercury	mercury.	⎪ der chemiſts were not
		⎪ acquainted with any of
		⎩ theſe ſalts.

ſilver	⎧ It is extremely probable that gold, ſilver,
gold	⎨ and platina only form nitrats, and cannot ſub-
platina	⎩ ſiſt in the ſtate of nitrites.

TABLE *of the Combinations of Azote, completely satura-*
ted with Oxygen, in the state of Nitric Acid, with
the Salifiable Bases, in the order of the affinity with
the Acid.

Bases.	Names of the resulting neutral salts.	
	New nomenclature.	*Old nomenclature.*
Barytes	Nitrat of barytes	{ Nitre, with a base of heavy earth.
Potash	potash	{ Nitre, saltpetre. Nitre with base of potash.
Soda	soda	{ Quadrangular nitre. Nitre with base of mineral alkali.
Lime	lime	{ Calcareous nitre. Nitre with calcareous base. Mother water of nitre, or saltpetre.
Magnesia	magnesia	{ Magnesian nitre. Nitre with base of magnesia.
Ammoniac	ammoniac	Ammoniacal nitre.
Argill	argill	{ Nitrous alum. Argillaceous nitre. Nitre with base of earth of alum.
Oxyd of zinc	zinc	Nitre of zinc.
iron	iron	{ Nitre of iron. Martial nitre. Nitrated iron.
manganese	manganese	Nitre of manganese.
cobalt	cobalt	Nitre of cobalt.
nickel	nickel	Nitre of nickel.
lead	lead	{ Saturnine nitre. Nitre of lead.
tin	tin	Nitre of tin.
copper	copper	{ Nitre of copper or of Venus.
bismuth	bismuth	Nitre of bismuth.
antimony	antimony	Nitre of antimony.
arsenic	arsenic	Arsenical nitre.
mercury	mercury	Mercurial nitre.
silver	silver	{ Nitre of silver or luna. Lunar caustic.
gold	gold	Nitre of gold.
platina	platina	Nitre of platina.

SECT.

SECT. XIII.—*Obfervations upon the Nitrous and Nitric Acids, and their Combinations.*

The nitrous and nitric acids are procured from a neutral falt long known in the arts under the name of *faltpetre*. This falt is extracted by lixiviation from the rubbifh of old buildings, from the earth of cellars, ftables, or barns, and in general of all inhabited places. In thefe earths the nitric acid is ufually combined with lime and magnefia, fometimes with potafh, and rarely with argill. As all thefe falts, excepting the nitrat of potafh, attract the moifture of the air, and confequently would be difficultly preferved, advantage is taken, in the manufactures of faltpetre and the royal refining-houfe, of the greater affinity of the nitric acid to potafh than thefe other bafes, by which means the lime, magnefia, and argill, are precipitated, and all thefe nitrats are reduced to the nitrat of potafh or faltpetre *.

The nitric acid is procured from this falt by diftillation, from three parts of pure faltpetre decompofed by one part of concentrated fulphuric

* Saltpetre is likewife procured in large quantities by lixiviating the natural foil in fome parts of Bengal, and of the Ruffian Ukrain.—E.

phuric acid, in a retort with Woulfe's apparatus, (Pl. IV. fig. 1.) having its bottles half filled with water, and all its joints carefully luted.
The nitrous acid paffes over in form of red vapours furcharged with nitrous gas, or, in other
words, not faturated with oxygen. Part of the
acid condenfes in the recipient in form of a
dark orange red liquid, while the reft combines
with the water in the bottles. During the diftillation, a large quantity of oxygen gas efcapes,
owing to the greater affinity of oxygen to caloric, in a high temperature, than to nitrous acid,
though in the ufual temperature of the atmofphere this affinity is reverfed. It is from the
difengagement of oxygen that the nitric acid of
the neutral falt is in this operation converted
into nitrous acid. It is brought back to the
ftate of nitric acid by heating over a gentle fire,
which drives off the fuperabundant nitrous gas,
and leaves the nitric acid much diluted with
water.

Nitric acid is procurable in a more concentrated ftate, and with much lefs lofs, by mixing
very dry clay with faltpetre. This mixture is
put into an earthern retort, and diftilled with a
ftrong fire. The clay combines with the potafh, for which it has great affinity, and the nitric acid paffes over, flightly impregnated with
nitrous gas. This is eafily difengaged by heating the acid gently in a retort, a fmall quantity

of

of nitrous gas paffes over into the recipient, and very pure concentrated nitric acid remains in the retort.

We have already feen that azote is the nitric radical. If to 20½ parts, by weight, of azote 43½ parts of oxygen be added, 64 parts of nitrous gas are formed ; and, if to this we join 36 additional parts of oxygen, 100 parts of nitric acid refult from the combination. Intermediate quantities of oxygen between thefe two extremes of oxygenation produce different fpecies of nitrous acid, or, in other words, nitric acid lefs or more impregnated with nitrous gas. I afcertained the above proportions by means of decompofition ; and, though I cannot anfwer for their abfolute accuracy, they cannot be far removed from truth. Mr Cavendifh, who firft fhowed by fynthetic experiments that azote is the bafe of nitric acid, gives the proportions of azote a little larger than I have done ; but, as it is not improbable that he produced the nitrous acid and not the nitric, that circumftance explains in fome degree the difference in the refults of our experiments.

As, in all experiments of a philofophical nature, the utmoft poffible degree of accuracy is required, we muft procure the nitric acid for experimental purpofes, from nitre which has been previoufly purified from all foreign matter. If, after diftillation, any fulphuric acid is fufpected

fpeƈted in the nitric acid, it is eafily feparated
by dropping in a little nitrat of barytes, fo long
as any precipitation takes place ; the fulphuric
acid, from its greater affinity, attraƈts the barytes, and forms with it an infoluble neutral
falt, which falls to the bottom. It may be purified in the fame manner from muriatic acid, by
dropping in a little nitrat of filver fo long as
any precipitation of muriat of filver is produced.
When thefe two precipitations are finifhed, diftill off about feven-eighths of the acid by a gentle heat, and what comes over is in the moſt
perfeƈt degree of purity.

The nitric acid is one of the moſt prone to
combination, and is at the fame time very eafily
decompofed. Almoſt all the fimple fubſtances,
with the exception of gold, filver, and platina,
rob it lefs or more of its oxygen ; fome of them
even decompofe it altogether. It was very anciently known, and its combinations have been
more ſtudied by chemiſts than thofe of any
other acid. Thefe combinations were named
nitres by Meſſrs Macquer and Beaumé ; but we
have changed their names to nitrats and nitrites,
according as they are formed by nitric or by
nitrous acid, and have added the fpecific name
of each particular bafe, to diftinguifh the feveral combinations from each other.

E e

TABLE *of the Combinations of Sulphuric Acid with the Salifiable Bafes, in the order of affinity.*

Names of the bafes.	Refulting compounds.	
	New nomenclature.	Old nomenclature.
Barytes	Sulphat of barytes	{ Heavy fpar. Vitriol of heavy earth.
Potafh	potafh	⎧ Vitriolated tartar. Sal ⎨ de duobus. Arcanum ⎩ duplicatam.
Soda	foda	Glauber's falt.
Lime	lime	{ Selenite, gypfum, cal- careous vitriol.
Magnefia	magnefia	{ Epfom falt, fedlitz falt, magnefian vitriol.
Ammoniac	ammoniac	{ Glauber's fecret fal ammoniac.
Argill	argill	Alum.
Oxyd of zinc	zinc	⎧ White vitriol, goflar ⎨ vitriol, white coperas, ⎩ vitriol of zinc.
iron	iron	⎧ Green coperas, green ⎨ vitriol, martial vitri- ⎩ ol, vitriol of iron.
manganefe	manganefe	Vitriol of manganefe.
cobalt	cobalt	Vitriol of cobalt.
nickel	nickel	Vitriol of nickel.
lead	lead	Vitriol of lead.
tin	tin	Vitriol of tin.
copper	copper	⎧ Blue coperas, blue vi- ⎨ triol, Roman vitriol, ⎩ vitriol of copper.
bifmuth	bifmuth	Vitriol of bifmuth.
antimony	antimony	Vitriol of antimony.
arfenic	arfenic	Vitriol of arfenic.
mercury	mercury	Vitriol of mercury.
filver	filver	Vitriol of filver.
gold	gold	Vitriol of gold.
platina	platina	Vitriol of platina.

SECT.

SECT. XIV.—*Obſervations upon Sulphuric Acid and its Combinations.*

For a long time this acid was procured by diſtillation from ſulphat of iron, in which ſulphuric acid and oxyd of iron are combined, according to the procefs defcribed by Bafil Valentine in the fifteenth century; but, in modern times, it is procured more oeconomically by the combuſtion of ſulphur in proper veſſels. Both to facilitate the combuſtion, and to aſſiſt the oxygenation of the ſulphur, a little powdered ſaltpetre, nitrat of potaſh, is mixed with it; the nitre is decompofed, and gives out its oxygen to the ſulphur, which contributes to its converſion into acid. Notwithſtanding this addition, the ſulphur will only continue to burn in clofe veſſels for a limited time; the combination ceafes, becaufe the oxygen is exhauſted, and the air of the veſſels reduced almoſt to pure azotic gas, and becaufe the acid itſelf remains long in the ſtate of vapour, and hinders the progrefs of combuſtion.

In the manufaċtories for making ſulphuric acid in the large way, the mixture of nitre and fulphur is burnt in large clofe built chambers lined with lead, having a little water at the bottom for facilitating the condenfation of the vapours,

pours. Afterwards, by diftillation in large retorts with a gentle heat, the water paffes over,
flightly impregnated with acid, and the fulphuric
acid remains behind in a concentrated ftate.   It
is then pellucid, without any flavour, and nearly double the weight of an equal bulk of water.
This procefs would be greatly facilitated, and
the combuftion much prolonged, by introducing frefh air into the chambers, by means of
feveral pairs of bellows directed towards the
flame of the fulphur, and by allowing the nitrous gas to efcape through long ferpentine ca-
nals, in contact with water, to abforb any fulphuric or fulphurous acid gas it might contain.

By one experiment, Mr Berthollet found that
69 parts of fulphur in combuftion, united with
31 parts of oxygen, to form 100 parts of fulphuric acid ; and, by another experiment, made
in a different manner, he calculates that 100
parts of fulphuric acid confifts of 72 parts fulphur, combined with 28 parts of oxygen, all
by weight.

This acid, in common with every other, can
only diffolve metals when they have been previoufly oxydated ; but moft of the metals are capable of decompofing a part of the acid, fo as
to carry off a fufficient quantity of oxygen, to
render themfelves foluble in the part of the acid
which remains undecompofed.   This happens
with filver, mercury, iron, and zinc, in boiling
concentrated

concentrated fulphuric acid ; they become firft oxydated by decompofing part of the acid, and then diffolve in the other part ; but they do not fufficiently difoxygenate the decompofed part of the acid to reconvert it into fulphur ; it is only reduced to the ftate of fulphurous acid, which, being volatilifed by the heat, flies off in form of fulphurous acid gas.

Silver, mercury, and all the other metals except iron and zinc, are infoluble in diluted fulphuric acid, becaufe they have not fufficient affinity with oxygen to draw it off from its combination either with the fulphur, the fulphurous acid, or the hydrogen ; but iron and zinc, being affifted by the action of the acid, decompofe the water, and become oxydated at its expence, without the help of heat.

TABLE

Table *of the Combinations of the Sulphurous Acid with the Salifiable Bafes, in the order of affinity.*

Names of the Bafes.	Names of the Neutral Salts.
Barytes	Sulphite of barytes.
Potafh	potafh.
Soda	foda.
Lime	lime.
Magnefia	magnefia.
Ammoniac	ammoniac.
Argill	argill.
Oxyd of zinc	zinc.
iron	iron.
manganefe	manganefe.
cobalt	cobalt.
nickel	nickel.
lead	lead.
tin	tin.
copper	copper.
bifmuth	bifmuth.
antimony	antimony.
arfenic	arfenic.
mercury	mercury.
filver	filver.
gold	gold.
platina	platina.

Sect.

Note.——The only one of thefe falts known to the old chemifts was the fulphite of potafh, under the name of *Stahl's fulphureous falt.* So that, before our new nomenclature, thefe compounds muft have been named *Stahl's fulphureous falt*, having bafe of fixed vegetable alkali, and fo of the reft.

In this Table we have followed Bergman's order of affinity of the fulphuric acid, which is the fame in regard to the earths and alkalies, but it is not certain if the order be the fame for the metallic oxyds.—A.

SECT. XV.—*Obfervations upon Sulphurous Acid,*
and its Combinations.

The fulphurous acid is formed by the union
of oxygen with fulphur by a leffer degree of o-
xygenation than the fulphuric acid. It is procurable either by burning fulphur flowly, or by
diftilling fulphuric acid from filver, antimony,
lead, mercury, or charcoal ; by which operation
a part of the oxygen quits the acid, and unites
to thefe oxydable bafes, and the acid paffes over
in the fulphurous ftate of oxygenation. This
acid, in the common preffure and temperature of the air, can only exift in form of gas ;
but it appears, from the experiments of Mr
Clouet, that, in a very low temperature, it condenfes, and becomes fluid. Water abforbs a
great deal more of this gas than of carbonic a-
cid gas, but much lefs than it does of muriatic
acid gas.

That the metals cannot be diffolved in acids
without being previoufly oxydated, or by procuring oxygen, for that purpofe, from the acids
during folution, is a general and well eftablifhed fact, which I have perhaps repeated too often. Hence, as fulphurous acid is already deprived of great part of the oxygen neceffary for
forming the fulphuric acid, it is more difpofed

to

to recover oxygen, than to furnish it to the greateft part of the metals ; and, for this reafon, it cannot diffolve them, unlefs previoufly oxyda-ted by other means. From the fame principle it is that the metallic oxyds diffolve without ef-fervefcence, and with great facility, in fulphu-rous acid. This acid, like the muriatic, has even the property of diffolving metallic oxyds furchar-ged with oxygn, and confequently infoluble in fulphuric acid, and in this way forms true ful-phats. Hence we might be led to conclude that there are no metallic fulphites, were it not that the phenomena which accompany the folution of iron, mercury, and fome other metals, convince us that thefe metallic fubftances are fufceptible of two degrees of oxydation, during their folu-tion in acids. Hence the neutral falt in which the metal is leaft oxydated muft be named *ful-phite*, and that in which it is fully oxydated muft be called *fulphat*. It is yet unknown whether this diftinction is applicable to any of the metal-lic fulphats, except thofe of iron and mercury.

TABLE

TABLE *of the Combinations of Phofphorous and Phofphoric Acids, with the Salifiable Bafes, in the Order of Affinity.*

Names of the Bafes.	Names of the Neutral Salts formed by	
	Phofphorous Acid,	*Phofphoric Acid.*
Lime	Phofphites of† lime	Phofphats of ‡ lime.
Barytes	barytes	barytes.
Magnefia	magnefia	magnefia.
Potafh	potafh	potafh.
Soda	foda	foda.
Ammoniac	ammoniac	ammoniac.
Argill	argill	argill.
Oxyds of *		
zinc	zinc	zinc.
iron	iron	iron.
manganefe	manganefe	manganefe.
cobalt	cobalt	cobalt.
nickel	nickel	nickel.
lead	lead	lead.
tin	tin	tin.
copper	copper	copper.
bifmuth	bifmuth	bifmuth.
antimony	antimony	antimony.
arfenic	arfenic	arfenic.
mercury	mercury	mercury.
filver	filver	filver.
gold	gold.	gold.
platina	platina	platina.

* The exiftence of metallic phofphites fuppofes that metals are fufceptible of folution in phofphoric acid at different degrees of oxygenation, which is not yet afcertained.—A.

† All the phofphites were unknown till lately, and confequently have not hitherto received names.—A.

† The greater part of the phofphats were only difcovered of late, and have not yet been named.—A.

F f

SECT. XVI.—*Obfervations upon Phofphorous and Phofporic Acids, and their Combinations.*

Under the article Phofphorus, Part II. Sect. X. we have already given a hiftory of the dif-covery of that fingular fubftance, with fome ob-fervations upon the mode of its exiftence in ve-getable and animal bodies. The beft method of obtaining this acid in a ftate of purity is by burning well purified phofphorus under bell-glaffes, moiftened on the infide with diftilled water; during combuftion it abforbs twice and a half its weight of oxygen; fo that 100 parts of phofphoric acid is compofed of $28\frac{1}{2}$ parts of phofphorus united to $71\frac{1}{2}$ parts of oxy-gen. This acid may be obtained concrete, in form of white flakes, which greedily attract the moifture of the air, by burning phofphorus in a dry glafs over mercury.

To obtain phofphorous acid, which is phof-phorus lefs oxygenated than in the ftate of phofphoric acid, the phofphorus muft be burnt by a very flow fpontaneous combuftion over a glafs-funnel leading into a cryftal phial; after a few days, the phofphorus is found oxy-genated, and the phofphorous acid, in propor-tion as it forms, has attracted moifture from the air, and dropped into the phial. The phofphorous

rous acid is readily changed into phofphoric acid by expofure for a long time to the free air ; it abforbs oxygen from the air, and becomes fully oxygenated.

As phofphorus has a fufficient affinity for oxygen to attract it from the nitric and muriatic acids, we may form phofphoric acid, by means of thefe acids, in a very fimple and cheap manner. Fill a tubulated receiver, half full of concentrated nitric acid, and heat it gently, then throw in fmall pieces of phofphorus through the tube, thefe are diffolved with effervefcence and red fumes of nitrous gas fly off; add phofphorus fo long as it will diffolve, and then increafe the fire under the retort to drive off the laft particles of nitric acid ; phofphoric acid, partly fluid and partly concrete, remains in the retort.

TABLE

TABLE *of the Combinations of Carbonic Acid, with the Salifiable Bases, in the Order of Affinity.*

Resulting Neutral Salts.

Names of Bases.	*New Nomenclature.*	*Old Nomenclature.*
Barytes	Carbonats of * barytes	{ Aërated or effervefcent heavy earth.
Lime	lime	{ Chalk, calcareous fpar, Aërated calcareous earth.
Potafh	potafh	{ Effervefcing or aërated fixed vegetable alkali, mephitis of potafh.
Soda	foda	{ Aërated or effervefcing fixed mineral alkali, mephitic foda.
Magnefia	magnefia	{ Aërated, effervefcing, mild, or mephitic magnefia.
Ammoniac	ammoniac	{ Aërated, effervefcing, mild, or mephitic volatile alkali.
Argill	argill	{ Aërated or effervefcing argillaceous earth, or earth of alum.
Oxyds of zinc	zinc	{ Zinc fpar, mephitic or aërated zinc.
iron	iron	{ Sparry iron-ore, mephitic or aërated iron.
manganefe	manganefe	Aërated manganefe.
cobalt	cobalt	Aërated cobalt.
nickel	nickel	Aërated nickel.
lead	lead	Sparry lead-ore, or aërated lead.
tin	tin	Aërated tin.
copper	copper	Aërated copper.
bifmuth	bifmuth	Aërated bifmuth.
antimony	antimony	Aërated antimony.
arfenic	arfenic	Aërated arfenic.
mercury	mercury	Aërated mercury.
filver	filver	Aërated filver.
gold	gold	Aërated gold.
platina	platina	Aërated platina.

* As thefe falts have only been underftood of late, they have not, properly fpeaking, any old names. Mr Morveau, in the Firft Volume of the Encyclopedia, calls them *Mephites*; Mr Bergman gives them the name of *aërated*; and Mr de Fourcroy, who calls the carbonic acid *chalky acid*, gives them the name of *chalks.*—A.

SECT. XVII.—*Obfervations upon Carbonic Acid, and its Combinations.*

Of all the known acids, the carbonic is the moſt abundant in nature; it exiſts ready form-ed in chalk, marble, and all the calcareous ſtones, in which it is neuturalized by a particu-lar earth called *lime.* To difengage it from this combination, nothing more is requiſite than to add ſome ſulphuric acid, or any other which has a ſtronger affinity for lime; a briſk effervef-cence enſues, which is produced by the difen-gagement of the carbonic acid which affumes the ſtate of gas immediately upon being ſet free. This gas, incapable of being condenſed into the ſolid or liquid form by any degree of cold or of preſſure hitherto known, unites to about its own bulk of water, and thereby forms a very weak acid. It may likewiſe be obtained in great a-bundance from faccharine matter in fermenta-tion, but is then contaminated by a ſmall por-tion of alkohol which it holds in ſolution.

As charcoal is the radical of this acid, we may form it artificially, by burning charcoal in oxygen gas, or by combining charcoal and me-tallic oxyds in proper proportions; the oxygen of the oxyd combines with the charcoal, form-

ing

ing carbonic acid gas, and the metal being left free, recovers its metallic or reguline form.

We are indebted for our firſt knowledge of this acid to Dr Black, before whoſe time its property of remaining always in the ſtate of gas had made it to elude the reſearches of chemiſtry.

It would be a moſt valuable diſcovery to ſociety, if we could decompoſe this gas by any cheap proceſs, as by that means we might obtain, for economical purpoſes, the immenſe ſtore of charcoal contained in calcareous earths, marbles, limeſtones, &c. This cannot be effected by ſingle affinity, becauſe, to decompoſe the carbonic acid, it requires a ſubſtance as combuſtible as charcoal itſelf, ſo that we ſhould only make an exchange of one combuſtible body for another not more valuable ; but it may poſſibly be accompliſhed by double affinity, ſince this proceſs is ſo readily performed by Nature, during vegetation, from the moſt common materials.

TABLE

TABLE *of the Combinations of Muriatic Acid, with the Salifiable Bafes, in the Order of Affinity.*

Names of the bafes.	Refulting Neutral Salts.	
	New nomenclature.	Old nomenclature.
Barytes.	Muriat of barytes	Sea-falt, having bafe of heavy earth.
Potafh	potafh	Febrifuge falt of Sylvius. Muriated vegetable fixed alkali.
Soda	foda	Sea-falt.
Lime	lime	Muriated lime. Oil of lime.
Magnefia	magnefia	Marine Epfom falt. Muriated magnefia.
Ammoniac	ammoniac	Sal ammoniac.
Argill	argill	Muriated alum, fea-falt with bafe of earth of alum.
Oxyd of zinc	zinc	Sea-falt of, or muriatic zinc.
iron	iron	Salt of iron, Martial fea-falt.
manganefe	manganefe	Sea-falt of manganefe.
cobalt	cobalt	Sea-falt of cobalt.
nickel	nickel	Sea-falt of nickel.
lead	lead	Horny-lead. Plumbum corneum.
tin	fmoaking of tin folid of tin	Smoaking liquor of Libavius. Solid butter of tin.
copper	copper	Sea-falt of copper.
bifmuth	bifmuth	Sea-falt of bifmuth.
antimony	antimony	Sea-falt of antimony.
arfenic	arfenic	Sea-falt of arfenic.
mercury	fweet of mercury	Sweet fublimate of mercury, calomel, aquila alba.
	corrofive of mercury	Corrofive fublimate of mercury.
filver	filver	Horny filver, argentum corneum, luna cornea.
gold	gold	Sea-falt of gold.
platina	platina	Sea-falt of platina.

TABLE

TABLE *Of the Combinations of Oxygenated Muriatic Acid, with the Salifiable Bases, in the Order of Affinity.*

Names of the Bases.	Names of the Neutral Salts by the new Nomenclature.
	Oxygenated muriat of
Barytes	barytes.
Potash	potash.
Soda	soda.
Lime	lime.
Magnesia	magnesia.
Argill	argill.
Oxyd of	
zinc	zinc.
iron	iron.
manganese	manganese.
cobalt	cobalt.
nickel	nickel.
lead	lead.
tin	tin.
copper	copper.
bismuth	bismuth.
antimony	antimony.
arsenic	arsenic.
mercury	mercury.
silver	silver.
gold	gold.
platina	platina.

This order of salts, entirely unknown to the ancient chemists, was discovered in 1786 by Mr Berthollet. —A.

SECT.

SECT. XIX.—*Observations upon Muriatic and Oxygenated Muriatic Acids, and their Combinations.*

Muriatic acid is very abundant in the mineral kingdom naturally combined with different falifiable bafes, efpecially with foda, lime, and magnefia. In fea-water, and the water of feveral lakes, it is combined with thefe three bafes, and in mines of rock-falt it is chiefly united to foda. This acid does not appear to have been hitherto decompofed in any chemical experiment ; fo that we have no idea whatever of the nature of its radical, and only conclude, from analogy with the other acids, that it contains oxygen as its acidifying principle. Mr Berthollet fufpects the radical to be of a metallic nature ; but, as Nature appears to form this acid daily, in inhabited places, by combining miafmata with aëriform fluids, this muft necelfarily fuppofe a metallic gas to exift in the atmofphere, which is certainly not impoffible, but cannot be admitted without proof.

The muriatic acid has only a moderate adherence to the falifiable bafes, and can readily be driven from its combination with thefe by fulphuric acid. Other acids, as the nitric, for inftance, may anfwer the fame purpofe ; but nitric acid being volatile, would mix, during di-

ftillation,

ftillation, with the muriatic. About one part of fulphuric acid is fufficient to decompofe two parts of decrepitated fea-falt. This operation is performed in a tubulated retort, having Woulfe's apparatus, (Pl. IV. Fig. 1.), adapted to it. When all the junctures are properly luted, the fea-falt is put into the retort through the tube, the fulphuric acid is poured on, and the opening immediately clofed with its ground cryftal ftopper. As the muriatic acid can only fubfift in the gaffeous form in the ordinary temperature, we could not condenfe it without the prefence of water. Hence the ufe of the water with which the bottles in Woulfe's apparatus are half filled; the muriatic acid gas, driven off from the fea-falt in the retort, combines with the water, and forms what the old chemifts called *fmoaking fpirit of falt*, or *Glauber's fpirit of fea-falt*, which we now name *muriatic acid*.

The acid obtained by the above procefs is ftill capable of combining with a farther dofe of oxygen, by being diftilled from the oxyds of manganefe, lead, or mercury, and the refulting acid, which we name *oxygenated muriatic acid*, can only, like the former, exift in the gaffeous form, and is abforbed, in a much fmaller quantity by water. When the impregnation of water with this gas is pufhed beyond a certain point, the fuperabundant acid precipitates to the bottom of the veffels in a concrete form. Mr Berthollet has fhown

fhown that this acid is capable of combining with a great number of the falifiable bafes ; the neutral falts which refult from this union are fufceptible of deflagrating with charcoal, and many of the metallic fubftances ; thefe deflagrations are very violent and dangerous, owing to the great quantity of caloric which the oxygen carries alongft with it into the compofition of oxygenated muriatic acid.

TABLE

TABLE *of the Combinations of Nitro-muriatic Acid with the Salifiable Bafes, in the Order of Affinity, fo far as is known.*

Names of the Bafes.	Names of the Neutral Salts.
Argill	Nitro-muriat of argill.
Ammoniac	ammoniac.
Oxyd of	
antimony	antimony.
filver	filver.
arfenic	arfenic.
Barytes	barytes.
Oxyd of	
bifmuth	bifmuth.
Lime	lime.
Oxyd of	
cobalt	cobalt.
copper	copper.
tin	tin.
iron	iron.
Magnefia	magnefia.
Oxyd of	
manganefe	manganefe.
mercury	mercury.
molybdena	molybdena
nickel	nickel.
gold	gold.
platina	platina.
lead	lead
Potafh	potafh.
Soda	foda.
Oxyd of	
tungftein	tungftein.
zinc	zinc.

Note.—Moft of thefe combinations, efpecially thofe with the earths and alkalies, have been little examined, and we are yet to learn whether they form a mixed falt in which the compound radical remains combined, or if the two acids feparate, to form two diftinct neutral falts.—A.

SECT. XX.—*Obfervations upon the Nitro-Muria-*
tic Acid, and its Combinations.

The nitro-muriatic acid, formerly called *a-*
qua regia, is formed by a mixture of nitric and
muriatic acids ; the radicals of thefe two acids
combine together, and form a compound bafe,
from which an acid is produced, having properties peculiar to itfelf, and diftinct from thofe of
all other acids, efpecially the property of diffolving gold and platina.

In diffolutions of metals in this acid, as in all
other acids, the metals are firft oxydated by attracting a part of the oxygen from the compound radical.   This occafions a difengagement of a particular fpecies of gas not hitherto
defcribed, which may be called *nitro-muriatic gas;*
it has a very difagreeable fmell, and is fatal to
animal life when refpired ; it attacks iron, and
caufes it to ruft ; it is abforbed in confiderable
quantity by water, which thereby acquires fome
flight characters of acidity.   I had occafion to
make thefe remarks during a courfe of experiments upon platina, in which I diffolved a confiderable quantity of that metal in nitro-muriatic
acid.

I at firft fufpected that, in the mixture of nitric and muriatic acids, the latter attracted a
part

part of the oxygen from the former, and became converted into oxygenated muriatic acid,
which gave it the property of diffolving gold;
but feveral facts remain inexplicable upon this
fuppofition. Were it fo, we muft be able to
difengage nitrous gas by heating this acid,
which however does not fenfibly happen. From
thefe confiderations, I am led to adopt the opinion of Mr Berthollet, and to confider nitro-
muriatic acid as a fingle acid, with a compound
bafe or radical.

TABLE

TABLE *of the Combinations of Fluoric Acid, with the Salifiable Bafes, in the Order of Affinity.*

Names of the Bafes.	*Names of the Neutral Salts.*
Lime	Fluat of lime.
Barytes	barytes.
Magnefia	magnefia.
Potafh	potafh.
Soda	foda.
Ammoniac	ammoniac.
Oxyd of	
zinc	zinc.
manganefe	manganefe.
iron	iron.
lead	lead.
tin	tin.
cobalt	cobalt.
copper	copper.
nickel	nickel.
arfenic	arfenic.
bifmuth	bifmuth.
mercury	mercury.
filver	filver.
gold	gold.
platina	platina.

And by the dry way,

Argill	Fluat of argill.

Note.—Thefe combinations were entirely unknown to the old chemifts, and confequently have no names in the old nomenclature.—A.

SECT.

SECT. XXI.—*Obfervations upon the Fluoric Acid,
and its Combinations.*

Fluoric exifts ready formed by Nature in the
fluoric fpars *, combined with calcareous earth,
fo as to form an infoluble neutral falt.   To obtain it difengaged from that combination, fluor
fpar, or fluat of lime, is put into a leaden retort, with a proper quantity of fulphuric acid, a
recipient likewife of lead, half full of water, is
adapted, and fire is applied to the retort.   The
fulphuric acid, from its greater affinity, expels
the fluoric acid which paffes over and is abforbed by the water in the receiver.   As fluoric a-
cid is naturally in the gaffeous form in the ordinary temperature, we can receive it in a pneumato-chemical apparatus over mercury.   We
are obliged to employ metallic veffels in this procefs, becaufe fluoric acid diffolves glafs and filicious earth, and even renders thefe bodies volatile,
carrying them over with itfelf in diftillation in
the gaffeous form.

We are indebted to Mr Margraff for our firft
acquaintance with this acid, though, as he could
never procure it free from combination with a
confiderable quantity of filicious earth, he was
                                          ignorant

* Commonly called *Derbyfhire fpars.*—E.

ignorant of its being an acid fui generis. The Duke de Liancourt, under the name of Mr Boulanger, confiderably increafed our knowledge of its properties ; and Mr Scheele feems to have exhaufted the fubject. The only thing remaining is to endeavour to difcover the nature of the fluoric radical, of which we cannot hitherto form any ideas, as the acid does not appear to have been decompofed in any experiment. It is only by means of compound affinity that experiments ought to be made with this view, with any probability of fuccefs.

H h   Table

TABLE *of the Combinations of Boracic Acid, with the Salifiable Bafes, in the Order of Affinity.*

Bafes.	Neutral Salts.
Lime	Borat of lime.
Barytes	barytes.
Magnefia	magnefia.
Potafh	potafh.
Soda	foda.
Ammoniac	ammoniac.
Oxyd of	
zinc	zinc.
iron	iron.
lead	lead.
tin	tin.
cobalt	cobalt.
copper	copper.
nickel	nickel.
mercury	mercury.
Argill	argill.

Note.—Moſt of thefe combinations were neither known nor named by the old chemiſts. The boracic acid was formerly called *fedative falt*, and its compounds *borax*, with bafe of fixed vegetable alkali, &c.—A.

SECT,

SECT. XXII.—*Obfervations upon Boracic Acid and its Combinations.*

This is a concrete acid, extracted from a falt procured from India called *borax* or *tincall.* Although borax has been very long employed in the arts, we have as yet very imperfect knowledge of its origin, and of the methods by which it is extracted and purified ; there is reafon to believe it to be a native falt, found in the earth in certain parts of the eaft, and in the water of fome lakes. The whole trade of borax is in the hands of the Dutch, who have been exclufively poffeffed of the art of purifying it till very lately, that Meffrs L'Eguillier of Paris have rivalled them in the manufacture ; but the procefs ftill remains a fecret to the world.

By chemical analyfis we learn that borax is a neutral falt with excefs of bafe, confifting of foda, partly faturated with a peculiar acid long called *Homberg's fedative falt,* now *the boracic acid.* This acid is found in an uncombined ftate in the waters of certain lakes. That of Cherchiais in Italy contains $94\frac{1}{2}$ grains in each pint of water.

To obtain boracic acid, diffolve fome borax in boiling water, filtrate the folution, and add fulphuric acid, or any other having greater affinity

nity to foda than the boracic acid; this latter acid is feparated, and is procured in a
cryftalline form by cooling. This acid was
long confidered as being formed during the
procefs by which it is obtained, and was confequently fuppofed to differ according to the
nature of the acid employed in feparating it
from the foda; but it is now univerfally ac-
knowledged that it is identically the fame acid,
in whatever way procured, provided it be properly purified from mixture of other acids, by
wafhing, and by repeated folution and criftallization. It is foluble both in water and alkohol,
and has the property of communicating a green
colour to the flame of that fpirit. This circumftance led to a fufpicion of its containing copper,
which is not confirmed by any decifive experiment. On the contrary, if it contain any of
that metal, it muft only be confidered as an accidental mixture. It combines with the falifiable bafes in the humid way; and though, in
this manner, it is incapable of diffolving any of
the metals directly, this combination is readily
affected by compound affinity.

The Table prefents its combinations in the
order of affinity in the humid way; but there
is a confiderable change in the order when we
operate via ficca; for, in that cafe, argill,
though the laft in our lift, muft be placed immediately after foda.

The

The boracic radical is hitherto unknown ; no experiments having as yet been able to decompofe the acid ; we conclude, from analogy with the other acids, that oxygen exifts in its compofition as the acidifying principle.

Table

Table *of the Combinations of Arfeniac Acid, with the Salifiable Bafes, in the Order of Affinity.*

Bafes.	Neutral Salts.
Lime	Arfeniat of lime.
Barytes	barytes.
Magnefia	magnefia.
Potafh	potafh.
Soda	foda.
Ammoniac	ammoniac.
Oxyd of	
zinc	zinc.
manganefe	manganefe.
iron	iron.
lead	lead.
tin	tin.
cobalt	cobalt.
copper	copper.
nickel	nickel.
bifmuth	bifmuth.
mercury	mercury.
antimony	antimony.
filver	filver.
gold	gold.
platina	platina.
Argill	argill.

Note.—This order of falts was entirely unknown to the antient chemifts. Mr Macquer, in 1746, difcovered the combinations of arfeniac acid with potafh and foda, to which he gave the name of *arfenical neutral falts.*—A.

Sect.

SECT. XXIII.—*Obfervations upon Arfeniac Acid,
and its Combinations.*

In the Collections of the Academy for 1746,
Mr Macquer fhows that, when a mixture of
white oxyd of arfenic and nitre are fubjected to
the action of a ftrong fire, a neutral falt is obtained, which he calls *neutral falt of arfenic.* At
that time, the caufe of this fingular phenomenon, in which a metal acts the part of an acid,
was quite unknown ; but more modern experiments teach that, during this procefs, the
arfenic becomes oxygenated, by carrying off the
oxygen of the nitric acid ; it is thus converted
into a real acid, and combines with the potafh.
There are other methods now known for oxygenating arfenic, and obtaining its acid free
from combination. The moft fimple and moft
effectual of thefe is as follows : Diffolve white
oxyd of arfenic in three parts, by weight, of
muriatic acid ; to this folution, in a boiling
ftate, add two parts of nitric acid, and evaporate to drynefs. In this procefs the nitric acid is
decompofed, its oxygen unites with the oxyd of
arfenic, and converts it into an acid, and the
nitrous radical flies off in the ftate of nitrous
gas ; whilft the muriatic acid is converted by
the heat into muriatic acid gas, and may be collected in proper veffels. The arfeniac acid is entirely

tirely freed from the other acids employed during
the procefs by heating it in a crucible till it begins to grow red; what remains is pure concrete arfeniac acid.

Mr Scheele's procefs, which was repeated
with great fuccefs by Mr Morveau, in the laboratory at Dijon, is as follows : Diftil muriatic a-
cid from the black oxyd of manganefe, this converts it into oxygenated muriatic acid, by carrying off the oxygen from the manganefe, receive this in a recipient containing white oxyd
of arfenic, covered by a little diftilled water ;
the arfenic decompofes the oxygenated muriatic
acid, by carrying off its fuperfaturation of oxygen, the arfenic is converted into arfeniac acid,
and the oxygenated muriatic acid is brought
back to the ftate of common muriatic acid.
The two acids are feparated by diftillation, with
a gentle heat increafed towards the end of the
operation, the muriatic acid paffes over, and the
arfeniac acid remains behind in a white concrete
form.

The arfeniac acid is confiderably lefs volatile
than white oxyd of arfenic ; it often contains
white oxyd of arfenic in folution, owing to
its not being fufficiently oxygenated ; this is
prevented by continuing to add nitrous acid, as
in the former procefs, till no more nitrous
gas is produced. From all thefe obfervations I would give the following definition of
                                        arfeniac

arſeniac acid. It is a white concrete metallic acid, formed by the combination of arſenic with oxygen, fixed in a red heat, ſoluble in water, and capable of combining with many of the ſaliſiable baſes.

SECT. XXIV.—*Obſervations upon Molybdic A-cid, and its Combinations with Acidifiable Ba-ſes* *.

Molybdena is a particular metallic body, capable of being oxygenated, ſo far as to become a true concrete acid †. For this purpoſe, one part ore of molybdena, which is a natural ſulphuret of that metal, is put into a retort, with five or ſix parts nitric acid, diluted with a quarter of its weight of water, and heat is applied to the retort ; the oxygen of the nitric acid acts both upon the molybdena and the ſulphur, converting the one into molybdic, and the other into ſulphuric acid ; pour on freſh quantities of nitric acid ſo long as any red fumes of nitrous

I i   gas

---

* I have not added the Table of theſe combinations, as the order of their affinity is entirely unknown ; they are called *molybdats of argil, antimony, potaſh*, &c.—E.,

† This acid was diſcovered by Mr Scheele, to whom chemiſtry is indebted for the diſcovery of ſeveral other acids.—A.

gas efcape; the molydbena is then oxygenated as far as is poffible, and is found at the bottom of the retort in a pulverulent form, refembling chalk. It muft be wafhed in warm water, to feparate any adhering particles of fulphuric a- cid; and, as it is hardly foluble, we lofe very little of it in this operation. All its combina- tions with falifiable bafes were unknown to the ancient chemifts.

TABLE

TABLE *of the Combinations of Tungſtic Acid with
the Salifiable Baſes.*

Baſes.	*Neutral Salts.*
Lime	Tungſtat of lime.
Barytes	barytes.
Magneſia	magneſia.
Potaſh	potaſh.
Soda	ſoda.
Ammoniac	ammoniac.
Argill	argill.
Oxyd of antimo-	
ny *, &c.	antimony †, &c.

SECT. XXV.—*Obſervations upon Tungſtic Acid,
and its Combinations.*

Tungſtein is a particular metal, the ore of
which has frequently been confounded with that
of tin. The ſpecific gravity of this ore is to water
as 6 to 1; in its form of criſtallization it re-
ſembles

---

* The combinations with metallic oxyds were ſet
down by Mr Lavoiſier in alphabetical order ; their order of affinity being unknown, I have omitted them, as
ſerving no purpoſe.—E.

† All theſe ſalts were unknown to the ancient chemiſts.—A.

fembles the garnet, and varies in colour from a pearl-white to yellow and reddifh ; it is found in feveral parts of Saxony and Bohemia. The mineral called *Wolfram*, which is frequent in the mines of Cornwal, is likewife an ore of this metal. In all thefe ores the metal is oxydated ; and, in fome of them, it appears even to be o-xygenated to the ftate of acid, being combined with lime into a true tungftat of lime.

To obtain the acid free, mix one part of ore of tungftein with four parts of carbonat of potafh, and melt the mixture in a crucible, then powder and pour on twelve parts of boiling water, add nitric acid, and the tungftic acid pre-cipitates in a concrete form. Afterwards, to infure the complete oxygenation of the metal, add more nitric acid, and evaporate to drynefs, repeating this operation fo long as red fumes of nitrous gas are produced. To procure tungftic acid perfectly pure, the fufion of the ore with carbonat of potafh muft be made in a crucible of platina, otherwife the earth of the common crucibles will mix with the products, and adul-terate the acid.

TABLE

TABLE *of the Combinations of Tartarous Acid,*
with the Salifiable Bafes, in the Order of
Affinity.

Bafes,	Neutral Salts.
Lime	Tartarite of lime.
Barytes	barytes.
Magnefia	magnefia.
Potafh	potafh.
Soda	foda.
Ammoniac	ammoniac.
Argill	argill.
Oxyd of	
zinc	zinc.
iron	iron.
manganefe	manganefe.
cobalt	cobalt.
nickel	nickel.
lead	lead.
tin	tin.
copper	copper.
bifmuth	bifmuth.
antimony	antimony.
arfenic	arfenic.
filver	filver.
mercury	mercury.
gold	gold.
platina	platina.

SECT.

SECT. XXVI.—*Obfervations upon Tartarous A-cid, and its Combinations.*

Tartar, or the concretion which fixes to the infide of veffels in which the fermentation of wine is completed, is a well known falt, com-pofed of a peculiar acid, united in confiderable excefs to potafh. Mr Scheele firft pointed out the method of obtaining this acid pure. Ha-ving obferved that it has a greater affinity to lime than to potafh, he directs us to proceed in the following manner. Diffolve purified tartar in boiling water, and add a fufficient quantity of lime till the acid be completely faturated. The tartarite of lime which is formed, being al-moft infoluble in cold water, falls to the bottom, and is feparated from the folution of potafh by decantation ; it is afterwards wafhed in cold water, and dried ; then pour on fome fulphuric acid, diluted with eight or nine parts of water, digeft for twelve hours in a gentle heat, fre-quently ftirring the mixture ; the fulphuric acid combines with the lime, and the tartarous acid is left free. A fmall quantity of gas, not hi-therto examined, is difengaged during this pro-cefs. At the end of twelve hours, having de-canted off the clear liquor, wafh the fulphat of lime in cold water, which add to the decanted liquor,

liquor, then evaporate the whole, and the tartarous acid is obtained in a concrete form. Two
pounds of purified tartar, by means of from
eight to ten ounces of fulphuric acid, yield a-
bout eleven ounces of tartarous acid.

As the combuftible radical exifts in excefs,
or as the acid from tartar is not fully faturated
with oxygen, we call it *tartarous acid*, and the
neutral falts formed by its combinations with falifiable bafes *tartarites*. The bafe of the tartarous acid is a carbono-hydrous or hydro-carbonous radical, lefs oxygenated than in the oxalic
acid; and it would appear, from the experiments of Mr Haffenfratz, that azote enters into
the compofition of the tartarous radical, even in
confiderable quantity. By oxygenating the tartarous acid, it is convertible into oxalic, malic,
and acetous acids; but it is probable the proportions of hydrogen and charcoal in the radical
are changed during thefe converfions, and that
the difference between thefe acids does not alone
confift in the different degrees of oxygenation.

The tartarous acid is fufceptible of two degrees of faturation in its combinations with the
fixed alkalies; by one of thefe a falt is formed
with excefs of acid, improperly called *cream of
tartar*, which in our new nomenclature is named *acidulous tartarite of potafh*; by a fecond or
equal degree of faturation a perfectly neutral
falt is formed, formerly called *vegetable falt*,
which

which we name *tartarite of potash*. With foda
this acid forms tartarite of foda, formerly called
fal de Seignette, or *fal polychrest of Rochell*.

SECT. XXVII.—*Obfervations upon Malic Acid,
and its Combinations with the Salifiable Bafes* *.

The malic acid exifts ready formed in the
four juice of ripe and unripe apples, and many
other fruits, and is obtained as follows : Satu-
rate the juice of apples with potafh or foda, and
add a proper proportion of acetite of lead diffolved in water ; a double decompofition takes
place, the malic acid combines with the oxyd
of lead and precipitates, being almoft infoluble,
and the acetite of potafh or foda remains in the
liquor. The malat of lead being feparated by
decantation, is wafhed with cold water, and fome
dilute fulphuric acid is added ; this unites with
the lead into an infoluble fulphat, and the malic
acid remains free in the liquor.

This acid, which is found mixed with citric
and tartarous acid in a great number of fruits,
is a kind of medium between oxalic and acetous

* I have omitted the Table, as the order of affinity
is unknown, and is given by Mr Lavoifier only in alphabetical order. All the combinations of malic acid
with falifiable bafes, which are named *malats*, were unknown to the ancient chemifts.—E.

tous acids being more oxygenated than the former, and lefs fo than the latter. From this circumftance, Mr Hermbftadt calls it *imperfect vinegar* ; but it differs likewife from acetous acid,
by having rather more charcoal, and lefs hydrogen, in the compofition of its radical.

When an acid much diluted has been ufed in
the foregoing procefs, the liquor contains oxalic
as well as malic acid, and probably a little tartarous, thefe are feparated by mixing lime-water with the acids, oxalat, tartarite, and malat
of lime are produced ; the two former, being
infoluble, are precipitated, and the malat of
lime remains diffolved ; from this the pure malic acid is feparated by the acetite of lead, and
afterwards by fulphuric acid, as directed above.

K k

TABLE *of the Combinations of Citric Acid, with the Salifiable Bases, in the Order of Affinity* *.

Bases.	Neutral Salts.
Barytes	Citrat of barytes.
Lime	lime.
Magnesia	magnesia.
Potash	potash.
Soda	soda.
Ammoniac	ammoniac.
Oxyd of	
zinc	zinc.
manganese	manganese.
iron	iron.
lead	lead.
cobalt	cobalt.
copper	copper.
arsenic	arsenic.
mercury	mercury.
antimony	antimony.
silver	silver.
gold	gold.
platina	platina.
Argill	argill.

* These combinations were unknown to the ancient chemists. The order of affinity of the salifiable bases with this acid was determined by Mr Bergman and by Mr de Breney of the Dijon Academy.—A.

SECT.

SECT. XXVIII.—*Obfervations upon Citric Acid, and its Combinations.*

The citric acid is procured by expreffion from lemons, and is found in the juices of many other fruits mixed with malic acid. To obtain it pure and concentrated, it is firft allowed to depurate from the mucous part of the fruit by long reft in a cool cellar, and is afterwards concentrated by expofing it to the temperature of 4 or 5 degrees below Zero, from 21° to 23° of Fahrenheit, the water is frozen, and the acid remains liquid, reduced to about an eighth part of its original bulk. A lower degree of cold would occafion the acid to be engaged amongft the ice, and render it difficultly feparable. This procefs was pointed out by Mr Georgius.

It is more eafily obtained by faturating the lemon-juice with lime, fo as to form a citrat of lime, which is infoluble in water ; wafh this falt, and pour on a proper quantity of fulphuric acid ; this forms a fulphat of lime, which precipitates and leaves the citric acid free in the liquor.

TABLE

TABLE *of the Combinations of Pyro-lignous Acid with the Salifiable Bafes, in the Order of Affinity* *.

Bafes.	Neutral Salts.
Lime	Pyro-mucite of lime.
Barytes	barytes.
Potafh	potafh.
Soda	foda.
Magnefia	magnefia.
Ammoniac	ammoniac.
Oxyd of	
zinc	zinc.
manganefe	manganefe.
iron	iron.
lead	lead
tin	tin.
cobalt	cobalt.
copper	copper.
nickel	nickel.
arfenic	arfenic.
bifmuth	bifmuth.
mercury	mercury.
antimony	antimony.
filver	filver.
gold	gold.
platina	platina.
Argill	argill.

* The above affinities were determined by Meffrs de Morveau and Eloi Bourfier de Clervaux. Thefe combinations were entirely unknown till lately.—A.

SECT.

SECT. XXIX.—*Obfervations upon Pyro-lignous Acid, and its Combinations.*

The ancient chemifts obferved that moft of the woods, efpecially the more heavy and compact ones, gave out a particular acid fpirit, by diftillation, in a naked fire; but, before Mr Goetling, who gives an account of his experiments upon this fubject in Crell's Chemical Journal for 1779, no one had ever made any inquiry into its nature and properties. This acid appears to be the fame, whatever be the wood it is procured from. When firft diftilled, it is of a brown colour, and confiderably impregnated with charcoal and oil; it is purified from thefe by a fecond diftillation. The pyro-lignous radical is chiefly compofed of hydrogen and charcoal.

SECT. XXX.—*Obfervations upon Pyro-tartarous Acid, and its Combinations with the Salifiable Bafes* *.

The name of *Pyro-tartarous acid* is given to a dilute empyreumatic acid obtained from purified

---

* The order of affinity of the falifiable bafes with this acid is hitherto unknown. Mr Lavoifier, from its fimila-

fied acidulous tartarite of potafh by diftillation in a naked fire. To obtain it, let a retort be half filled with powdered tartar, adapt a tubulated recipient, having a bent tube communicating with a bell-glafs in a pneumato-chemical apparatus ; by gradually raifing the fire under the retort, we obtain the pyro-tartarous acid mixed with oil, which is feparated by means of a funnel. A vaft quantity of carbonic acid gas is difengaged during the diftillation. The acid obtained by the above procefs is much contaminated with oil, which ought to be feparated from it. Some authors advife to do this by a fecond diftillation ; but the Dijon academicians inform us, that this is attended with great danger from explofions which take place during the procefs.

TABLE

fimilarity to pyro-lignous acid, fuppofes the order to be the fame in both ; but, as this is not afcertained by experiment, the table is omitted. All thefe combinations, called *Pyro-tartarites,* were unknown till lately.—E.

TABLE *of the Combinations of Pyro-mucous Acid, with the Salifiable Bases, in the Order of Affinity* *.

Bases.	Neutral Salts.
Potaſh	Pyro-mucite of potaſh.
Soda	ſoda.
Barytes	barytes.
Lime	lime.
Magneſia	magneſia.
Ammoniac	ammoniac.
Argill	argill.
Oxyd of	
zinc	zinc.
manganeſe	manganeſe.
iron	iron.
lead	lead.
tin	tin.
cobalt	cobalt.
copper	copper.
nickel	nickel.
arſenic	arſenic.
biſmuth	biſmuth.
antimony	antimony.

* All theſe combinations were unknown to the ancient chemiſts.—A.

SECT. XXXI.—*Obſervations upon Pyro-mucous Acid, and its Combinations.*

This acid is obtained by diſtillation in a naked fire from ſugar, and all the ſaccharine bodies ; and, as theſe ſubſtances ſwell greatly in the fire, it is neceſſary to leave ſeven-eighths of the retort empty. It is of a yellow colour, verging to red, and leaves a mark upon the ſkin, which will not remove but alongſt with the epidermis. It may be procured leſs-coloured, by means of a ſecond diſtillation, and is concentrated by freezing, as is directed for the citric acid. It is chiefly compoſed of water and oil ſlightly oxygenated, and is convertible into oxalic and malic acids by farther oxygenation with the nitric acid.

It has been pretended that a large quantity of gas is diſengaged during the diſtillation of this acid, which is not the caſe if it be conducted ſlowly, by means of moderate heat.

TABLE *of the Combinations of the Oxalic Acid,*
with the Salifiable Bafes, in the Order of Affi-
nity *.

Bafes.	Neutral Salts.
Lime	Oxalat of lime.
Barytes	barytes.
Magnefia	magnefia.
Potafh	potafh.
Soda	foda.
Ammoniac	ammoniac.
Argill	argill.
Oxyd of	
zinc	zinc.
iron	iron.
manganefe	manganefe.
cobalt	cobalt.
nickel	nickel.
lead	lead.
copper	copper.
bifmuth	bifmuth.
antimony	antimony.
arfenic	arfenic.
mercury	mercury.
filver	filver.
gold	gold.
platina	platina.

All unknown to the ancient chemifts.—A.

L l

SECT. XXXII.—*Obfervations upon Oxalic Acid,
and its Combinations.*

The oxalic acid is moftly prepared in Swit-
zerland and Germany from the expreffed juice
of forrel, from which it criftallizes by being left
long at reft ; in this ftate it is partly faturated
with potafh, forming a true acidulous oxalat of
potafh, or falt with excefs of acid. To obtain
it pure, it muft be formed artificially by oxygenating fugar, which feems to be the true oxalic
radical. Upon one part of fugar pour fix or
eight parts of nitric acid, and apply a gentle
heat ; a confiderable effervefcence takes place,
and a great quantity of nitrous gas is difengaged ; the nitric acid is decompofed, and its oxygen unites to the fugar : By allowing the liquor
to ftand at reft, criftals of pure oxalic acid are
formed, which muft be dried upon blotting paper, to feparate any remaining portions of nitric acid ; and, to enfure the purity of the acid,
diffolve the criftals in diftilled water, and criftallize them afrefh.

From the liquor remaining after the firft criftallization of the oxalic acid we may obtain malic acid by refrigeration : This acid is more oxygenated than the oxalic ; and, by a further oxygenation,

TABLE of the Combinations of Acetous Acid with the Satifiable Bafes in the Order of Affinity.

Bafes.	Neutral falts.	Names of the refulting neutral falts according to the old nomenclature.
Barytes	Acetite of barytes	Unknown to the ancients. Difcovered by Mr de Morveau, who calls it *barotic acite*.
Potafh	—— potafh	Secret terra foliata tartari of Muller. Arcanum tartari of Bafil Valentin and Paracelfus. Purgative magiftery of tartar of Schroëder. Effential falt of wine of Zwelfer. Regenerated tartar of Tachenius. Diuretic falt of Sylvius and Wilfon.
Soda	—— foda	Foliated earth with bafe of mineral alkali. Mineral or cryftallifable foliated earth. Mineral acetous falt.
Lime	—— lime	Salt of chalk, coral, or crabs eyes; mentioned by Hartman.
Magnefia	—— magnefia	Firft mentioned by Mr Wenzel.
Ammoniac	—— ammoniac	Spiritus Mindereri. Ammoniacal acetous falt.
Oxyd of zinc	—— zinc	Known to Glauber, Schwedemberg, Refpour, Pott, de Laffone, and Wenzel, but not named.
—— manganefe	—— manganefe	Unknown to the ancients.
—— iron	—— iron	Martial vinegar. Defcribed by Monnet, Wenzel, and the Duke d'Ayen.
—— lead	—— lead	Sugar, vinegar, and falt of lead or Saturn.
—— tin	—— tin	Known to Lemery, Margraf, Monnet, Weflendorf, and Wenzel, but not named.
—— cobalt	—— cobalt	Sympathetic ink of Mr Cadet.
—— copper	—— copper	Verdigris, cryftals of verditer, verditer, diftilled verdigris, cryftals of Venus or of copper.
—— nickel	—— nickel	Unknown to the ancients.
—— arfenic	—— arfenic	Arfenico-acetous fuming liquor, liquid phofphorus of Mr Cadet.
—— bifmuth	—— bifmuth	Sugar of bifmuth of Mr Geoffroi. Known to Gellert, Pott, Weflendorf, Bergman, and de Morveau.
—— mercury	—— mercury	Mercurial foliated earth, Keyfer's famous antivenereal remedy. Mentioned by Gebaver in 1748; known to Helot, Margraff, Baumé, Bergman, and de Morveau.
—— antimony	—— antimony	Unknown.
—— filver	—— filver	Defcribed by Margraff, Monnet, and Wenzel; unknown to the ancients.
—— gold	—— gold	Little known, mentioned by Schroëder and Juncker.
—— platina	—— platina	Unknown.
Argill	—— argill	According to Mr Wenzel, vinegar diffolves only a very fmall proportion of argill.

genation, the fugar is convertible into acetous acid, or vinegar.

The oxalic acid, combined with a fmall quan-tity of foda or potafh, has the property, like the tartarous acid, of entering into a number of combinations without fuffering decompofition : Thefe combinations form triple falts, or neutral falts with double bafes, which ought to have proper names. The falt of forrel, which is pot-afh having oxalic acid combined in excefs, is named acidulous oxalat of potafh in our new nomenclature.

The acid procured from forrel has been known to chemifts for more than a century, being mentioned by Mr Duclos in the Memoirs of the Academy for 1688, and was pretty accu-rately defcribed by Boerhaave; but Mr Scheele firft fhowed that it contained potafh, and de-monftrated its identity with the acid formed by the oxygenation of fugar.

SECT. XXXIII.—*Obfervations upon Acetous Acid, and its Combinations.*

This acid is compofed of charcoal and hy-drogen united together, and brought to the ftate of an acid by the addition of oxygen; it is confequently formed by the fame elements with the

the tartarous oxalic, citric, malic acids, and
others, but the elements exift in different proportions in each of thefe; and it would appear
that the acetous acid is in a higher ftate of oxygenation than thefe other acids. I have fome
reafon to believe that the acetous radical contains a fmall portion of azote; and, as this element is not contained in the radicals of any vegetable acid except the tartarous, this circumftance is one of the caufes of difference. The
acetous acid, or vinegar, is produced by expofing wine to a gentle heat, with the addition of
fome ferment: This is ufually the ley, or mother, which has feparated from other vinegar
during fermentation, or fome fimilar matter.
The fpiritous part of the wine, which confifts
of charcoal and hydrogen, is oxygenated, and
converted into vinegar: This operation can only take place with free accefs of air, and is always attended by a diminution of the air employed in confequence of the abforption of oxygen; wherefore, it ought always to be carried
on in veffels only half filled with the vinous liquor fubmitted to the acetous fermentation.
The acid formed during this procefs is very volatile, is mixed with a large proportion of water, and with many foreign fubftances; and, to
obtain it pure, it is diftilled in ftone or glafs
veffels by a gentle fire. The acid which paffes
over in diftillation is fomewhat changed by the
procefs,

procefs, and is not exactly of the fame nature
with what remains in the alembic, but feems
lefs oxygenated : This circumftance has not
been formerly obferved by chemifts.

Diftillation is not fufficient for depriving this
acid of all its unneceffary water ; and, for this
purpofe, the beft way is by expofing it to a degree of cold from 4° to 6° below the freezing
point, from 19° to 23° of Fahrenheit ; by this
means the aqueous part becomes frozen, and
leaves the acid in a liquid ftate, and confiderably concentrated. In the ufual temperature of
the air, this acid can only exift in the gaffeous
form, and can only be retained by combination
with a large proportion of water. There are
other chemical proceffes for obtaining the acetous acid, which confift in oxygenating the tartarous, oxalic, or malic acids, by means of nitric
acid ; but there is reafon to believe the proportions of the elements of the radical are changed
during this procefs. Mr Haffenfratz is at prefent engaged in repeating the experiments by
which thefe converfions are faid to be produced.

The combinations of acetous acid with the
various falifiable bafes are very readily formed ;
but moft of the refulting neutral falts are not
criftallizable, whereas thofe produced by the
tartarous and oxalic acids are, in general, hardly foluble. Tartarite and oxalat of lime are

not

not foluble in any fenfible degree: The malats are a medium between the oxalats and acetites, with refpect to folubility, and the malic acid is in the middle degree of faturation between the oxalic and acetous acids. With this, as with all the acids, the metals require to be oxydated previous to folution.

The ancient chemifts knew hardly any of the falts formed by the combinations of acetous acid with the falifiable bafes, except the acetites of potafh, foda, ammoniac, copper, and lead. Mr Cadet difcovered the acetite of arfenic *; Mr Wenzel, the Dijon academicians Mr de Laffone, and Mr Prouft, made us acquainted with the properties of the other acetites. From the property which acetite of potafh poffeffes, of giving out ammoniac in diftillation, there is fome reafon to fuppofe, that, befides charcoal and hydrogen, the acetous radical contains a fmall proportion of azote, though it is not impoffible but the above production of ammoniac may be occafioned by the decompofition of the potafh.

TABLE

---

* Savans Etrangers, Vol. III.

TABLE *of the Combinations of Acetic Acid with the Salifiable Bases, in the order of affinity.*

Bases.	*Neutral Salts.*
Barytes	Acetat of barytes.
Potash	potash.
Soda	soda.
Lime	lime.
Magnesia	magnesia.
Ammoniac	ammoniac.
Oxyd of zinc	zinc.
manganese	manganese.
iron	iron.
lead	lead.
tin	tin.
cobalt	cobalt.
copper	copper.
nickel	nickel.
arsenic	arsenic.
bismuth	bismuth.
mercury	mercury.
antimony	antimony.
silver	silver.
gold	gold.
platina	platina.
Argill	argill.

SECT,

Note.—All these salts were unknown to the ancients; and even those chemists who are most verfant in modern difcoveries, are yet at a lofe whether the greater part of the falts produced by the oxygenated acetic radical belong properly to the clafs of acetites, or to that of acetats.—A.

SECT. XXXIV.—*Obfervations upon Acetic Acid,
and its Combinations.*

We have given to radical vinegar the name
of acetic acid, from fuppofing that it confifts of
the fame radical with that of the acetous acid,
but more highly faturated with oxygen. According to this idea, acetic acid is the higheft
degree of oxygenation of which the hydro-carbonous radical is fufceptible; but, although this
circumftance be extremely probable, it requires
to be confirmed by farther, and more decifive
experiments, before it be adopted as an abfolute chemical truth. We procure this acid as
follows: Upon three parts acetite of potafh or
of copper, pour one part of concentrated fulphuric acid, and, by diftillation, a very highly
concentrated vinegar is obtained, which we call
acetic acid, formerly named radical vinegar. It
is not hitherto rigoroufly proved that this acid
is more highly oxygenated than the acetous
acid, nor that the difference between them may
not confift in a different proportion between the
elements of the radical or bafe.

TABLE

TABLE *of the Combinations of Succinic Acid with the Salifiable Bases, in the order of Affinity.*

Bases.	Neutral Salts.
Barytes	Succinat of barytes.
Lime	lime.
Potaſh	potaſh.
Soda	ſoda.
Ammoniac	ammoniac.
Magneſia	magneſia.
Argill	argill.
Oxyd of zinc	zinc.
iron	iron.
manganeſe	manganeſe.
cobalt	cobalt.
nickel	nickel.
lead	lead.
tin	tin.
copper	copper.
biſmuth	biſmuth.
antimony	antimony.
arſenic	arſenic.
mercury	mercury.
ſilver	ſilver.
gold	gold.
platina	platina.

M m   SECT.

Note.—All the ſuccinats were unknown to the ancient chemiſts.—A.

Sect. XXXV.—*Obfervations upon Succinic Acid, and its Combinations.*

The fuccinic acid is drawn from amber by fublimation in a gentle heat, and rifes in a concrete form into the neck of the fubliming veffel. The operation muft not be pufhed too far, or by too ftrong a fire, otherwife the oil of the amber rifes alongft with the acid. The falt is dried upon blotting paper, and purified by repeated folution and criftallization.

This acid is foluble in twenty-four times its weight of cold water, and in a much fmaller quantity of hot water. It poffeffes the qualities of an acid in a very fmall degree, and only affects the blue vegetable colours very flightly. The affinities of this acid, with the falifiable bafes, are taken from Mr de Morveau, who is the firft chemift that has endeavoured to afcertain them.

Sect.

SECT. XXXVI.—*Obfervations upon Benzoic Acid, and its Combinations with Salifiable Bafes* *.

This acid was known to the ancient chemifts under the name of Flowers of Benjamin, or of Benzoin, and was procured, by fublimation, from the gum or refin called Benzoin : The means of procuring it, *via humida*, was difcovered by Mr Geoffroy, and perfeɛted by Mr Scheele. Upon benzoin, reduced to powder, pour ftrong lime-water, having rather an excefs of lime ; keep the mixture continually ftirring, and, after half an hour's digeftion, pour off the liquor, and ufe frefh portions of lime-water in the fame manner, fo long as there is any appearance of neutralization. Join all the decanted liquors, and evaporate, as far as poffible, without occafioning criftallization, and, when the liquor is cold, drop in muriatic acid till no more precipitate is formed. By the former part of the procefs a benzoat of lime is formed, and, by the latter, the muriatic acid combines with the lime, forming muriat of lime, which remains

---

* Thefe combinations are called Benzoats of Lime, Potafh, Zinc, &c. ; but, as the order of affinity is unknown, the alphabetical table is omitted, as unneceffary.— E.

mains diffolved, while the benzoic acid, being infoluble, precipitates in a concrete ftate.

SECT. XXXVII.—*Obfervations upon Camphoric Acid, and its Combinations with Salifiable Bafes* *.

Camphor is a concrete effential oil, obtained, by fublimation, from a fpecies of laurus which grows in China and Japan. By diftilling nitric acid eight times from camphor, Mr Kofegarten converted it into an acid analogous to the oxalic; but, as it differs from that acid in fome circumftances, we have thought neceffary to give it a particular name, till its nature be more completely afcertained by farther experiment.

As camphor is a carbono-hydrous or hydro-carbonous radical, it is eafily conceived, that, by oxygenation, it fhould form oxalic, malic, and feveral other vegetable acids : This conjecture is rendered not improbable by the experiments of Mr Kofegarten ; and the principal phenomena exhibited in the combinations of camphoric acid with the falifiable bafes, being

very

* Thefe combinations, which were all unknown to the ancients, are called Camphorats. The table is omitted, as being only in alphabetical order.—E.

very fimilar to thofe of the oxalic and malic acids, lead me to believe that it confifts of a mixture of thefe two acids.

SECT. XXXVIII.—*Obfervations upon Gallic Acid, and its Combinations with Salifiable Bafes* *.

The Gallic acid, formerly called Principle of Aftringency, is obtained from gall nuts, either by infufion or decoction with water, or by di-ftillation with a very gentle heat. This acid has only been attended to within thefe few years. The Committee of the Dijon Academy have followed it through all its combinations, and give the beft account of it hitherto produced. Its acid properties are very weak; it reddens the tincture of turnfol, decompofes fulphurets, and unites to all the metals when they have been previoufly diffolved in fome other acid. Iron, by this combination, is precipitated of a very deep blue or violet colour. The radical of this acid, if it deferves the name of one, is hitherto entirely unknown; it is contained in
oak

* Thefe combinations, which are called Gallats, were all unknown to the ancients; and the order of their affinity is not hitherto eftablifhed.—A.

oak willow, marſh iris, the ſtrawberry, nym-
phea, Peruvian bark, the flowers and bark of
pomgranate, and in many other woods and
barks.

SECT. XXXIX.—*Obſervations upon Laſtic Acid,
and its Combinations with Salifiable Baſes* *.

The only accurate knowledge we have of this
acid is from the works of Mr Scheele. It is
contained in whey, united to a ſmall quantity
of earth, and is obtained as follows : Reduce
whey to one eighth part of its bulk by evaporation, and filtrate, to ſeparate all its cheeſy
matter; then add as much lime as is neceſſary
to combine with the acid; the lime is afterwards
diſengaged by the addition of oxalic acid, which
combines with it into an inſoluble neutral ſalt.
When the oxalat of lime has been ſeparated by
decantation, evaporate the remaining liquor to
the conſiſtence of honey; the laſtic acid is diſſolved by alkohol, which does not unite with
the ſugar of milk and other foreign matters;
                                                    theſe

---

* Theſe combinations are called Laſtats; they were
all unknown to the ancient chemiſts, and their affinities have not yet been aſcertained.—A.

thefe are feparated by filtration from the alkohol and acid; and the alkohol being evaporated,
or diftilled off, leaves the lactic acid behind.

This acid unites with all the falifiable bafes
forming falts which do not criftallize; and it
feems confiderably to refemble the acetous acid.

TABLE

TABLE *of the Combinations of Saccholactic Acid with the Salifiable Bases, in the Order of Affinity.*

Bafes,	Neutral Salts.
Lime	Saccholat of lime.
Barytes	barytes.
Magnefia	magnefia.
Potafh	potafh.
Soda	foda.
Ammoniac	ammoniac.
Argill	argill.
Oxyd of zinc	zinc.
manganefe	manganefe.
iron	iron.
lead	lead.
tin	tin.
cobalt	cobalt.
copper	copper.
nickel	nickel.
arfenic	arfenic.
bifmuth	bifmuth.
mercury	mercury.
antimony	antimony.
filver	filver.

SECT.

Note.—All thefe were unknown to the ancient che-mifts.—A.

SECT. XL.—*Obfervations upon Saccholactic Acid,
and its Combinations.*

A fpecies of fugar may be extracted, by evaporation, from whey, which has long been
known in pharmacy, and which has a confiderable refemblance to that procured from fugar
canes. This faccharine matter, like ordinary
fugar, may be oxygenated by means of nitric
acid : For this purpofe, feveral portions of nitric acid are diftilled from it ; the remaining liquid is evaporated, and fet to criftallize, by
which means criftals of oxalic acid are procured; at the fame time a very fine white powder
precipitates, which is the faccholactic acid difcovered by Scheele. It is fufceptible of combining with the alkalies, ammoniac, the earths,
and even with the metals: Its action upon the
latter is hitherto but little known, except that,
with them, it forms difficultly foluble falts. The
order of affinity in the table is taken from Bergman.

N n   TABLE

TABLE *of the Combinations of Formic Acid, with the Salifiable Bases, in the Order of Affinity.*

Bases.	Neutral Salts.
Barytes	Formiat of barytes,
Potafh	potafh.
Soda	foda.
Lime	lime.
Magnefia	magnefia.
Ammoniac	ammoniac.
Oxyd of	
zinc	zinc.
manganefe	manganefe.
iron	iron.
lead	lead.
tin	tin.
cobalt	cobalt.
copper	copper.
nickel	nickel.
bifmuth	bifmuth.
filver	filver.
Argill	argill.

SECT.

Note.—All unknown to the ancient chemifts.—A.

SECT. XLI.—*Obfervations upon Formic Acid, and its Combinations.*

This acid was firſt obtained by diſtillation from ants, in the laſt century, by Samuel Fiſh-er. The fubjeƈt was treated of by Margraff in 1749, and by Meſſrs Ardwiſſon and Ochrn of Leipſic in 1777. The formic acid is drawn from a large ſpecies of red ants, *formica rufa, Lin.* which form large ant hills in woody places. It is procured, either by diſtilling the ants with a gentle heat in a glaſs retort or an alembic; or, after having waſhed the ants in cold water, and dried them upon a cloth, by pouring on boiling water, which diſſolves the acid ; or the acid may be procured by gentle expreſſion from the infeƈts, in which caſe it is ſtronger than in any of the former ways. To obtain it pure, we muſt reƈtify, by means of diſtillation, which fe-parates it from the uncombined oily and charry matter ; and it may be concentrated by freezing, in the manner recommended for treating the acetous acid.

SECT. XLII.—*Obfervations upon Bombic Acid, and its Combinations with Acidifiable Bafes* *.

The juices of the filk worm feem to affume an acid quality when that infect changes from a larva to a chryfalis. At the moment of its efcape from the latter to the butterfly form, it emits a reddifh liquor which reddens blue pa-per, and which was firft attentively obferved by Mr Chauffier of the Dijon academy, who ob-tains the acid by infufing filk worm chryfalids in alkohol, which diffolves their acid without being charged with any of the gummy parts of the infect; and, by evaporating the alkohol, the acid remains tollerably pure. The properties and affinities of this acid are not hitherto afcer-tained with any precifion; and we have reafon to believe that analogous acids may be procu-red from other infects. The radical of this acid is probably, like that of the other acids from the animal kingdom, compofed of charcoal, hy-drogen, and azote, with the addition, perhaps, of phofphorus.

TABLE

* Thefe combinations named Bombats were un-known to the ancient chemifts; and the affinities of the falifiable bafes with the bombic acid are hitherto undetermined.—A.

TABLE *of the Combinations of the Sebacic Acid, with the Salifiable Bases, in the Order of Affinity.*

Bases.	Neutral Salts.
Barytes	Sebat of barytes.
Potafh	potafh.
Soda	foda.
Lime	lime.
Magnefia	magnefia.
Ammoniac	ammoniac.
Argill	argill.
Oxyd of	
zinc	zinc.
manganefe	manganefe.
iron	iron.
lead	lead
tin	tin.
cobalt	cobalt.
copper	copper.
nickel	nickel.
arfenic	arfenic.
bifmuth	bifmuth.
mercury	mercury.
antimony	antimony.
filver	filver.

SECT.

Note.—All thefe were unknown to the ancient chemifts.—A.

SECT. XLIII.—*Obfervations upon Sebacid Acid, and its Combinations.*

To obtain the febacic acid, let fome fuet be melted in a fkillet over the fire, alongft with fome quick-lime in fine powder, and conftantly ftirred, raifing the fire towards the end of the operation, and taking care to avoid the vapours, which are very offenfive. By this procefs the febacic acid unites with the lime into a febat of lime, which is difficultly foluble in water; it is, however, feparated from the fatty matters with which it is mixed by folution in a large quantity of boiling water. From this the neutral falt is feparated by evaporation; and, to render it pure, is calcined, rediffolved, and again criftallized. After this we pour on a proper quantity of fulphuric acid, and the febacic acid paffes over by diftillation.

SECT.

SECT. XLIV.—*Obfervations upon the Lithic Acid, and its Combinations with the Salifiable Bafes* *.

From the later experiments of Bergman and Scheele, the urinary calculus appears to be a fpeciés of falt with an earthy bafis; it is flightly acidulous, and requires a large quantity of water for folution, three grains being fcarcely foluble in a thoufand grains of boiling water, and the greater part again criftallizes when cold. To this concrete acid, which Mr de Morveau calls Lithiafic Acid, we give the name of Lithic Acid, the nature and properties of which are hitherto very little known. There is fome appearance that it is an acidulous neutral falt, or acid combined in excefs with a falifiable bafe; and I have reafon to believe that it really is an acidulous phofphat of lime; if fo, it muft be excluded from the clafs of peculiar acids.

TABLE

---

* All the combinations of this acid, fhould it finally turn out to be one, were unknown to the ancient chemifts, and its affinities with the falifiable bafes have not been hitherto determined.—A.

TABLE *of the Combinations of the Pruſſic Acid with the Salifiable Baſes, in the order of affinity.*

Baſes.	Neutral Salts.
Potaſh	Pruſſiat of potaſh.
Soda	soda.
Ammoniac	ammoniac.
Lime	lime.
Barytes	barytes.
Magneſia	magneſia.
Oxyd of zinc	zinc.
iron	iron.
manganeſe	manganeſe.
cobalt	cobalt.
nickel	nickel.
lead	lead.
tin	tin.
copper	copper.
biſmuth	biſmuth.
antimony	antimony.
arſenic	arſenic.
ſilver	ſilver.
mercury	mercury.
gold	gold.
platina	platina.

Obſer-

Note.——All theſe were unknown to former chemiſts.—A.

Obfervations upon the Pruffic Acid, and its Combinations.

As the experiments which have been made hitherto upon this acid feem ftill to leave a confiderable degree of uncertainty with regard to its nature, I fhall not enlarge upon its properties, and the means of procuring it pure and diffengaged from combination. It combines with iron, to which it communicates a blue colour, and is equally fufceptible of entering into combination with moft of the other metals, which are precipitated from it by the alkalies, ammoniac, and lime, in confequence of greater affinity. The Pruffic radical, from the experiments of Scheele, and efpecially from thofe of Mr Berthollet, feems compofed of charcoal and azote ; hence it is an acid with a double bafe. The phofphorus which has been found combined with it appears, from the experiments of Mr Haffenfratz, to be only accidental.

Although this acid combines with alkalies, earths, and metals, in the fame way with other acids, it poffeffes only fome of the properties we have been in ufe to attribute to acids, and it may confequently be improperly ranked here in

O o   the

the clafs of acids ; but, as I have already obferved, it is difficult to form a decided opinion
upon the nature of this fubftance until the fubject has been farther elucidated by a greater
number of experiments.

PART

PART   III.

Defcription of the Inftruments and Operations of Chemiftry.

———————

INTRODUCTION.

IN the two former parts of this work I defign-
edly avoided being particular in defcribing
the manual operations of chemiftry, becaufe I
had found from experience, that, in a work appropriated to reafoning, minute defcriptions of
proceffes and of plates interrupt the chain of
ideas, and render the attention neceffary both
difficult and tedious to the reader. On the
other hand, if I had confined myfelf to the fummary defcriptions hitherto given, beginners
could have only acquired very vague conceptions of practical chemiftry from my work, and
muft have wanted both confidence and intereft
in operations they could neither repeat nor
thoroughly

thoroughly comprehend. This want could not
have been fupplied from books ; for, befides
that there are not any which defcribe the modern inftruments and experiments fufficiently at
large, any work that could have been confulted
would have prefented thefe things under a very
different order of arrangement, and in a different chemical language, which muft greatly
tend to injure the main objeƈ of my performance.

Influenced by thefe motives, I determined to
referve, for a third part of my work, a fummary
defcription of all the inftruments and manipula-
tions relative to elementary chemiftry. I confidered it as better placed at the end, rather
than at the beginning of the book, becaufe I
muft have been obliged to fuppofe the reader
acquainted with circumftances which a begin-
ner cannot know, and muft therefore have read
the elementary part to become acquainted with.
The whole of this third part may therefore be
confidered as refembling the explanations of
plates which are ufually placed at the end of
academic memoirs, that they may not interrupt
the conneƈion of the text by lengthened defcription. Though I have taken great pains to
render this part clear and methodical, and have
not omitted any effential inftrument or apparatus, I am far from pretending by it to fet afide
the neceffity of attendance upon leƈures and laboratories,

boratories, for fuch as wifh to acquire accurate
knowledge of the fcience of chemiftry. Thefe
fhould familiarife themfelves to the employment
of apparatus, and to the performance of experiments by actual experience. *Nihil eft in intel-
lectu quod non prius fuerit in fenfu,* the motto
which the celebrated Rouelle caufed to be painted in large characters in a confpicuous part of
his laboratory, is an important truth never to
be loft fight of either by teachers or ftudents of
chemiftry.

Chemical operations may be naturally divided
into feveral claffes, according to the purpofes
they are intended for performing. Some may
be confidered as purely mechanical, fuch as the
determination of the weight and bulk of bodies,
trituration, levigation, fearching, wafhing, filtration, &c. Others may be confidered as real
chemical operations, becaufe they are performed by means of chemical powers and agents ;
fuch are folution, fufion, &c. Some of thefe
are intended for feparating the elements of bodies from each other, fome for reuniting thefe
elements together ; and fome, as combuftion,
produce both thefe effects during the fame procefs.

Without rigoroufly endeavouring to follow
the above method, I mean to give a detail of
the chemical operations in fuch order of arrangement as feemed beft calculated for conveying

veying inſtruction. I ſhall be more particular
in deſcribing the apparatus connected with modern chemiſtry, becauſe theſe are hitherto little
known by men who have devoted much of their
time to chemiſtry, and even by many profeſſors
of the ſcience.

C H A P.

C H A P. I.

*Of the Inftruments neceffary for determining the
Abfolute and Specific Gravities of Solid and Liquid Bodies.*

THE beft method hitherto known for determining the quantities of fubftances fubmitted to chemical experiment, or refulting from
them, is by means of an accurately conftructed
beam and fcales, with properly regulated weights,
which well known operation is called *weighing*.
The denomination and quantity of the weights
ufed as an unit or ftandard for this purpofe are
extremely arbitrary, and vary not only in different kingdoms, but even in different provinces
of the fame kingdom, and in different cities of
the fame province. This variation is of infinite
confequence to be well underftood in commerce
and in the arts ; but, in chemiftry, it is of no
moment what particular denomination of weight
be employed, provided the refults of experiments be expreffed in convenient fractions of
the fame denomination. For this purpofe, until all the weights ufed in fociety be reduced to
the fame ftandard, it will be fufficient for chemifts in different parts to ufe the common

pound

pound of their own country as the unit or standard, and to exprefs all its fractional parts in decimals, inftead of the arbitrary divifions now in ufe. By this means the chemifts of all countries will be thoroughly underftood by each other, as, although the abfolute weights of the ingredients and products cannot be known, they will readily, and without calculation, be able to determine the relative proportions of thefe to each other with the utmoft accuracy ; fo that in this way we fhall be poffeffed of an univerfal language for this part of chemiftry.

With this view I have long projected to have the pound divided into decimal fractions, and I have of late fucceeded through the affiftance of Mr Fourche balance-maker at Paris, who has executed it for me with great accuracy and judgment. I recommend to all who carry on experiments to procure fimilar divifions of the pound, which they will find both eafy and fimple in its application, with a very fmall knowledge of decimal fractions *.

As

* Mr Lavoifier gives, in this part of his work, very accurate directions for reducing the common fubdivifions of the French pound into decimal fractions, and *vice verfa*, by means of tables fubjoined to this 3d part. As thefe inftructions, and the table, would be ufelefs to the Britifh chemift, from the difference between the fubdivifions of the French and Troy pounds, I have omitted them, but have fubjoined in the appendix accurate rules for converting the one into the other.—E.

As the ufefulnefs and accuracy of chemiftry depends entirely upon the determination of the weights of the ingredients and products both before and after experiments, too much preci-fion cannot be employed in this part of the fub-ject; and, for this purpofe, we muft be provided with good inftruments. As we are often obli-ged, in chemical proceffes, to afcertain, within a grain or lefs, the tare or weight of large and heavy inftruments, we muft have beams made with peculiar nicenefs by accurate workmen, and thefe muft always be kept apart from the laboratory in fome place where the vapours of acids, or other corrofive liquors, cannot have accefs, otherwife the fteel will ruft, and the ac-curacy of the balance be deftroyed. I have three fets, of different fizes, made by Mr Fon-tin with the utmoft nicety, and, excepting thofe made by Mr Ramfden of London, I do not think any can compare with them for precifion and fenfibility. The largeft of thefe is about three feet long in the beam for large weights, up to fifteen or twenty pounds; the fecond, for weights of eighteen or twenty ounces, is exact to a tenth part of a grain; and the fmalleft, calculated only for weighing about one gros, is fenfibly affected by the five hundredth part of a grain.

Befides thefe nicer balances, which are only ufed for experiments of refearch, we muft have

P p   others

others of lefs value for the ordinary purpofes of the laboratory. A large iron balance, capable of weighing forty or fifty pounds within half a dram, one of a middle fize, which may afcertain eight or ten pounds, within ten or twelve grains, and a fmall one, by which about a pound may be determined, within one grain.

We muft likewife be provided with weights divided into their feveral fractions, both vulgar and decimal, with the utmoft nicety, and verified by means of repeated and accurate trials in the niceft fcales ; and it requires fome experience, and to be accurately acquainted with the different weights, to be able to ufe them properly. The beft way of precifely afcertaining the weight of any particular fubftance is to weigh it twice, once with the decimal divifions of the pound, and another time with the common fubdivifions or vulgar fractions, and, by comparing thefe, we attain the utmoft accuracy.

By the fpecific gravity of any fubftance is underftood the quotient of its abfolute weight divided by its magnitude, or, what is the fame, the weight of a determinate bulk of any body. The weight of a determinate magnitude of water has been generally affumed as unity for this purpofe ; and we exprefs the fpecific gravity of gold, fulphuric acid, &c. by faying, that gold is nineteen times, and fulphuric acid twice the weight of water, and fo of other bodies.

It

It is the more convenient to affume water as unity in fpecific gravities, that thofe fubftances whofe fpecific gravity we wifh to determine, are moft commonly weighed in water for that purpofe. Thus, if we wifh to determine the fpecific gravity of gold flattened under the hammer, and fuppofing the piece of gold to weigh 8 *oz.* 4 *gros* 2½ *grs.* in the air *, it is fufpended by means of a fine metallic wire under the fcale of a hydroftatic balance, fo as to be entirely immerfed in water, and again weighed. The piece of gold in Mr Briffon's experiment loft by this means 3 *gros* 37 *grs.*; and, as it is evident that the weight loft by a body weighed in water is precifely equal to the weight of the water difplaced, or to that of an equal volume of water, we may conclude, that, in equal magnitudes, gold weighs 4893½ *grs.* and water 253 *grs.* which, reduced to unity, gives 1.0000 as the fpecific gravity of water, and 19.3617 for that of gold. We may operate in the fame manner with all folid fubftances. We have rarely any occafion, in chemiftry, to determine the fpecific gravity of folid bodies, unlefs when operating upon alloys or metallic glaffes; but we have very frequent neceffity to afcertain that of fluids, as it is often the only means of judging of their purity or degree of concentration.

This

* Vide Mr Briffon's Effay upon Specific Gravity, p. 5.—A.

This object may be very fully accomplished with the hydrostatic balance, by weighing a solid body; such, for example, as a little ball of rock cristal suspended by a very fine gold wire, first in the air, and afterwards in the fluid whose specific gravity we wish to discover. The weight lost by the cristal, when weighed in the liquor, is equal to that of an equal bulk of the liquid. By repeating this operation successively in water and different fluids, we can very readily ascertain, by a simple and easy calculation, the relative specific gravities of these fluids, either with respect to each other or to water. This method is not, however, sufficiently exact, or, at least, is rather troublesome, from its extreme delicacy, when used for liquids differing but little in specific gravity from water; such, for instance, as mineral waters, or any other water containing very small portions of salt in solution.

In some operations of this nature, which have not hitherto been made public, I employed an instrument of great sensibility for this purpose with great advantage. It consists of a hollow cylinder, *A b c f*, Pl. vii. fig. 6. of brass, or rather of silver, loaded at its bottom, b c f, with tin, as represented swimming in a jug of water, *l m n o*. To the upper part of the cylinder is attached a stalk of silver wire, not more than three fourths of a line diameter, surmounted by

a

a little cup *d*, intended for containing weights;
upon the ftalk a mark is made at *g*, the ufe of
which we fhall prefently explain. This cylinder may be made of any fize; but, to be accurate, ought at leaft to difplace four pounds of
water. The weight of tin with which this inftrument is loaded ought to be fuch as will make
it remain almoft in equilibrium in diftilled water, and fhould not require more than half a
dram, or a dram at moft, to make it fink to *g*.

We muft firft determine, with great precifion, the exact weight of the inftrument, and
the number of additional grains requifite for
making it fink, in diftilled water of a determinate temperature, to the mark: We then perform the fame experiment upon all the fluids
of which we wifh to afcertain the fpecific gravity, and, by means of calculation, reduce the
obferved differences to a common ftandard of
cubic feet, pints or pounds, or of decimal fractions, comparing them with water. This method, joined to experiments with certain reagents *, is one of the beft for determining the
quality of waters, and is even capable of pointing out differences which efcape the moft accurate chemical analyfis. I fhall, at fome future
period,

---

* For the ufe of thefe reagents fee Bergman's excellent treatife upon the analyfis of mineral waters, in his
Chemical and Phyfical Effays.—E.

period, give an account of a very extenſive ſet of experiments which I have made upon this ſubjeĉt.

Theſe metallic hydrometers are only to be uſed for determining the ſpecific gravities of ſuch waters as contain only neutral ſalts or alkaline ſubſtances; and they may be conſtruĉted with different degrees of ballaſt for alkohol and other ſpiritous liquors. When the ſpecific gravities of acid liquors are to be aſcertained, we muſt uſe a glaſs hydrometer, as repreſented Pl. vii. fig. 14 †. This conſiſts of a hollow cylinder of glaſs, *a b c f*, hermetically ſealed at its lower end, and drawn out at the upper into a capillary tube *a*, ending in the little cup or baſon *d*. This inſtrument is ballaſted with more or leſs mercury, at the bottom of the cylinder introduced through the tube, in proportion to the weight of the liquor intended to be examined: We may introduce a ſmall graduated ſlip of paper into the tube *a d*; and, though theſe degrees do not exaĉtly correſpond to the fractions of grains in the different liquors, they may be rendered very uſeful in calculation.

What is ſaid in this chapter may ſuffice, without farther enlargement, for indicating the means

---

† Three or four years ago, I have ſeen ſimilar glaſs hydrometers, made for Dr Black by B. Knie, a very ingenious artiſt of this city.—E.

means of afcertaining the abfolute and fpecific
gravities of folids and fluids, as the neceffary
inftruments are generally known, and may eafily be procured : But, as the inftruments I have
ufed for meafuring the gaffes are not any where
defcribed, I fhall give a more detailed account
of thefe in the following chapter.

CHAP.

C H A P.   II,

*Of Gazometry, or the Meafurement of the Weight
and Volume of Aëriform Subftances.*

S E C T.   I.

Defcription of the Pneumato-chemical Apparatus.

THE French chemifts have of late applied
the name of *pneumato-chemical apparatus*
to the very fimple and ingenious contrivance,
invented by Dr Prieftley, which is now indifpenfibly neceffary to every laboratory. This confifts of a wooden trough, of larger or fmaller
dimenfions as is thought convenient, lined with
plate-lead or tinned copper, as reprefented in
perfpective, Pl. V. In Fig. 1. the fame trough or
ciftern is fuppofed to have two of its fides cut a-
way, to fhow its interior conftruction more diftinctly. In this apparatus, we diftinguifh between the fhelf ABCD Fig. 1. and 2. and the
bottom or body of the ciftern FGHI Fig. 2.
The

The jars or bell-glaffes are filled with water in this deep part, and, being turned with their mouths downwards, are afterwards fet upon the fhelf ABCD, as fhown Plate X. Fig. 1. F. The upper parts of the fides of the ciftern above the level of the fhelf are called the *rim* or *borders*.

The ciftern ought to be filled with water, fo as to ftand at leaft an inch and a half deep upon the fhelf, and it fhould be of fuch dimenfions as to admit of at leaft one foot of water in every direction in the well. This fize is fufficient for ordinary occafions ; but it is often convenient, and even neceffary, to have more room ; I would therefore advife fuch as intend to employ themfelves ufefully in chemical experiments, to have this apparatus made of confiderable magnitude, where their place of operating will allow. The well of my principal ciftern holds four cubical feet of water, and its fhelf has a furface of fourteen fquare feet ; yet, in fpite of this fize, which I at firft thought immoderate, I am often ftraitened for room.

In laboratories, where a confiderable number of experiments are performed, it is neceffary to have feveral leffer cifterns, befides the large one, which may be called the *general magazine ;* and even fome portable ones, which may be moved when neceffary, near a furnace, or wherever they may be wanted. There are likewife fome operations which dirty the water of the appara-

Q q   tus,

tus, and therefore require to be carried on in cifterns by themfelves.

It were doubtlefs confiderably cheaper to ufe cifterns, or iron-bound tubs, of wood fimply dove-tailed, inftead of being lined with lead or copper; and in my firft experiments I ufed them made in that way; but I foon difcovered their inconvenience. If the water be not always kept at the fame level, fuch of the dovetails as are left dry fhrink, and, when more water is added, it efcapes through the joints, and runs out.

We employ criftal jars or bell glaffes, Pl. V. Fig. 9. A. for containing the gaffes in this apparatus; and, for tranfporting thefe, when full of gas, from one ciftern to another, or for keeping them in referve when the ciftern is too full, we make ufe of a flat difh BC, furrounded by a ftanding up rim or border, with two handles DE for carrying it by.

After feveral trials of different materials, I have found marble the beft fubftance for conftructing the mercurial pneumato-chemical apparatus, as it is perfectly impenetrable by mercury, and is not liable, like wood, to feparate at the junctures, or to allow the mercury to efcape through chinks; neither does it run the rifk of breaking, like glafs, ftone-ware, or porcelain. Take a block of marble BCDE, Plate V. Fig. 3. and 4. about two feet long, 15 or 18 inches broad,

broad, and ten inches thick, and caufe it to be hollowed out as at *m n* Fig. 5. about four inches deep, as a refervoir for the mercury ; and, to be able more conveniently to fill the jars, cut the gutter T V, Fig. 3. 4. and 5. at leaft four inches deeper ; and, as this trench may fome-times prove troublefome, it is made capable of being covered at pleafure by thin boards, which flip into the grooves *x y,* Fig. 5. I have two marble cifterns upon this conftruction, of dif-ferent fizes, by which I can always employ one of them as a refervoir of mercury, which it pre-ferves with more fafety than any other veffel, being neither fubject to overturn, nor to any other accident. We operate with mercury in this ap-paratus exactly as with water in the one before defcribed; but the bell-glaffes muft be of fmaller diameter, and much ftronger; or we may ufe glafs tubes, having their mouths widened, as in Fig. 7. ; thefe are called *eudiometers* by the glafs-men who fell them. One of the bell-glaffes is repre-fented Fig. 5. A. ftanding in its place, and what is called a jar is engraved Fig. 6.

The mercurial pneumato-chemical apparatus is neceffary in all experiments wherein the dif-engaged gaffes are capable of being abforbed by water, as is frequently the cafe, efpecially in all combinations, excepting thofe of metals, in fer-mentation, &c.

S E C T.

S E C T.   II.

Of the Gazometer.

I give the name of *gazometer* to an inftrument which I invented, and caufed conftruct, for the purpofe of a kind of bellows, which might fur- nifh an uniform and continued ftream of oxy- gen gas in experiments of fufion. Mr Meuf- nier and I have fince made very confiderable corrections and additions, having converted it into what may be called an *univerfal inftrument*, without which it is hardly poffible to perform moft of the very exact experiments. The name we have given the inftrument indicates its in- tention for meafuring the volume or quantity of gas fubmitted to it for examination.

It confifts of a ftrong iron beam, DE, Pl. VIII. Fig. 1. three feet long, having at each end, D and E, a fegment of a circle, likewife ftrongly conftructed of iron, and very firmly joined. In- ftead of being poifed as in ordinary balances, this beam refts, by means of a cylindrical axis of polifhed fteel, F, Fig. 9. upon two large moveable brafs friction-wheels, by which the re- fiftance to its motion from friction is confider- ably diminifhed, being converted into friction

of

of the fecond order. As an additional precaution, the parts of thefe wheels which fupport
the axis of the beam are covered with plates of
polifhed rock-criftal. The whole of this machinery is fixed to the top of the folid column of
wood BC, Fig. 1. To one extremity D of the
beam, a fcale P for holding weights is fufpended by a flat chain, which applies to the curvature of the arc nD*o*, in a groove made for the
purpofe. To the other extremity E of the beam
is applied another flat chain, *i k m*, fo conftructed, as to be incapable of lengthening or
fhortening, by being lefs or more charged with
weight; to this chain, an iron trivet, with three
branches, *a i*, *c i*, and *h i*, is ftrongly fixed at *i*,
and thefe branches fupport a large inverted jar
A, of hammered copper, of about 18 inches diameter, and 20 inches deep. The whole of
this machine is reprefented in perfpective, Pl.
VIII. Fig. 1. and Pl. IX. Fig. 2. and 4. give
perpendicular fections, which fhow its interior
ftructure.

Round the bottom of the jar, on its outfide,
is fixed (Pl. IX. Fig. 2.) a border divided into
compartments 1, 2, 3, 4, &c. intended to receive leaden weights feparately reprefented 1,
2, 3, Fig. 3. Thefe are intended for increafing the weight of the jar when a confiderable
preffure is requifite, as will be afterwards explained, though fuch neceffity feldom occurs.

The

The cylindrical jar **A** is entirely open below, *de*, Pl. IX. Fig. 4.; but is clofed above with a copper lid, *a b c*, open at *b f*, and capable of being fhut by the cock *g*. This lid, as may be feen by infpecting the figures, is placed a few inches within the top of the jar to prevent the jar from being ever entirely immerfed in the water, and covered over. Were I to have this inftrument made over again, I fhould caufe the lid to be confiderably more flattened, fo as to be almoft level. This jar or refervoir of air is contained in the cylindrical copper veffel, LMNO, Pl. VIII. Fig. 1. filled with water.

In the middle of the cylindrical veffel LMNO, Pl. IX. Fig. 4. are placed two tubes *st*, *xy*, which are made to approach each other at their upper extremities *t y*; thefe are made of fuch a length as to rife a little above the upper edge LM of the veffel LMNO, and when the jar *abcde* touches the bottom NO, their upper ends enter about half an inch into the conical hollow *b*, leading to the ftop-cock *g*.

The bottom of the veffel LMNO is reprefented Pl. IX. Fig. 3. in the middle of which a fmall hollow femifpherical cap is foldered, which may be confidered as the broad end of a funnel reverfed; the two tubes *st*, *xy*, Fig. 4. are adapted to this cap at *s* and *x*, and by this means communicate with the tubes *mm*, *nn*, *oo*, *pp*, Fig. 3. which are fixed horizontally upon the

bottom

bottom of the veffel, and all of which terminate in, and are united by, the fpherical cap *sx*. Three of thefe tubes are continued out of the veffel, as in Pl. VIII. Fig. 1. The firft marked in that figure 1, 2, 3, is inferted at its extremity 3, by means of an intermediate ftop-cock 4, to the jar V. which ftands upon the fhelf of a fmall pneumato-chemical apparatus GHIK, the infide of which is fhown Pl. IX. Fig. 1. The fecond tube is applied againft the outfide of the veffel LMNO from 6 to 7, is continued at 8, 9, 10, and at 11 is engaged below the jar V. The former of thefe tubes is intended for conveying gas into the machine, and the latter for conducting fmall quantities for trials under jars. The gas is made either to flow into or out of the machine, according to the degree of preffure it receives ; and this preffure is varied at pleafure, by loading the fcale P lefs or more, by means of weights. When gas is to be introduced into the machine, the preffure is taken off, or even rendered negative ; but, when gas is to be expelled, a preffure is made with fuch degree of force as is found neceffary.

The third tube 12, 13, 14, 15, is intended for conveying air or gas to any neceffary place or apparatus for combuftions, combinations, or any other experiment in which it is required.

To explain the ufe of the fourth tube, I muft enter into fome difcuffions. Suppofe the veffel

fel LMNO, Pl. VIII. Fig. 1. full of water, and
the jar A partly filled with gas, and partly with
water ; it is evident that the weights in the bafon P may be fo adjufted, as to occafion an exact equilibrium between the weight of the bafon
and of the jar, fo that the external air fhall not
tend to enter into the jar, nor the gas to efcape
from it ; and in this cafe the water will ftand
exactly at the fame level both within and without the jar.   On the contrary, if the weight in
the bafon P be diminifhed, the jar will then
prefs downwards from its own gravity, and the
water will ftand lower within the jar than it
does without ; in this cafe, the included air or
gas will fuffer a degree of compreffion above
that experienced by the external air, exactly
proportioned to the weight of a column of water, equal to the difference of the external and
internal furfaces of the water.   From thefe reflections, Mr Meufnier contrived a method of
determining the exact degree of preffure to
which the gas contained in the jar is at any
time expofed.   For this purpofe, he employs a
double glafs fyphon 19, 20, 21, 22, 23, firmly
cemented at 19 and 23.   The extremity 19 of
this fyphon communicates freely with the water
in the external veffel of the machine, and the
extremity 23 communicates with the fourth
tube at the bottom of the cylindrical veffel, and
confequently, by means of the perpendicular
tube

tube *st*, Pl. IX. Fig. 4. with the air contained
in the jar. He likewife cements, at 16, Pl. VIII.
Fig. 1. another glafs tube 16, 17, 18, which
communicates at 16 with the water in the exte-
rior veffel LMNO, and, at its upper end 18, is
open to the external air.

By thefe feveral contrivances, it is evident
that the water muft ftand in the tube 16, 17,
18, at the fame level with that in the ciftern
LMNO ; and, on the contrary, that, in the
branch 19, 20, 21, it muft ftand higher or lower, according as the air in the jar is fubjected to
a greater or leffer preffure than the external air.
To afcertain thefe differences, a brafs fcale divided into inches and lines is fixed between thefe
two tubes. It is readily conceived that, as air,
and all other elaftic fluids, muft increafe in
weight by compreffion, it is neceffary to know
their degree of condenfation to be enabled to
calculate their quantities, and to convert the
meafure of their volumes into correfpondent
weights ; and this object is intended to be ful-
filled by the contrivance now defcribed.

But, to determine the fpecific gravity of air
or of gaffes, and to afcertain their weight in a
known volume, it is neceffary to know their
temperature, as well as the degree of preffure
under which they fubfift ; and this is accomplifhed by means of a fmall thermometer, ftrongly cemented into a brafs collet, which fcrews

R r   into

into the lid of the jar A. This thermometer
is reprefented feparately, Pl. VIII. Fig. 10. and
in its place 24, 25, Fig. 1. and Pl. IX. Fig. 4.
The bulb is in the infide of the jar A, and its
graduated ftalk rifes on the outfide of the lid.

The practice of gazometry would ftill have
laboured under great difficulties, without farther precautions than thofe above defcribed.
When the jar A finks in the water of the ciftern
LMNO, it muft lofe a weight equal to that of
the water which it difplaces; and confequently
the compreffion which it makes upon the contained air or gas muft be proportionally diminifhed. Hence the gas furnifhed, during experiments from the machine, will not have the fame
denfity towards the end that it had at the beginning, as its fpecific gravity is continually dimi-
nifhing. This difference may, it is true, be determined by calculation; but this would have
occafioned fuch mathematical inveftigations as
muft have rendered the ufe of this apparatus
both troublefome and difficult. Mr Meufnier has remedied this inconvenience by the
following contrivance. A fquare rod of iron,
26, 27, Pl. VIII. Fig. 1. is raifed perpendicular
to the middle of the beam DE. This rod paffes
through a hollow box of brafs 28, which opens,
and may be filled with lead; and this box is
made to flide alongft the rod, by means of a
toothed pinion playing in a rack, fo as to raife

or

or lower the box, and to fix it at fuch places as is judged proper.

When the lever or beam DE ftands horizontal, this box gravitates to neither fide; but, when the jar A finks into the ciftern LMNO, fo as to make the beam incline to that fide, it is evident the loaded box 28, which then paffes beyond the center of fufpenfion, muft gravitate to the fide of the jar, and augment its preffure upon the included air. This is increafed in proportion as the box is raifed towards 27, becaufe the fame weight exerts a greater power in proportion to the length of the lever by which it acts. Hence, by moving the box 28 alongft the rod 26, 27, we can augment or diminifh the correction it is intended to make upon the preffure of the jar ; and both experience and calculation fhow that this may be made to compenfate very exactly for the lofs of weight in the jar at all degrees of preffure.

I have not hitherto explained the moft important part of the ufe of this machine, which is the manner of employing it for afcertaining the quantities of the air or gas furnifhed during experiments. To determine this with the moft rigorous precifion, and likewife the quantity fupplied to the machine from experiments, we fixed to the arc which terminates the arm of the beam E, Pl. VIII. Fig. 1. the brafs fector *l m*, divided into degrees and half degrees, which

which confequently moves in common with the beam; and the lowering of this end of the beam is meafured by the fixed index 29, 30, which has a Nonius giving hundredth parts of a degree at its extremity 30.

The whole particulars of the different parts of the above defcribed machine are reprefented in Plate VIII. as follow.

Fig. 2. Is the flat chain invented by Mr Vaucanfon, and employed for fufpending the fcale or bafon P, Fig. 1; but, as this lengthens or fhortens according as it is more or lefs loaded, it would not have anfwered for fufpending the jar A, Fig. 1.

Fig. 5. Is the chain *i k m*, which in Fig. 1. fuftains the jar A. This is entirely formed of plates of polifhed iron interlaced into each other, and held together by iron pins. This chain does not lengthen in any fenfible degree, by any weight it is capable of fupporting.

Fig. 6. The trivet, or three branched ftirrup, by which the jar A is hung to the balance, with the fcrew by which it is fixed in an accurately vertical pofition.

Fig. 3. The iron rod 26, 27, which is fixed perpendicular to the center of the beam, with its box 28.

Fig. 7. & 8. The friction-wheels, with the plates of rock-criftal Z, as points of contact

by

by which the friction of the axis of the lever of the balance is avoided.

Fig. 4. The piece of metal which fupports the axis of the friction-wheels.

Fig. 9. The middle of the lever or beam, with the axis upon which it moves.

Fig. 10. The thermometer for determining the temperature of the air or gas contained in the jar.

When this gazometer is to be ufed, the cif-tern or external veffel, LMNO, Pl. VIII. Fig. 1. is to be filled with water to a determinate height, which fhould be the fame in all experiments. The level of the water fhould be taken when the beam of the balance ftands horizontal; this level, when the jar is at the bottom of the cif-tern, is increafed by all the water which it dif-places, and is diminifhed in proportion as the jar rifes to its higheft elevation. We next en-deavour, by repeated trials, to difcover at what elevation the box 28 muft be fixed, to render the preffure equal in all fituations of the beam. I fhould have faid nearly, becaufe this correc-tion is not abfolutely rigorous ; and differences of a quarter, or even of half a line, are not of any confequence. This height of the box 28 is not the fame for every degree of preffure, but varies according as this is of one, two, three, or more inches. All thefe fhould be regiftered with great order and precifion.

We

We next take a bottle which holds eight or ten pints, the capacity of which is very accurately determined by weighing the water it is capable of containing. This bottle is turned bottom upwards, full of water, in the ciftern of the pneumato chemical apparatus GHIK, Fig. 1. and is fet on its mouth upon the fhelf of the apparatus, inftead of the glafs jar V, having the extremity 11 of the tube 7, 8, 9, 10, 11, inferted into its mouth. The machine is fixed at zero of preffure, and the degree marked by the index 30 upon the fector *m l* is accurately obferved; then, by opening the ftopcock 8, and preffing a little upon the jar A, as much air is forced into the bottle as fills it entirely. The degree marked by the index upon the fector is now obferved, and we calculate what number of cubical inches correfpond to each degree. We then fill a fecond and third bottle, and fo on, in the fame manner, with the fame precautions, and even repeat the operation feveral times with bottles of different fizes, till at laft, by accurate attention, we afcertain the exact gage or capacity of the jar A, in all its parts; but it is better to have it formed at firft accurately cylindrical, by which we avoid thefe calculations and eftimates.

The inftrument I have been defcribing was conftructed with great accuracy and uncommon fkill by Mr Meignie junior, engineer and phyfi-

cal

cal inftrument-maker. It is a moft valuable inftrument, from the great number of purpofes to
which it is applicable ; and, indeed, there are
many experiments which are almoft impoffible
to be performed without it. It becomes expenfive, becaufe, in many experiments, fuch as
the formation of water and of nitric acid, it is
abfolutely neceffary to employ two of the fame
machines. In the prefent advanced ftate of chemiftry, very expenfive and complicated inftruments are become indifpenfibly neceffary for
afcertaining the analyfis and fynthefis of bodies
with the requifite precifion as to quantity and
proportion ; it is certainly proper to endeavour
to fimplify thefe, and to render them lefs coftly ; but this ought by no means to be attempted at the expence of their conveniency of application, and much lefs of their accuracy.

S E C T.   III.

Some other methods of meafuring the volume of
Gaffes.

The gazometer defcribed in the foregoing
fection is too coftly and too complicated for being generally ufed in laboratories for meafuring
the gaffes, and is not even applicable to every
circumftance

circumftance of this kind. In numerous feries of experiments, more fimple and more readily applicable methods muft be employed. For this purpofe I fhall defcribe the means I ufed before I was in poffeffion of a gazometer, and which I ftill ufe in preference to it in the ordinary courfe of my experiments.

Suppofe that, after an experiment, there is a refiduum of gas, neither abforbable by alkali nor water, contained in the upper part of the jar AEF, Pl. IV. Fig. 3. ftanding on the fhelf of a pneumato-chemical apparatus, of which we wifh to afcertain the quantity, we muft firft mark the height to which the mercury or water rifes in the jar with great exactnefs, by means of flips of paper pafted in feveral parts round the jar. If we have been operating in mercury, we begin by difplacing the mercury from the jar, by introducing water in its ftead. This is readily done by filling a bottle quite full of water; having ftopped it with your finger, turn it up, and introduce its mouth below the edge of the jar; then, turning down its body again, the mercury, by its gravity, falls into the bottle, and the water rifes in the jar, and takes the place occupied by the mercury. When this is accomplifhed, pour fo much water into the ciftern ABCD as will ftand about an inch over the furface of the mercury; then pafs the difh BC, Pl. V. Fig. 9. under the jar, and carry it to the

water

water ciftern, Fig. 1. and 2. We here exchange the gas into another jar, which has been previoufly graduated in the manner to be afterwards defcribed; and we thus judge of the quantity or volume of the gas by means of the degrees which it occupies in the graduated jar.

There is another method of determining the volume of gas, which may either be fubftituted in place of the one above defcribed, or may be ufefully employed as a correction or proof of that method. After the air or gas is exchanged from the firft jar, marked with flips of paper, into the graduated jar, turn up the mouth of the marked jar, and fill it with water exactly to the marks EF, Pl. IV. Fig. 3. and by weighing the water we determine the volume of the air or gas it contained, allowing one cubical foot, or 1728 cubical inches, of water for each 70 pounds, French weight.

The manner of graduating jars for this purpofe is very eafy, and we ought to be provided with feveral of different fizes, and even feveral of each fize, in cafe of accidents. Take a tall, narrow, and ftrong glafs jar, and, having filled it with water in the ciftern, Pl. V. Fig. 1. place it upon the fhelf ABCD; we ought always to ufe the fame place for this operation, that the level of the fhelf may be always exactly fimilar, by which almoft the only error to which this procefs is liable will be avoided. Then take a nar-

S s   row

row mouthed phial which holds exactly 6 *oz.*
3 *gros* 61 *grs.* of water, which correfponds to
10 cubical inches. If you have not one exactly of this dimenfion, choofe one a little larger,
and diminifh its capacity to the fize requifite,
by dropping in a little melted wax and rofin.
This bottle ferves the purpofe of a ftandard for
gaging the jars. Make the air contained in
this bottle pafs into the jar, and mark exactly
the place to which the water has defcended;
add another meafure of air, and again mark
the place of the water, and fo on, till all the
water be difplaced. It is of great confequence
that, during the courfe of this operation, the
bottle and jar be kept at the fame temperature
with the water in the ciftern; and, for this reafon, we muft avoid keeping the hands upon either as much as poffible; or, if we fufpect they
have been heated, we muft cool them by means
of the water in the ciftern. The height of the
barometer and thermometer during this experiment is of no confequence.

When the marks have been thus afcertained
upon the jar for every ten cubical inches, we
engrave a fcale upon one of its fides, by means
of a diamond pencil. Glafs tubes are graduated in the fame manner for ufing in the mercurial apparatus, only they muft be divided into cubical inches, and tenths of a cubical inch.
The bottle ufed for gaging thefe muft hold

8 *oz.*

8 *oz.* 6 *gros* 25 *grs.* of mercury, which exactly correfponds to a cubical inch of that metal.

The method of determining the volume of air or gas, by means of a graduated jar, has the advantage of not requiring any correction for the difference of height between the furface of the water within the jar, and in the ciftern ; but it requires corrections with refpect to the height of the barometer and thermometer. But, when we afcertain the volume of air by weighing the water which the jar is capable of containing, up to the marks EF, it is neceffary to make a farther correction, for the difference between the furface of the water in the ciftern, and the height to which it rifes within the jar. This will be explained in the fifth fection of this chapter.

S E C T. IV.

Of the method of Separating the different Gaffes from each other.

As experiments often produce two, three, or more fpecies of gas, it is neceffary to be able to feparate thefe from each other, that we may afcertain the quantity and fpecies of each. Suppofe that under the jar A, Pl. IV. Fig. 3. is

contained

contained a quantity of different gaſſes mixed together, and ſtanding over mercury, we begin by marking with ſlips of paper, as before directed, the height at which the mercury ſtands within the glaſs ; then introduce about a cubical inch of water into the jar, which will ſwim over the ſurface of the mercury : If the mixture of gas contains any muriatic or ſulphurous acid gas, a rapid and conſiderable abſorption will inſtantly take place, from the ſtrong tendency theſe two gaſſes have, eſpecially the former, to combine with, or be abſorbed by water. If the water only produces a ſlight abſorption of gas hardly equal to its own bulk, we conclude, that the mixture neither contains muriatic acid, ſulphuric acid, or ammoniacal gas, but that it contains carbonic acid gas, of which water only abſorbs about its own bulk. To aſcertain this conjecture, introduce ſome ſolution of cauſtic alkali, and the carbonic acid gas will be gradually abſorbed in the courſe of a few hours ; it combines with the cauſtic alkali or potaſh, and the remaining gas is left almoſt perfectly free from any ſenſible reſiduum of carbonic acid gas.

After each experiment of this kind, we muſt carefully mark the height at which the mercury ſtands within the jar, by ſlips of paper paſted on, and varniſhed over when dry, that they may not be waſhed off when placed in the water apparatus.

paratus. It is likewife neceffary to regifter the difference between the furface of the mercury in the ciftern and that in the jar, and the height of the barometer and thermometer, at the end of each experiment.

When all the gas or gaffes abforbable by water and potafh are abforbed, water is admitted into the jar to difplace the mercury ; and, as is defcribed in the preceding fection, the mercury in the ciftern is to be covered by one or two inches of water. After this, the jar is to be tranfported by means of the flat difh BC, Pl. V. Fig. 9. into the water apparatus ; and the quantity of gas remaining is to be afcertained by changing it into a graduated jar. After this, fmall trials of it are to be made by experiments in little jars, to afcertain nearly the nature of the gas in queftion. For inftance, into a fmall jar full of the gas, Fig. 8. Pl. V. a lighted taper is introduced ; if the taper is not immediately extinguifhed, we conclude the gas to contain oxygen gas ; and, in proportion to the brightnefs of the flame, we may judge if it contain lefs or more oxygen gas than atmofpheric air contains. If, on the contrary, the taper be inftantly extinguifhed, we have ftrong reafon to prefume that the refiduum is chiefly compofed of azotic gas. If, upon the approach of the taper, the gas takes fire and burns quietly at the furface with a white flame, we conclude it to be

pure

pure hydrogen gas ; if this flame is blue, we
judge it confifts of carbonated hydrogen gas ;
and, if it takes fire with a fudden deflagration,
that it is a mixture of oxygen and hydrogen
gas. If, again, upon mixing a portion of the
refiduum with oxygen gas, red fumes are produced, we conclude that it contains nitrous gas.

Thefe preliminary trials give fome general
knowledge of the properties of the gas, and
nature of the mixture, but are not fufficient to
determine the proportions and quantities of the
feveral gaffes of which it is compofed. For this
purpofe all the methods of analyfis muft be employed ; and, to direct thefe properly, it is of
great ufe to have a previous approximation by
the above methods. Suppofe, for inftance, we
know that the refiduum confifts of oxygen and
azotic gas mixed together, put a determinate
quantity, 100 parts, into a graduated tube of
ten or twelve lines diameter, introduce a folution
of fulphuret of potafh in contact with the gas,
and leave them together for fome days ; the fulphuret abforbs the whole oxygen gas, and leaves
the azotic gas pure.

If it is known to contain hydrogen gas, a determinate quantity is introduced into Volta's
eudiometer alongft with a known proportion of
hydrogen gas ; thefe are deflagrated together by
means of the electrical fpark ; frefh portions of
oxygen gas are fucceffively added, till no farther

ther deflagration takes place, and till the greateft poffible diminution is produced. By this
procefs water is formed, which is immediately
abforbed by the water of the apparatus ; but, if
the hydrogen gas contain charcoal, carbonic
acid is formed at the fame time, which is not
abforbed fo quickly ; the quantity of this is
readily afcertained by affifting its abforption, by
means of agitation. If the refiduum contains
nitrous gas, by adding oxygen gas, with which
it combines into nitric acid, we can very nearly
afcertain its quantity, from the diminution produced by this mixture.

I confine myfelf to thefe general examples,
which are fufficient to give an idea of this kind
of operations ; a whole volume would not ferve
to explain every poffible cafe. It is neceffary to
become familiar with the analyfis of gaffes by
long experience ; we muft even acknowledge
that they moftly poffefs fuch powerful affinities
to each other, that we are not always certain of
having feparated them completely. In thefe
cafes, we muft vary our experiments in every
poffible point of view, add new agents to the
combination, and keep out others, and continue
our trials, till we are certain of the truth and
exactitude of our conclufions.

S E C T.

S E C T. V.

Of the neceſſary corrections upon the volume of the Gaſſes, according to the preſſure of the Atmoſphere.

All elaſtic fluids are compreſſible or condenſible in proportion to the weight with which they are loaded. Perhaps this law, which is aſcertained by general experience, may ſuffer ſome irregularity when theſe fluids are under a degree of condenſation almoſt ſufficient to reduce them to the liquid ſtate, or when either in a ſtate of extreme rarefaction or condenſation; but we ſeldom approach either of theſe limits with moſt of the gaſſes which we ſubmit to our experiments. I underſtand this propoſition of gaſſes being compreſſible, in proportion to their ſuperincumbent weights, as follows:

A barometer, which is an inſtrument generally known, is, properly ſpeaking, a ſpecies of ſyphon, ABCD, Pl. XII. Fig. 16. whoſe leg AB is filled with mercury, whilſt the leg CD is full of air. If we ſuppoſe the branch CD indefinitely continued till it equals the height of our atmoſphere, we can readily conceive that the barometer is, in reality, a ſort of balance, in which

a

column of mercury ſtands in equilibrium with a column of air of the ſame weight. But it is unneceſſary to prolongate the branch CD to ſuch a height, as it is evident that the barometer being immerſed in air, the column of mercury AB will be equally in equilibrium with a column of air of the ſame diameter, though the leg CD be cut off at C, and the part CD be taken away altogether.

The medium height of mercury in equilibrium with the weight of a column of air, from the higheſt part of the atmoſphere to the ſurface of the earth is about twenty-eight French inches in the lower parts of the city of Paris; or, in other words, the air at the ſurface of the earth at Paris is uſually preſſed upon by a weight equal to that of a column of mercury twenty-eight inches in height. I muſt be underſtood in this way in the ſeveral parts of this publication when talking of the different gaſſes, as, for inſtance, when the cubical foot of oxygen gas is ſaid to weigh 1 *oz.* 4 *gros*, under 28 inches preſſure. The height of this column of mercury, ſupported by the preſſure of the air, diminiſhes in proportion as we are elevated above the ſurface of the earth, or rather above the level of the ſea, becauſe the mercury can only form an equilibrium with the column of air which is above it, and is not in the ſmalleſt

T t   degree

degree affected by the air which is below its level.

In what ratio does the mercury in the baro-meter defcend in proportion to its elevation? or, what is the fame thing, according to what law or ratio do the feveral ftrata of the atmo-fphere decreafe in denfity? This queftion, which has exercifed the ingenuity of natural philofophers during laft century, is confiderably elucidated by the following experiment.

If we take the glafs fyphon ABCDE, Pl. XII. Fig. 17. fhut at E, and open at A, and intro-duce a few drops of mercury, fo as to intercept the communication of air between the leg AB and the leg BE, it is evident that the air con-tained in BCDE is preffed upon, in common with the whole furrounding air, by a weight or column of air equal to 28 inches of mercury. But, if we pour 28 inches of mercury into the leg AB, it is plain the air in the branch BCDE will now be preffed upon by a weight equal to twice 28 inches of mercury, or twice the weight of the atmofphere; and experience fhows, that, in this cafe, the included air, inftead of filling the tube from B to E, only occupies from C to E, or exactly one half of the fpace it filled be-fore. If to this firft column of mercury we add two other portions of 28 inches each, in the branch AB, the air in the branch BCDE will be preffed upon by four times the weight of the atmofphere,

atmofphere, or four times the weight of 28 inches of mercury, and it will then only fill the fpace from D to E, or exactly one quarter of the fpace it occupied at the commencement of the experiment. From thefe experiments, which may be infinitely varied, has been deduced as a general law of nature, which feems applicable to all permanently elaftic fluids, that they di- minifh in volume in proportion to the weights with which they are preffed upon ; or, in other words, " *the volume of all elaftic fluids is in the* " *inverfe ratio of the weight by which they are* " *compreffed.*"

The experiments which have been made for meafuring the heights of mountains by means of the barometer, confirm the truth of thefe deductions ; and, even fuppofing them in fome degree inaccurate, thefe differences are fo ex- tremely fmall, that they may be reckoned as nullities in chemical experiments. When this law of the compreffion of elaftic fluids is once well underftood, it becomes eafily applicable to the corrections neceffary in pneumato che- mical experiments upon the volume of gas, in relation to its preffure. Thefe corrections are of two kinds, the one relative to the varia- tions of the barometer, and the other for the column of water or mercury contained in the jars. I fhall endeavour to explain thefe by ex- amples, beginning with the moft fimple cafe.

Suppofe

Suppofe that 100 cubical inches of oxygen gas are obtained at 10° (54.5°) of the thermo-meter, and at 28 inches 6 lines of the barome-ter, it is required to know what volume the 100 cubical inches of gas would occupy, under the preffure of 28 inches *, and what is the exact weight of the 100 inches of oxygen gas ? Let the unknown volume, or the number of inches this gas would occupy at 28 inches of the baro-meter, be expreffed by x ; and, fince the vo-lumes are in the inverfe ratio of their fuperin-cumbent weights, we have the following ftate-ment : 100 cubical inches is to x inverfely as 28.5 inches of preffure is to 28.0 inches ; or directly $28 : 28.5 :: 100 : x = 101.786$ — cubi-cal inches, at 28 inches barometrical preffure ; that is to fay, the fame gas or air which at 28.5 inches of the barometer occupies 100 cubical inches of volume, will occupy 101.786 cubical inches when the barometer is at 28 inches. It is equally eafy to calculate the weight of this gas, occupying 100 cubical inches, under 28.5 inches of barometrical preffure ; for, as it cor-refponds

---

* According to the proportion of 114 to 107, given between the French and Englifh foot, 28 inches of the French barometer are equal to 29.83 inches of the Englifh. Directions will be found in the appendix for converting all the French weights and meafures ufed in this work into correfponding Englifh denominations. —E.

refponds to 101.786 cubical inches at the preffure of 28, and as, at this preffure, and at 10°
(54.5°) of temperature, each cubical inch of
oxygen gas weighs half a grain, it follows, that
100 cubical inches, under 28.5 barometrical
preffure, muft weigh 50.893 grains. This conclufion might have been formed more directly,
as, fince the volume of elaftic fluids is in the
inverfe ratio of their compreffion, their weights
muft be in the direct ratio of the fame compreffion : Hence, fince 100 cubical inches weigh
50 grains, under the preffure of 28 inches, we
have the following ftatement to determine the
weight of 100 cubical inches of the fame gas as
28.5 barometrical preffure, $28 : 50 :: 28.5 : x$,
the unknown quantity, $= 50.893$.

The following cafe is more complicated :
Suppofe the jar A, Pl. XII. Fig. 18. to contain
a quantity of gas in its upper part ACD, the
reft of the jar below CD being full of mercury,
and the whole ftanding in the mercurial bafon
or refervoir GHIK, filled with mercury up to
EF, and that the difference between the furface
CD of the mercury in the jar, and EF, that in
the ciftern, is fix inches, while the barometer
ftands at 27.5 inches. It is evident from thefe
data, that the air contained in ACD is preffed
upon by the weight of the atmofphere, diminifhed by the weight of the column of mercury CE,
or by $27.5 - 6 = 21.5$ inches of barometrical
preffure.

preffure. This air is therefore lefs compreffed than the atmofphere at the mean height of the barometer, and confequently occupies more fpace than it would occupy at the mean pref-fure, the difference being exactly proportional to the difference between the compreffing weights. If, then, upon meafuring the fpace ACD, it is found to be 120 cubical inches, it muft be reduced to the volume which it would occupy under the mean preffure of 28 inches. This is done by the following ftatement: 120 : x, the unknown volume, :: 21.5 : 28 in-verfely; this gives $x = \dfrac{120 \times 21.5}{28} = 92.143$ cu-bical inches.

In thefe calculations we may either reduce the height of the mercury in the barometer, and the difference of level in the jar and bafon, into lines or decimal fractions of the inch; but I prefer the latter, as it is more readily calcula-ted. As, in thefe operations, which frequently recur, it is of great ufe to have means of abbre-viation, I have given a table in the appendix for reducing lines and fractions of lines into de-cimal fractions of the inch.

In experiments performed in the water-appa-ratus, we muft make fimilar corrections to pro-cure rigoroufly exact refults, by taking into ac-count, and making allowances for the difference of height of the water within the jar above the furface of the water in the ciftern. But, as the

preffure

preffure of the atmofphere is expreffed in inches
and lines of the mercurial barometer, and, as
homogeneous quantities only can be calculated
together, we muft reduce the obferved inches
and lines of water into correfpondent heights of
the mercury. I have given a table in the appendix for this converfion, upon the fuppofition
that mercury is 13.5681 times heavier than water.

S E C T.   VI.

Of Corrections relative to the Degrees of the Ther-
mometer.

In afcertaining the weight of gaffes, befides
reducing them to a mean of barometrical preffure, as directed in the preceding fection, we
muft likewife reduce them to a ftandard thermometrical temperature ; becaufe, all elaftic
fluids being expanded by heat, and condenfed
by cold, their weight in any determinate volume is thereby liable to confiderable alterations. As the temperature of 10° (54.5°) is a
medium between the heat of fummer and the
cold of winter, being the temperature of fub-
terraneous places, and that which is moft eafily
approached to at all feafons, I have chofen that
degree as a mean to which I reduce air or gas
in this fpecies of calculation.

Mr

Mr de Luc found that atmofpheric air was increafed $\frac{1}{215}$ part of its bulk, by each degree of a mercurial thermometer, divided into 81 degrees, between the freezing and boiling points ; this gives $\frac{1}{211}$ part for each degree of Reaumur's thermometer, which is divided into 80 degrees between thefe two points. The experiments of Mr Monge feem to make this dilatation lefs for hydrogen gas, which he thinks is only dilated $\frac{1}{180}$. We have not any exact experiments hitherto publifhed refpecting the ratio of dilatation of the other gaffes ; but, from the trials which have been made, their dilatation feems to differ little from that of atmofpheric air. Hence I may take for granted, till farther experiments give us better information upon this fubject, that atmofpherical air is dilated $\frac{1}{210}$ part, and hydrogen gas $\frac{1}{190}$ part for each degree of the thermometer ; but, as there is ftill great uncertainty upon this point, we ought always to operate in a temperature as near as poffible to the ftandard of 10°, (54.5°) by this means any errors in correcting the weight or volume of gaffes by reducing them to the common ftandard, will become of little moment.

The calculation for this correction is extremely eafy. Divide the obferved volume of air by 210, and multiply the quotient by the degrees of temperature above or below 10°
(54.5°).

(54.5°). This correction is negative when the actual temperature is above the standard, and positive when below. By the use of logarithmical tables this calculation is much facilitated *.

S E C T.   VII.

Example for calculating the Corrections relative to the Variations of Pressure and Temperature.

C A S E.

In the jar A, Pl. IV. Fig. 3. standing in a water apparatus, is contained 353 cubical inches of air; the surface of the water within the jar at EF is $4\frac{1}{2}$ inches above the water in the cistern, the barometer is at 27 inches $9\frac{1}{2}$ lines, and the thermometer at 15° (65.75°). Having burnt a quantity of phosphorus in the air, by which concrete phosphoric acid is produced, the air after the combustion occupies 295 cubical

U u   inches,

---

* When Fahrenheit's thermometer is employed, the dilatation by each degree must be smaller, in the proportion of 1 to 2.25, because each degree of Reaumur's scale contains 2.25 degrees of Fahrenheit; hence we must divide by 472 5, and finish the rest of the calculation as above.—E.

inches, the water within the jar ſtands 7 inches above that in the ciſtern, the barometer is at 27 inches 9¼ lines, and the thermometer at 16° (68°). It is required from theſe data to determine the actual volume of air before and after combuſtion, and the quantity abſorbed during the proceſs.

Calculation before Combuſtion.

The air in the jar before combuſtion was 353 cubical inches, but it was only under a barometrical preſſure of 27 inches 9½ lines ; which, reduced to decimal fractions by Tab. I. of the Appendix, gives 27.79167 inches ; and from this we muſt deduct the difference of 4½ inches of water, which, by Tab. II. correſponds to 0.33166 inches of the barometer ; hence the real preſſure of the air in the jar is 27.46001. As the volume of elaſtic fluids diminiſh in the inverſe ratio of the compreſſing weights, we have the following ſtatement to reduce the 353 inches to the volume the air would occupy at 28 inches barometrical preſſure.

353 : x, the unknown volume, :: 27.46001 : 28.

Hence, $x = \dfrac{353 \times 27.46001}{28} = 346,192$ cubical inches, which is the volume the ſame quantity of air would have occupied at 28 inches of the barometer.

The

The 210th part of this corrected volume is 1.65, which, for the five degrees of temperature above the standard gives 8.255 cubical inches; and, as this correction is subtractive, the real corrected volume of the air before combustion is 337.942 inches.

Calculation after Combustion.

By a similar calculation upon the volume of air after combustion, we find its barometrical pressure $27.77083 - 0.51593 = 27.25490$. Hence, to have the volume of air under the pressure of 28 inches, $295 : x :: 27.77083 : 28$ inversely; or, $x = \dfrac{295 \times 27.25490}{28} = 287.150$. The 210th part of this corrected volume is 1.368, which, multiplied by 6 degrees of thermometrical difference, gives the subtractive correction for temperature 8.208, leaving the actual corrected volume of air after combustion 278.942 inches.

Result.

The corrected volume before combustion 337.942

Ditto remaining after combustion . 278.942

Volume absorbed during combustion   59.000.

SECT.

S E C T. VIII.

Method of determining the Abfolute **Gravity** *of the different Gaffes.*

Take a large balloon A, Pl.V. Fig. 10. capable of holding 17 or 18 pints, or about half a cubical foot, having the brafs cap *bcde* ftrongly cemented to its neck, and to which the tube and ftop-cock *f g* is fixed by a tight fcrew. This apparatus is connected by the double fcrew reprefented feparately at Fig. 12. to the jar BCD, Fig. 10. which muft be fome pints larger in dimenfions than the balloon. This jar is open at top, and is furnifhed with the brafs cap *h i*, and ftop-cock *l m*. One of thefe ftop-cocks is reprefented feparately at Fig. 11.

We firft determine the exact capacity of the balloon by filling it with water, and weighing it both full and empty. When emptied of water, it is dried with a cloth introduced through its neck *d e*, and the laft remains of moifture are removed by exhaufting it once or twice in an air-pump.

When the weight of any gas is to be afcertained, this apparatus is ufed as follows : Fix the balloon A to the plate of an air-pump by means of the fcrew of the ftop-cock *f g*, which is

left

left open ; the balloon is to be exhaufted as completely as poffible, obferving carefully the degree of exhauftion by means of the barometer
attached to the air-pump.   When the vacuum is
formed, the ftop-cock *f g* is fhut, and the weight
of the balloon determined with the moft fcrupulous exactitude.   It is then fixed to the jar
BCD, which we fuppofe placed in water in
the fhelf of the pneumato chemical apparatus
Fig. 1. ; the jar is to be filled with the gas we
mean to weigh, and then, by opening the ftop-
cocks *f g* and *l m*, the gas afcends into the balloon, whilft the water of the ciftern rifes at the
fame time into the jar.   To avoid very troublefome corrections, it is neceffary, during this firft
part of the operation, to fink the jar in the ciftern till the furfaces of the water within the jar
and without exactly correfpond.   The ftop-
cocks are again fhut, and the balloon being un-
fcrewed from its connection with the jar, is to
be carefully weighed ; the difference between
this weight and that of the exhaufted balloon is
the precife weight of the air or gas contained in
the balloon. Multiply this weight by 1728, the
number of cubical inches in a cubical foot, and
divide the product by the number of cubical
inches contained in the balloon, the quotient is
the weight of a cubical foot of the gas or air
fubmitted to experiment.

Exact

Exact account muſt be kept of the barometrical height and temperature of the thermometer during the above experiment; and from
theſe the reſulting weight of a cubical foot is
eaſily corrected to the ſtandard of 28 inches and
10°, as directed in the preceding ſection. The
ſmall portion of air remaining in the balloon after forming the vacuum muſt likewiſe be attended to, which is eaſily determined by the
barometer attached to the air-pump. If that barometer, for inſtance, remains at the hundredth
part of the height it ſtood at before the vacuum
was formed, we conclude that one hundredth
part of the air originally contained remained in
the balloon, and conſequently that only $\frac{99}{100}$ of
gas was introduced from the jar into the balloon.

C H A P.

C H A P. III.

Defcription of the Calorimeter, or Apparatus for meafuring Caloric.

THE calorimeter, or apparatus for meafuring the relative quantities of heat contained in bodies, was defcribed by Mr de la
Place and me in the Memoirs of the Academy
for 1780, p. 355. and from that Eflay the materials of this chapter are extracted.

If, after having cooled any body to the freezing point, it be expofed in an atmofphere of
25° (88.25°), the body will gradually become
heated, from the furface inwards, till at laft it
acquire the fame temperature with the furrounding air. But, if a piece of ice be placed
in the fame fituation, the circumftances are
quite different; it does not approach in the
fmalleft degree towards the temperature of the
circumambient air, but remains conftantly at
Zero (32°), or the temperature of melting ice,
till the laft portion of ice be completely melted.

This phenomenon is readily explained; as,
to melt ice, or reduce it to water, it requires to
be combined with a certain portion of caloric;
the

the whole caloric attracted from the surround-
ing bodies, is arrested or fixed at the surface or
external layer of ice which it is employed to diffolve, and combines with it to form water; the
next quantity of caloric combines with the fecond layer to diffolve it into water, and fo on
fucceffively till the whole ice be diffolved or
converted into water by combination with caloric, the very laft atom ftill remaining at its former temperature, becaufe the caloric has never
penetrated fo far as long as any intermediate ice
remained to melt.

Upon thefe principles, if we conceive a hollow fphere of ice at the temperature of Zero
(32°) placed in an atmofphere 10° (54.5°), and
containing a fubftance at any degree of temperature above freezing, it follows, 1ft, That the
heat of the external atmofphere cannot penetrate
into the internal hollow of the fphere of ice;
2dly, That the heat of the body placed in the
hollow of the fphere cannot penetrate outwards
beyond it, but will be ftopped at the internal
furface, and continually employed to melt fucceffive layers of ice, until the temperature of
the body be reduced to Zero (32°), by having
all its fuperabundant caloric above that temperature carried off by the ice. If the whole water, formed within the fphere of ice during the
reduction of the temperature of the included
body to Zero, be carefully collected, the weight
of

of the water will be exactly proportional to the quantity of caloric lost by the body in passing from its original temperature to that of melting ice; for it is evident that a double quantity of caloric would have melted twice the quantity of ice; hence the quantity of ice melted is a very exact measure of the quantity of caloric employed to produce that effect, and consequently of the quantity lost by the only substance that could possibly have supplied it.

I have made this supposition of what would take place in a hollow sphere of ice, for the purpose of more readily explaining the method used in this species of experiment, which was first conceived by Mr de la Place. It would be difficult to procure such spheres of ice, and inconvenient to make use of them when got; but, by means of the following apparatus, we have remedied that defect. I acknowledge the name of Calorimeter, which I have given it, as derived partly from Greek and partly from Latin, is in some degree open to criticism; but, in matters of science, a slight deviation from strict etymology, for the sake of giving distinctness of idea, is excusable; and I could not derive the name entirely from Greek without approaching too near to the names of known instruments employed for other purposes.

The calorimeter is represented in Pl. VI. It is shown in perspective at Fig. 1. and its interior

structure

ftructure is engraved in Fig. 2. and 3.; the former being a horizontal, and the latter a perpendicular fection. Its capacity or cavity is divided into three parts, which, for better diftinction, I fhall name the interior, middle, and external cavities. The interior cavity *ffff*, Fig. 4.
into which the fubftances fubmitted to experiment are put, is compofed of a grating or cage
of iron wire, fupported by feveral iron bars; its
opening or mouth LM, is covered by the lid
HG, of the fame materials. The middle cavity *b b b b*, Fig. 2. and 3. is intended to contain
the ice which furrounds the interior cavity, and
which is to be melted by the caloric of the fubftance employed in the experiment. The ice is
fupported by the grate *m m* at the bottom of the
cavity, under which is placed the fieve *n n*.
Thefe two are reprefented feparately in Fig. 5.
and 6.

In proportion as the ice contained in the
middle cavity is melted, by the caloric difengaged from the body placed in the interior cavity,
the water runs through the grate and fieve, and
falls through the conical funnel *c c d*, Fig. 3. and
tube *x y*, into the receiver F, Fig. 1. This water may be retained or let out at pleafure, by
means of the ftop-cock *u*. The external cavity
a a a a, Fig. 2. and 3. is filled with ice, to prevent any effect upon the ice in the middle cavity from the heat of the furrounding air, and
the

the water produced from it is carried off through the pipe ST, which fhuts by means of the ftop-cock *r*. The whole machine is covered by the lid FF, Fig. 7. made of tin painted with oil co-lour, to prevent ruft.

When this machine is to be employed, the middle cavity *b b b b*, Fig. 2. and 3. the lid GH, Fig. 4. of the interior cavity, the external cavi-ty *a a a a*, Fig. 2. and 3. and the general lid FF, Fig. 7. are all filled with pounded ice, well rammed, fo that no void fpaces remain, and the ice of the middle cavity is allowed to drain. The machine is then opened, and the fubftance fubmitted to experiment being placed in the in-terior cavity, it is inftantly clofed. After wait-ing till the included body is completely cooled to the freezing point, and the whole melted ice has drained from the middle cavity, the water collected in the veffel F, Fig. 1. is accurately weighed. The weight of the water produced during the experiment is an exact meafure of the caloric difengaged during the cooling of the included body, as this fubftance is evidently in a fimilar fituation with the one formerly men-tioned as included in a hollow fphere of ice; the whole caloric difengaged is ftopped by the ice in the middle cavity, and that ice is prefer-ved from being affected by any other heat by means of the ice contained in the general lid, Fig. 7. and in the external cavity. Experiments

of

of this kind laſt from fifteen to twenty hours; they are ſometimes accelerated by covering up the ſubſtance in the interior cavity with well drained ice, which haſtens its cooling.

The ſubſtances to be operated upon are placed in the thin iron bucket, Fig. 8. the cover of which has an opening fitted with a cork, into which a ſmall thermometer is fixed. When we uſe acids, or other fluids capable of injuring the metal of the inſtruments, they are contained in the matras, Fig. 10. which has a ſimilar thermometer in a cork fitted to its mouth, and which ſtands in the interior cavity upon the ſmall cylindrical ſupport RS, Fig. 10.

It is abſolutely requiſite that there be no communication between the external and middle cavities of the calorimeter, otherwiſe the ice melted by the influence of the ſurrounding air, in the external cavity, would mix with the water produced from the ice of the middle cavity, which would no longer be a meaſure of the caloric loſt by the ſubſtance ſubmitted to experiment.

When the temperature of the atmoſphere is only a few degrees above the freezing point, its heat can hardly reach the middle cavity, being arreſted by the ice of the cover, Fig. 7. and of the external cavity ; but, if the temperature of the air be under the degree of freezing, it might cool the ice contained in the middle cavity, by

cauſing

caufing the ice in the external cavity to fall, in the firft place, below zero (32°). It is therefore effential that this experiment be carried on in a temperature fomewhat above freezing: Hence, in time of froft, the calorimeter muft be kept in an apartment carefully heated. It is likewife neceffary that the ice employed be not under zero (32°); for which purpofe it muft be pounded, and fpread out thin for fome time, in a place of a higher temperature.

The ice of the interior cavity always retains a certain quantity of water adhering to its furface, which may be fuppofed to belong to the refult of the experiment; but as, at the beginning of each experiment, the ice is already faturated with as much water as it can contain, if any of the water produced by the caloric fhould remain attached to the ice, it is evident, that very nearly an equal quantity of what adhered to it before the experiment muft have run down into the veffel F in its ftead; for the inner furface of the ice in the middle cavity is very little changed during the experiment.

By any contrivance that could be devifed, we could not prevent the accefs of the external air into the interior cavity when the atmofphere was 9° or 10° (52° or 54°) above zero. The air confined in the cavity being in that cafe fpecifically heavier than the external air, efcapes downwards through the pipe x y, Fig. 3, and is
replaced

replaced by the warmer external air, which, giving out its caloric to the ice, becomes heavier, and finks in its turn; thus a current of air is formed through the machine, which is the more rapid in proportion as the external air exceeds the internal in temperature. This current of warm air muſt melt a part of the ice, and injure the accuracy of the experiment: We may, in a great degree, guard againſt this ſource of error by keeping the ſtop-cock u continually ſhut; but it is better to operate only when the temperature of the external air does not exceed 3°, or at moſt 4°, (39° to 41°); for we have obſerved, that, in this caſe, the melting of the interior ice by the atmoſpheric air is perfectly infenſible; ſo that we may anſwer for the accuracy of our experiments upon the ſpecific heat of bodies to a fortieth part.

We have cauſed make two of the above deſcribed machines; one, which is intended for ſuch experiments as do not require the interior air to be renewed, is preciſely formed according to the deſcription here given; the other, which anſwers for experiments upon combuſtion, reſpiration, &c. in which freſh quantities of air are indiſpenſibly neceſſary, differs from the former in having two ſmall tubes in the two lids, by which a current of atmoſpheric air may be blown into the interior cavity of the machine.

It

It is extremely eafy, with this apparatus, to determine the phenomena which occur in operations where caloric is either difengaged or abforbed. If we wifh, for inftance, to afcertain the quantity of caloric which is difengaged from a folid body in cooling a certain number of degrees, let its temperature be raifed to 80° (212°); it is then placed in the interior cavity *ffff*, Fig. 2. and 3. of the calorimeter, and allowed to remain till we are certain that its temperature is reduced to zero (32°); the water produced by melting the ice during its cooling is collected, and carefully weighed ; and this weight, divided by the volume of the body fubmitted to experiment, multiplied into the degrees of temperature which it had above zero at the commencement of the experiment, gives the proportion of what the Englifh philofophers call fpecific heat.

Fluids are contained in proper veffels, whofe fpecific heat has been previoufly afcertained, and operated upon in the machine in the fame manner as directed for folids, taking care to deduct, from the quantity of water melted during the experiment, the proportion which belongs to the containing veffel.

If the quantity of caloric difengaged during the combination of different fubftances is to be determined, thefe fubftances are to be previoufly reduced to the freezing degree by keeping
them

them a fufficient time furrounded with pounded
ice; the mixture is then to be made in the inner cavity of the calorimeter, in a proper veffel likewife reduced to zero (32°); and they are
kept inclofed till the temperature of the combination has returned to the fame degree: The
quantity of water produced is a meafure of the
caloric difengaged during the combination.

To determine the quantity of caloric difengaged during combuftion, and during animal
refpiration, the combuftible bodies are burnt,
or the animals are made to breathe in the interior cavity, and the water produced is carefully
collected. Guinea pigs, which refift the effects
of cold extremely well, are well adapted for this
experiment. As the continual renewal of air is
abfolutely neceffary in fuch experiments, we
blow frefh air into the interior cavity of the calorimeter by means of a pipe deftined for that
purpofe, and allow it to efcape through another
pipe of the fame kind; and that the heat of this
air may not produce errors in the refults of the
experiments, the tube which conveys it into the
machine is made to pafs through pounded ice,
that it may be reduced to zero (32°) before
it arrives at the calorimeter. The air which
efcapes muft likewife be made to pafs through
a tube furrounded with ice, included in the interior cavity of the machine, and the water
which is produced muft make a part of what is
collected,

collected, becaufe the caloric difengaged from this air is part of the product of the experiment.

It is fomewhat more difficult to determine the fpecific caloric contained in the different gaffes, on account of their fmall degree of denfity ; for, if they are only placed in the calorimeter in veffels like other fluids, the quantity of ice melted is fo fmall, that the refult of the experiment becomes at beft very uncertain. For this fpecies of experiment we have contrived to make the air pafs through two metallic worms, or fpiral tubes ; one of thefe, through which the air paffes, and becomes heated in its way to the calorimeter, is contained in a veffel full of boiling water, and the other, through which the air circulates within the calorimeter to difengage its caloric, is placed in the interior cavity, *ffff*, of that machine. By means of a fmall thermometer placed at one end of the fecond worm, the temperature of the air, as it enters the calorimeter, is determined, and its temperature in getting out of the interior cavity is found by another thermometer placed at the other end of the worm. By this contrivance we are enabled to afcertain the quantity of ice melted by determinate quantities of air or gas, while lofing a certain number of degrees of temperature, and, confequently, to determine their feveral degrees of fpecific caloric. The

Y y

fame apparatus, with fome particular precautions, may be employed to afcertain the quantity of caloric difengaged by the condenfation of the vapours of different liquids.

The various experiments which may be made with the calorimeter do not afford abfolute conclufions, but only give us the meafure of relative quantities; we have therefore to fix a unit, or ftandard point, from whence to form a fcale of the feveral refults. The quantity of caloric neceffary to melt a pound of ice has been chofen as this unit; and, as it requires a pound of water of the temperature of 60° (167°) to melt a pound of ice, the quantity of caloric expreffed by our unit or ftandard point is what raifes a pound of water from zero (32°) to 60° (167°). When this unit is once determined, we have only to exprefs the quantities of caloric difengaged from different bodies by cooling a certain number of degrees, in analogous values: The following is an eafy mode of calculation for this purpofe, applied to one of our earlieft experiments.

We took 7 *lib.* 11 *oz.* 2 *gros* 36 *grs.* of plate-iron, cut into narrow flips, and rolled up, or expreffing the quantity in decimals, 7.7070319. Thefe, being heated in a bath of boiling water to about 78° (207.5°), were quickly introduced into the interior cavity of the calorimeter: At

the

the end of eleven hours, when the whole quantity of water melted from the ice had thoroughly drained off, we found that 1.109795 pounds
of ice were melted. Hence, the caloric difengaged from the iron by cooling 78° (175.5°)
having melted 1.109795 pounds of ice, how
much would have been melted by cooling 60°
(135°)? This queftion gives the following ftate-
ment in direct proportion, 78:1.109795::60:x=
0.85369. Dividing this quantity by the weight
of the whole iron employed, viz. 7.7070319, the
quotient 0.110770 is the quantity of ice which
would have been melted by one pound of iron
whilft cooling through 60° (135°) of temperature.

Fluid fubftances, fuch as fulphuric and nitric
acids, &c. are contained in a matras, Pl. VI.
Fig. 9. having a thermometer adapted to the
cork, with its bulb immerfed in the liquid. The
matras is placed in a bath of boiling water, and
when, from the thermometer, we judge the liquid is raifed to a proper temperature, the matras is placed in the calorimeter. The calculation of the products, to determine the fpecific
caloric of thefe fluids, is made as above directed, taking care to deduct from the water obtained the quantity which would have been
produced by the matras alone, which muft be
afcertained by a previous experiment. The
table

table of the refults obtained by thefe experiments is omitted, becaufe not yet fufficiently
complete, different circumftances having occafioned the feries to be interrupted; it is not,
however, loft fight of; and we are lefs or more
employed upon the fubject every winter.

C H A P.

C H A P. IV.

Of Mechanical Operations for Division of Bodies.

S E C T. I.

Of Trituration, Levigation, and Pulverization.

THESE are, properly fpeaking, only preliminary mechanical operations for dividing and feparating the particles of bodies, and reducing them into very fine powder. Thefe operations can never reduce fubftances into their primary, or elementary and ultimate particles; they do not even deftroy the aggregation of bodies; for every particle, after the moft accurate trituration, forms a fmall whole, refembling the original mafs from which it was divided. The real chemical operations, on the contrary, fuch as folution, deftroy the aggregation of bodies, and feparate their conftituent and integrant particles from each other.

Brittle

Brittle fubftances are reduced to powder by means of peftles and mortars. Thefe are of brafs or iron, Pl. I. Fig. 1.; of marble or granite, Fig. 2.; of lignum vitae, Fig. 3.; of glafs, Fig. 4.; of agate, Fig. 5.; or of porcellain, Fig. 6. The peftles for each of thefe are reprefented in the plate, immediately below the mortars to which they refpectively belong, and are made of hammered iron or brafs, of wood, glafs, porcellain, marble, granite, or agate, according to the nature of the fubftances they are intended to triturate. In every laboratory, it is requifite to have an affortment of thefe utenfils, of various fizes and kinds : Thofe of porcellain and glafs can only be ufed for rubbing fubftances to powder, by a dexterous ufe of the peftle round the fides of the mortar, as it would be eafily broken by reiterated blows of the peftle.

The bottom of mortars ought to be in the form of a hollow fphere, and their fides fhould have fuch a degree of inclination as to make the fubftances they contain fall back to the bottom when the peftle is lifted, but not fo perpendicular as to collect them too much together, otherwife too large a quantity would get below the peftle, and prevent its operation. For this reafon, likewife, too large a quantity of the fubftance to be powdered ought not to be put into the mortar at one time ; and we muft from

time

time to time get rid of the particles already reduced to powder, by means of fieves to be afterwards defcribed.

The moft ufual method of levigation is by
means of a flat table ABCD, Pl. 1. Fig. 7. of
porphyry, or other ftone of fimilar hardnefs, upon which the fubftance to be reduced to powder
is fpread, and is then bruifed and rubbed by a
muller M, of the fame hard materials, the bottom of which is made a fmall portion of a large
fphere ; and, as the muller tends continually to
drive the fubftances towards the fides of the
table, a thin flexible knife, or fpatula of iron,
horn, wood, or ivory, is ufed for bringing them
back to the middle of the ftone.

In large works, this operation is performed
by means of large rollers of hard ftone, which
turn upon each other, either horizontally, in the
way of corn-mills, or by one vertical roller
turning upon a flat ftone.   In the above operations, it is often requifite to moiften the fubftances a little, to prevent the fine powder from
flying off.

There are many bodies which cannot be reduced to powder by any of the foregoing methods ; fuch are fibrous fubftances, as woods ;
fuch as are tough and elaftic, as the horns of a-
nimals, elaftic gum, &c. and the malleable metals which flatten under the peftle, inftead of
being reduced to powder. For reducing the
woods

woods to powder, rafps, as Pl. 1. Fig. 8. are employed; files of a finer kind are ufed for horn,
and ftill finer, Pl. 1. Fig. 9. and 10. for metals.

Some of the metals, though not brittle e-
nough to powder under the peftle, are too foft
to be filed, as they clog the file, and prevent its
operation. Zinc is one of thefe, but it may be
powdered when hot in a heated iron mortar, or
it may be rendered brittle, by alloying it with
a fmall quantity of mercury. One or other of
thefe methods is ufed by fire-work makers for
producing a blue flame by means of zinc. Metals may be reduced into grains, by pouring
them when melted into water, which ferves very well when they are not wanted in fine powder.

Fruits, potatoes, &c. of a pulpy and fibrous
nature may be reduced to pulp by means of the
grater, Pl. 1. Fig. 11.

The choice of the different fubftances of
which thefe inftruments are made is a matter of
importance; brafs or copper are unfit for operations upon fubftances to be ufed as food or in
pharmacy; and marble or metallic inftruments
muft not be ufed for acid fubftances; hence
mortars of very hard wood, and thofe of porcelain, granite, or glafs, are of great utility in
many operations.

S E C T.

S E C T.   II.

Of Sifting and Washing Powdered Substances.

None of the mechanical operations employed
for reducing bodies to powder is capable of producing it of an equal degree of finenefs throughout; the powder obtained by the longeft and
moft accurate trituration being ftill an affem-
blage of particles of various fizes. The coarfer
of thefe are removed, fo as only to leave the
finer and more homogeneous particles by means
of fieves, Pl. I. Fig. 12. 13. 14. 15. of different
finenefles, adapted to the particular purpofes they
are intended for ; all the powdered matter
which is larger than the inteftices of the fieve
remains behind, and is again fubmitted to the
peftle, while the finer pafs through. The fieve
Fig. 12. is made of hair-cloth, or of filk gauze ;
and the one reprefented Fig. 13. is of parch-
ment pierced with round holes of a proper fize ;
this latter is employed in the manufacture of
gun-powder. When very fubtile or valuable
materials are to be fifted, which are eafily
difperfed, or when the finer parts of the powder
may be hurtful, a compound fieve, Fig. 15. is
made ufe of, which confifts of the fieve ABCD,
with a lid EF, and receiver GH ; thefe three

Z z   parts

parts are reprefented as joined together for ufe, Fig. 14.

There is a method of procuring powders of an uniform finenefs, confiderably more accurate than the fieve; but it can only be ufed with fuch fubftances as are not acted upon by water. The powdered fubftance is mixed and agitated with water, or other convenient fluid ; the liquor is allowed to fettle for a few moments, and is then decanted off; the coarfeft powder remains at the bottom of the veffel, and the finer paffes over with the liquid. By repeated decantations in this manner, various fediments are obtained of different degrees of finenefs; the laft fediment, or that which remains longeft fufpended in the liquor, being the fineft. This procefs may likewife be ufed with advantage for feparating fubftances of different degrees of fpecific gravity, though of the fame finenefs; this laft is chiefly employed in mining, for feparating the heavier metallic ores from the lighter earthy matters with which they are mixed.

In chemical laboratories, pans and jugs of glafs or earthen ware are employed for this operation ; fometimes, for decanting the liquor without difturbing the fediment, the glafs fyphon ABCHI, Pl. II. Fig. 11. is ufed, which may be fupported by means of the perforated board DE, at the proper depth in the veffel FG, to draw off all the liquor required into the receiver

receiver LM. The principles and application of this ufeful inftrument are fo well known as to need no explanation.

SECT. III.

Of Filtration.

A filtre is a fpecies of very fine fieve, which is permeable to the particles of fluids, but through which the particles of the fineft pow-dered folids are incapable of paffing; hence its ufe in feparating fine powders from fufpenfion in fluids. In pharmacy, very clofe and fine woollen cloths are chiefly ufed for this opera-tion; thefe are commonly formed in a conical fhape, Pl. II. Fig. 2. which has the advantage of uniting all the liquor which drains through into a point A, where it may be readily collect-ed in a narrow mouthed veffel. In large phar-maceutical laboratories, this filtring bag is ftreached upon a wooden ftand, Pl. II. Fig. 1.

For the purpofes of chemiftry, as it is requi-fite to have the filtres perfectly clean, unfized paper is fubftituted inftead of cloth or flannel; through this fubftance, no folid body, however finely it be powdered, can penetrate, and fluids percolate through it with the greateft readinefs.

As

As paper breaks eafily when wet, various methods of fupporting it are ufed according to circumftances. When a large quantity of fluid is
to be filtrated, the paper is fupported by the
frame of wood, Pl. II. Fig. 3. ABCD, having a
piece of coarfe cloth ftretched over it, by means
of iron-hooks. This cloth muft be well cleaned
each time it is ufed, or even new cloth muft be
employed, if there is reafon to fufpect its being
impregnated with any thing which can injure
the fubfequent operations. In ordinary operations, where moderate quantities of fluid are to
be filtrated, different kinds of glafs funnels are
ufed for fupporting the paper, as reprefented
Pl. II. Fig. 5. 6. and 7. When feveral filtrations
muft be carried on at once, the board or fhelf
AB, Fig. 9. fupported upon ftands C and D,
and pierced with round holes, is very convenient for containing the funnels.

Some liquors are fo thick and clammy, as
not to be able to penetrate through paper without fome previous preparation, fuch as clarifica-
tion by means of white of eggs, which being
mixed with the liquor, coagulates when brought
to boil, and, entangling the greater part of the
impurities of the liquor, rifes with them to the
furface in the ftate of fcum. Spiritous liquors
may be clarified in the fame manner by means
of ifinglafs diffolved in water, which coagulates

by

by the action of the alkohol without the affiftance of heat.

As moft of the acids are produced by diftillation, and are confequently clear, we have rarely
any occafion to filtrate them ; but if, at any
time, concentrated acids require this operation,
it is impoffible to employ paper, which would
be corroded and deftroyed by the acid. For
this purpofe, pounded glafs, or rather quartz or
rock-criftal, broke in pieces and grofsly powdered, anfwers very well ; a few of the larger
pieces are put in the neck of the funnel ; thefe
are covered with the fmaller pieces, the finer
powder is placed over all, and the acid is poured
on at top. For the ordinary purpofes of fociety, river-water is frequently filtrated by means
of clean wafhed fand, to feparate its impurities.

S E C T.   IV.

Of Decantation.

This operation is often fubftituted inftead of
filtration for feparating folid particles which are
diffufed through liquors. Thefe are allowed to
fettle in conical veffels, ABCDE, Pl. II. Fig. 10.
the diffufed matters gradually fubfide, and the

clear

clear fluid is gently poured off. If the fediment be extremely light, and apt to mix again with the fluid by the flighteft motion, the fyphon, Fig. 11. is ufed, inftead of decantation, for drawing off the clear fluid.

In experiments, where the weight of the precipitate muft be rigoroufly afcertained, decantation is preferable to filtration, providing the precipitate be feveral times wafhed in a confiderable proportion of water. The weight of the precipitate may indeed be afcertained, by carefully weighing the filtre before and after the operation ; but, when the quantity of precipitate is fmall, the different proportions of moifture retained by the paper, in a greater or leffer degree of exficcation, may prove a material fource of error, which ought carefully to be guarded againft.

C H A P.

C H A P.   V.

Of Chemical Means for separating the Particles of Bodies from each other, without Decomposition, and for uniting them again.

I have already shown that there are two methods of dividing the particles of bodies, the *mechanical* and *chemical*. The former only separates a solid mass into a great number of smaller masses; and for these purposes various species of forces are employed, according to circumstances, such as the strength of man or of animals, the weight of water applied through the means of hydraulic engines, the expansive power of steam, the force of the wind, &c. By all these mechanical powers, we can never reduce substances into powder beyond a certain degree of fineness; and the smallest particle produced in this way, though it seems very minute to our organs, is still in fact a mountain, when compared with the ultimate elementary particles of the pulverized substance.

The chemical agents, on the contrary, divide bodies into their primitive particles. If, for instance, a neutral salt be acted upon by these, it is divided, as far as is possible, without ceasing to be a neutral salt. In this Chapter, I mean to

give

give examples of this kind of divifion of bodies, to which I fhall add fome account of the rela-tive operations.

S E C T.   I.

Of the Solution of Salts.

In chemical language, the terms of *folution* and *diffolution* have long been confounded, and have very improperly been indifcriminately em-ployed for expreffing both the divifion of the particles of a falt in a fluid, fuch as water, and the divifion of a metal in an acid.   A few re-flections upon the effects of thefe two opera-tions will fuffice to fhow that they ought not to be confounded together.   In the folution of falts, the faline particles are only feparated from each other, whilft neither the falt nor the water are at all decompofed ; we are able to recover both the one and the other in the fame quantity as before the operation.   The fame thing takes place in the folution of refins in alkohol.   Du-ring metallic diffolutions, on the contrary, a decompofition, either of the acid, or of the wa-ter which dilutes it, always takes place ; the metal combines with oxygen, and is changed into an oxyd, and a gaffeous fubftance is difen-gaged ; fo that in reality none of the fubftan-

ces

ces employed remain, after the operation, in the fame ftate they were in before. This article is entirely confined to the confideration of folution.

To underftand properly what takes place during the folution of falts, it is neceffary to know, that, in moft of thefe operations, two diftinct effects are complicated together, viz. folution by water, and folution by caloric ; and, as the explanation of moft of the phenomena of folution depends upon the diftinction of thefe two circumftances, I fhall enlarge a little upon their nature.

Nitrat of potafh, ufually called nitre or faltpetre, contains very little water of criftallization, perhaps even none at all ; yet this falt liquifies in a degree of heat very little fuperior to that of boiling water. This liquifaction cannot therefore be produced by means of the water of criftallization, but in confequence of the falt being very fufible in its nature, and from its paffing from the folid to the liquid ftate of aggregation, when but a little raifed above the temperature of boiling water. All falts are in this manner fufceptible of being liquified by caloric, but in higher or lower degrees of temperature. Some of thefe, as the acetites of potafh and foda, liquify with a very moderate heat, whilft others, as fulphat of potafh, lime, &c. require the ftrongeft fires we are capable of producing. This liqui-

3 A

faction

faction of falts by caloric produces exactly the fame phenomena with the melting of ice; it is accomplished in each falt by a determinate degree of heat, which remains invariably the fame during the whole time of the liquifaction. Caloric is employed, and becomes fixed during the melting of the falt, and is, on the contrary, difengaged when the falt coagulates. Thefe are general phenomena which univerfally occur during the paffage of every fpecies of fubftance from the folid to the fluid ftate of aggregation, and from fluid to folid.

Thefe phenomena arifing from folution by caloric are always lefs or more conjoined with thofe which take place during folutions in water. We cannot pour water upon a falt, on purpofe to diffolve it, without employing a compound folvent, both water and caloric; hence we may diftinguifh feveral different cafes of folution, according to the nature and mode of exiftence of each falt. If, for inftance, a falt be difficultly foluble in water, and readily fo by caloric, it evidently follows, that this falt will be difficultly foluble in cold water, and confiderably in hot water; fuch is nitrat of potafh, and more efpecially oxygenated muriat of potafh. If another falt be little foluble both in water and caloric, the difference of its folubility in cold and warm water will be very inconfiderable; fulphat of lime is of this kind. From thefe confi-

derations,

derations, it follows, that there is a neceſſary
relation between the following circumſtances;
the ſolubility of a ſalt in cold water, its ſolubility
in boiling water, and the degree of temperature
at which the ſame ſalt liquifies by caloric, unaſſiſt-
ed by water; and that the difference of ſolubility
in hot and cold water is ſo much greater in proportion to its ready ſolution in caloric, or in
proportion to its ſuſceptibility of liquifying in a
low degree of temperature.

The above is a general view of ſolution; but,
for want of particular facts, and ſufficiently exact experiments, it is ſtill nothing more than an
approximation towards a particular theory. The
means of compleating this part of chemical ſcience is extremely ſimple; we have only to aſcertain how much of each ſalt is diſſolved by a
certain quantity of water at different degrees of
temperature; and as, by the experiments publiſhed by Mr de la Place and me, the quantity
of caloric contained in a pound of water at each
degree of the thermometer is accurately known,
it will be very eaſy to determine, by ſimple experiments, the proportion of water and caloric
required for ſolution by each ſalt, what quantity
of caloric is abſorbed by each at the moment of
liquifaction, and how much is diſengaged at the
moment of criſtallization.   Hence the reaſon
why ſalts are more rapidly ſoluble in hot than
in cold water is perfectly evident.   In all ſolu-
tions

tions of falts caloric is employed; when that is
furnifhed intermediately from the furrounding
bodies, it can only arrive flowly to the falt;
whereas this is greatly accelerated when the requifite caloric exifts ready combined with the
water of folution.

In general, the fpecific gravity of water is
augmented by holding falts in folution; but
there are fome exceptions to the rule. Some
time hence, the quantities of radical, of oxygen,
and of bafe, which conftitute each neutral falt,
the quantity of water and caloric neceffary for
folution, the increafed fpecific gravity communicated to water, and the figure of the elementary particles of the criftals, will all be accurately known. From thefe all the circumftances
and phenomena of criftallization will be explained, and by thefe means this part of chemiftry will be compleated. Mr Seguin has formed
the plan of a thorough inveftigation of this
kind, which he is extremely capable of execu-
ting.

The folution of falts in water requires no particular apparatus; fmall glafs phials of different
fizes, Pl. II. Fig. 16. and 17. pans of earthern
ware, A, Fig. 1. and 2. long-necked matraffes,
Fig. 14. and pans or bafons of copper or of filver, Fig. 13. and 15. anfwer very well for thefe
operations.

SECT.

S E C T.   II.

Of Lixiviation.

This is an operation ufed in chemiſtry and manufaétures for feparating fubſtances which are foluble in water from fuch as are infoluble. The large vat or tub, Pl. II. Fig. 12. having a hole D near its bottom, containing a wooden fpiget and foffet or metallic ſtop-cock DE, is generally ufed for this purpofe. A thin ſtratum of ſtraw is placed at the bottom of the tub; over this, the fubſtance to be lixiviated is laid and covered by a cloth, then hot or cold water, according to the degree of folubility of the fa-line matter, is poured on. When the water is fuppofed to have diffolved all the faline parts, it is let off by the ſtop-cock ; and, as fome of the water charged with falt neceffarily adheres to the ſtraw and infoluble matters, feveral freſh quantities of water are poured on. The ſtraw ferves to fecure a proper paffage for the water, and may be compared to the ſtraws or glafs rods ufed in filtrating, to keep the paper from touching the fides of the funnel. The cloth which is laid over the matters under lixiviation prevents the water from making a hollow in

thefe

thefe fubftances where it is poured on, through which it might efcape without acting upon the whole mafs.

This operation is lefs or more imitated in chemical experiments ; but as in thefe, efpecially with analytical views, greater exactnefs is required, particular precautions muft be employed, fo as not to leave any faline or foluble part in the refiduum. More water muft be employed than in ordinary lixiviations, and the fubftances ought to be previoufly ftirred up in the water before the clear liquor is drawn off, otherwife the whole mafs might not be equally lixiviated, and fome parts might even efcape altogether from the action of the water. We muft likewife employ frefh portions of water in confiderable quantity, until it comes off entirely free from falt, which we may afcertain by means of the hydrometer formerly defcribed.

In experiments with fmall quantities, this operation is conveniently performed in jugs or matraffes of glafs, and by filtrating the liquor through paper in a glafs funnel. When the fubftance is in larger quantity, it may be lixiviated in a kettle of boiling water, and filtrated through paper fupported by cloth in the wooden frame, Pl. II. Fig. 3. and 4. ; and in operations in the large way, the tub already mentioned muft be ufed.

S E C T.

S E C T. III.

Of Evaporation.

This operation is ufed for feparating two fubftances from each other, of which one at leaft
muft be fluid, and whofe degrees of volatility
are confiderably different. By this means we
obtain a falt, which has been diffolved in water,
in its concrete form ; the water, by heating, becomes combined with caloric, which renders it
volatile, while the particles of the falt being
brought nearer to each other, and within the
fphere of their mutual attraction, unite into the
folid ftate.

As it was long thought that the air had great
influence upon the quantity of fluid evaporated,
it will be proper to point out the errors which
this opinion has produced. There certainly is
a conftant flow evaporation from fluids expofed
to the free air ; and, though this fpecies of evaporation may be confidered in fome degree as a
folution in air, yet caloric has confiderable influence in producing it, as is evident from the
refrigeration which always accompanies this procefs ; hence we may confider this gradual evaporation as a compound folution made partly in

air

air, and partly in caloric. But the evaporation
which takes place from a fluid kept continually
boiling, is quite different in its nature, and in it
the evaporation produced by the action of the
air is exceedingly inconfiderable in comparifon
with that which is occafioned by caloric. This
latter fpecies may be termed *vaporization* rather
than *evaporation*. This procefs is not accelerated in proportion to the extent of evaporating
furface, but in proportion to the quantities of
caloric which combine with the fluid. Too free
a current of cold air is often hurtful to this procefs, as it tends to carry off caloric from the water, and confequently retards its converfion into
vapour. Hence there is no inconvenience produced by covering, in a certain degree, the veffels in which liquids are evaporated by continual
boiling, provided the covering body be of fuch
a nature as does not ftrongly draw off the caloric, or, to ufe an expreffion of Dr Franklin's,
provided it be a bad conductor of heat. In this
cafe, the vapours efcape through fuch opening
as is left, and at leaft as much is evaporated,
frequently more than when free accefs is allowed to the external air.

As during evaporation the fluid carried off by
caloric is entirely loft, being facrificed for the
fake of the fixed fubftances with which it was
combined, this procefs is only employed where
the fluid is of fmall value, as water, for inftance.

But,

But, when the fluid is of more confequence, we have recourfe to diftillation, in which procefs we preferve both the fixed fubftance and the volatile fluid. The veffels employed for evaporation are bafons or pans of copper, filver, or lead, Pl. II. Fig. 13. and 15. or capfules of glafs, porcellain, or ftone ware, Pl. II. A, Fig. 1. and 2. Pl. III. Fig. 3 and 4. The beft utenfils for this purpofe are made of the bottoms of glafs retorts and matraffes, as their equal thinnefs renders them more fit than any other kind of glafs veffel for bearing a brifk fire and fudden alterations of heat and cold without breaking.

As the method of cutting thefe glafs veffels is no where defcribed in books, I fhall here give a defcription of it, that they may be made by chemifts for themfelves out of fpoiled retorts, matraffes, and recipients, at a much cheaper rate than any which can be procured from glafs manufacturers. The inftrument, Pl. III. Fig. 5. confifting of an iron ring AC, fixed to the rod AB, having a wooden handle D, is employed as follows: Make the ring red hot in the fire, and put it upon the matrafs G, Fig. 6. which is to be cut; when the glafs is fufficiently heated, throw on a little cold water, and it will generally break exactly at the circular line heated by the ring.

Small flafks or phials of thin glafs are exceeding good veffels for evaporating fmall quantities

3 B

of fluid; they are very cheap, and ſtand the fire remarkably. One or more of theſe may be placed upon a ſecond grate above the furnace, Pl. III. Fig. 2. where they will only experience a gentle heat. By this means a great number of experiments may be carried on at one time. A glaſs retort, placed in a ſand bath, and covered with a dome of baked earth, Pl. III. Fig. 1. anſwers pretty well for evaporations; but in this way it is always conſiderably ſlower, and is even liable to accidents; as the ſand heats unequally, and the glaſs cannot dilate in the ſame unequal manner, the retort is very liable to break. Sometimes the ſand ſerves exactly the office of the iron ring formerly mentioned; for, if a ſingle drop of vapour, condenſed into liquid, happens to fall upon the heated part of the veſſel, it breaks circularly at that place. When a very intenſe fire is neceſſary, earthen crucibles may be uſed; but we generally uſe the word *evaporation* to expreſs what is produced by the temperature of boiling water, or not much higher.

S E C T

SECT. IV.

Of Criſtallization.

In this procefs the integrant parts of a folid body, feparated from each other by the inter-vention of a fluid, are made to exert the mutual attraction of aggregation, fo as to coalefce and reproduce a folid mafs. When the particles of a body are only feparated by caloric, and the fubſtance is thereby retained in the liquid ſtate, all that is neceſſary for making it criſtallize, is to remove a part of the caloric which is lodged between its particles, or, in other words, to cool it. If this refrigeration be flow, and the body be at the fame time left at reſt, its particles af-fume a regular arrangement, and criſtallization, properly fo called, takes place ; but, if the re-frigeration is made rapidly, or if the liquor be agitated at the moment of its paſſage to the con-crete ſtate, the criſtallization is irregular and confufed.

The fame phenomena occur with watery folu-tions, or rather in thofe made partly in water, and partly by caloric. So long as there remains a fufficiency of water and caloric to keep the particles of the body afunder beyond the fphere

of

of their mutual attraction, the falt remains in
the fluid ftate ; but, whenever either caloric or
water is not prefent in fufficient quantity, and
the attraction of the particles for each other becomes fuperior to the power which keeps them
afunder, the falt recovers its concrete form, and
the criftals produced are the more regular in
proportion as the evaporation has been flower
and more tranquilly performed.

All the phenomena we formerly mentioned
as taking place during the folution of falts, occur in a contrary fenfe during their criftallization. Caloric is difengaged at the inftant of
their affuming the folid ftate, which furnifhes
an additional proof of falt being held in folution by the compound action of water and caloric. Hence, to caufe falts to criftallize which
readily liquify by means of caloric, it is not fufficient to carry off the water which held them
in folution, but the caloric united to them muft
likewife be removed. Nitrat of potafh, oxygenated muriat of potafh, alum, fulphat of foda,
&c. are examples of this circumftance, as, to
make thefe falts criftallize, refrigeration muft be
added to evaporation. Such falts, on the contrary, as require little caloric for being kept in
folution, and which, from that circumftance,
are nearly equally foluble in cold and warm
water, are criftallizable by fimply carrying off
the water which holds them in folution, and
even

even recover their folid ftate in boiling water; fuch are fulphat of lime, muriat of potaſh and of foda, and feveral others.

The art of refining faltpetre depends upon thefe properties of falts, and upon their different degrees of folubility in hot and cold water. This falt, as produced in the manufactories by the firſt operation, is compofed of many different falts; fome are deliquefcent, and not fufceptible of being criftallized, fuch as the nitrat and muriat of lime; others are almoſt equally foluble in hot and cold water, as the muriats of potaſh and of foda; and, laſtly, the faltpetre, or nitrat of potaſh, is greatly more foluble in hot than it is in cold water. The operation is begun, by pouring upon this mixture of falts as much water as will hold even the leaſt foluble, the muriats of foda and of potaſh, in folution; fo long as it is hot, this quantity readily diſſolves all the faltpetre, but, upon cooling, the greater part of this falt criftallizes, leaving about a fixth part remaining diſſolved, and mixed with the nitrat of lime and the two muriats. The nitre obtained by this procefs is ſtill fomewhat impregnated with other falts, becaufe it has been criftallized from water in which thefe abound: It is completely purified from thefe by a fecond folution in a fmall quantity of boiling water, and fecond criftallization. The water remaining after thefe criftallizations of nitre is ſtill loaded with a mix-

ture

ture of faltpetre, and other falts; by farther evaporation, crude faltpetre, or rough-petre, as the
workmen call it, is procured from it, and this
is purified by two frefh folutions and criftallizations.

The deliquefcent earthy falts which do not
contain the nitric acid are rejected in this manufacture; but thofe which confift of that acid
neutralized by an earthy bafe are diffolved in
water, the earth is precipitated by means of potafh, and allowed to fubfide; the clear liquor is
then decanted, evaporated, and allowed to criftallize. The above management for refining
faltpetre may ferve as a general rule for feparating falts from each other which happen to
be mixed together. The nature of each muft
be confidered, the proportion in which each diffolves in given quantities of water, and the different folubility of each in hot and cold water.
If to thefe we add the property which fome falts
poffefs, of being foluble in alkohol, or in a mixture of alkohol and water, we have many re-
fources for feparating falts from each other by
means of criftallization, though it muft be allowed that it is extremely difficult to render this
feparation perfectly complete.

The veffels ufed for criftallization are pans
of earthen ware, A, Pl. II. Fig. 1. and 2. and
large flat difhes, Pl. III. Fig. 7. When a faline
folution is to be expofed to a flow evaporation

in

in the heat of the atmofphere, with free accefs
of air, veffels of fome depth, Pl. III. Fig. 3.
muft be employed, that there may be a confiderable body of liquid; by this means the criftals produced are of confiderable fize, and remarkably regular in their figure.

Every fpecies of falt criftallizes in a peculiar
form, and even each falt varies in the form of
its criftals according to circumftances, which
take place during criftallization. We muft not
from thence conclude that the faline particles of
each fpecies are indeterminate in their figures:
The primative particles of all bodies, efpecially
of falts, are perfectly conftant in their fpecific
forms; but the criftals which form in our experiments are compofed of congeries of minute
particles, which, though perfectly equal in fize
and fhape, may affume very diffimilar arrangements, and confequently produce a vaft variety
of regular forms, which have not the fmalleft
apparent refemblance to each other, nor to the
original criftal. This fubject has been very ably
treated by the Abbe Haüy, in feveral memoirs
prefented to the Academy, and in his work upon the ftructure of criftals: It is only neceffary
to extend generally to the clafs of falts the principles he has particularly applied to fome crifta-
lized ftones.

SECT.

S E C T.   V.

Of Simple Diftillation.

As diftillation has two diftinct objects to ac-
complish, it is divifible into fimple and compound; and, in this fection, I mean to confine
myfelf entirely to the former. When two bodies, of which one is more volatile than the
other, or has more affinity to caloric, are fubmitted to diftillation, our intention is to feparate them from each other: The more volatile
fubftance affumes the form of gas, and is afterwards condenfed by refrigeration in proper veffels. In this cafe diftillation, like evaporation,
becomes a fpecies of mechanical operation,
which feparates two fubftances from each other
without decompofing or altering the nature of
either. In evaporation, our only object is to
preferve the fixed body, without paying any regard to the volatile matter; whereas, in diftillation, our principal attention is generally paid
to the volatile fubftance, unlefs when we intend
to preferve both the one and the other. Hence,
fimple diftillation is nothing more than evaporation produced in clofe veffels.

The moft fimple diftilling veffel is a fpecies
of bottle or matrafs, A, Pl. III. Fig. 8. which has

been

been bent from its original form BC to BD, and which is then called a retort; when ufed, it is placed either in a reverberatory furnace, Pl. XIII. Fig. 2. or in a fand bath under a dome of baked earth, Pl. III. Fig. 1. To receive and condenfe the products, we adapt a recipient, E, Pl. III. Fig. 9. which is luted to the retort. Sometimes, more efpecially in pharmaceutical operations, the glafs or ftone ware cucurbit, A, with its capital B, Pl. III. Fig. 12. or the glafs alembic and capital, Fig. 13. of one piece, is employed. This latter is managed by means of a tubulated opening T, fitted with a ground ftopper of criftal; the capital, both of the cucurbit and alembic, has a furrow or trench, r r, intended for conveying the condenfed liquor into the beak RS, by which it runs out. As, in almoft all diftillations, expanfive vapours are produced, which might burft the veffels employed, we are under the neceffity of having a fmall hole, T, Fig. 9. in the balloon or recipient, through which thefe may find vent; hence, in this way of diftilling, all the products which are permanently aëriform are entirely loft, and even fuch as difficultly lofe that ftate have not fufficient fpace to condenfe in the balloon: This apparatus is not, therefore, proper for experiments of inveftigation, and can only be admitted in the ordinary operations of the laboratory or in pharmacy. In the article appropriated for com-

3 C   pound

pound diftillation, I fhall explain the various methods which have been contrived for preferving the whole products from bodies in this procefs.

As glafs or earthen veffels are very brittle, and do not readily bear fudden alterations of heat and cold, every well regulated laboratory ought to have one or more alembics of metal for diftilling water, fpiritous liquors, effential oils, &c. This apparatus confifts of a cucurbit and capital of tinned copper or brafs, Pl. III. Fig. 15. and 16. which, when judged proper, may be placed in the water bath, D, Fig. 17. In diftillations, efpecially of fpiritous liquors, the capital muft be furnifhed with a refrigetory, SS, Fig. 16. kept continually filled with cold water; when the water becomes heated, it is let off by the ftop-cock, R, and renewed with a frefh fupply of cold water. As the fluid diftilled is converted into gas by means of caloric furnifhed by the fire of the furnace, it is evident that it could not condenfe, and, confequently, that no diftillation, properly fpeaking, could take place, unlefs it is made to depofit in the capital all the caloric it received in the cucurbit; with this view, the fides of the capital muft always be preferved at a lower temperature than is neceffary for keeping the diftilling fubftance in the ftate of gas, and the water in the refrigetory is intended for this purpofe.

Water

Water is converted into gas by the temperature
of 80° (212°), alkohol by 67° (182.75°), ether
by 32° (104°); hence thefe fubftances cannot
be diftilled, or, rather, they will fly off in the
ftate of gas, unlefs the temperature of the refrigetory be kept under thefe refpective degrees.

In the diftillation of fpiritous, and other expanfive liquors, the above defcribed refrigetory
is not fufficient for condenfing all the vapours
which arife; in this cafe, therefore, inftead of
receiving the diftilled liquor immediately from
the beak, TU, of the capital into a recipient,
a worm is interpofed between them. This inftrument is reprefented Pl. III. Fig. 18. contained in a worm tub of tinned copper, it confifts
of a metallic tube bent into a confiderable number of fpiral revolutions. The veffel which contains the worm is kept full of cold water, which
is renewed as it grows warm. This contrivance
is employed in all diftilleries of fpirits, without
the intervention of a capital and refrigetory,
properly fo called. The one reprefented in the
plate is furnifhed with two worms, one of them
being particularly appropriated to diftillations
of odoriferous fubftances.

In fome fimple diftillations it is neceffary to
interpofe an adopter between the retort and receiver, as fhown Pl. III. Fig. 11. This may

serve

ferve two different purpofes, either to feparate
two products of different degrees of volatility,
or to remove the receiver to a greater diftance
from the furnace, that it may be lefs heated.
But thefe, and feveral other more complicated
inftruments of ancient contrivance, are far from
producing the accuracy requifite in modern
chemiftry, as will be readily perceived when I
come to treat of compound diftillation.

S E C T.   VI.

Of Sublimation.

This term is applied to the diftillation of fubftances which condenfe in a concrete or folid
form, fuch as the fublimation of fulphur, and of
muriat of ammoniac, or fal ammoniac. Thefe
operations may be conveniently performed in
the ordinary diftilling veffels already defcribed,
though, in the fublimation of fulphur, a fpecies of veffels, named Alludels, have been ufually employed. Thefe are veffels of ftone or
porcelain ware, which adjuft to each other over
a cucurbit containing the fulphur to be fublim-
ed.   One of the beft fubliming veffels, for fubftances which are not very volatile, is a flafk,

or

or phial of glafs, funk about two thirds into a fand bath; but in this way we are apt to lofe a part of the products. When thefe are wifhed to be entirely preferved, we muft have recourfe to the pneumato-chemical diftilling apparatus, to be defcribed in the following chapter.

CHAP.

C H A P. VI.

Of Pneumato-chemical Diſtillations, Metallic Diſſo-
lutions, and ſome other operations which require
very complicated inſtruments.

S E C T. I.

Of Compound and Pneumato-chemical Diſtillations.

IN the preceding chapter, I have only treated
of diſtillation as a ſimple operation, by which
two ſubſtances, differing in degrees of volatility,
may be ſeparated from each other ; but diſtillation often actually decompoſes the ſubſtances
ſubmitted to its action, and becomes one of the
moſt complicated operations in chemiſtry. In
every diſtillation, the ſubſtance diſtilled muſt be
brought to the ſtate of gas, in the cucurbit or
retort, by combination with caloric : In ſimple
diſtillation, this caloric is given out in the re-
frigeratory or in the worm, and the ſubſtance
again recovers its liquid or ſolid form, but the
ſubſtances ſubmitted to compound diſtillation

are

are abfolutely decompounded ; one part, as for
inftance the charcoal they contain, remains fixed in the retort, and all the reft of the elements
are reduced to gaffes of different kinds. Some
of thefe are fufceptible of being condenfed, and
of recovering their folid or liquid forms, whilft
others are permanently aëriform ; one part of
thefe are abforbable by water, fome by the alkalies, and others are not fufceptible of being
abforbed at all. An ordinary diftilling apparatus, fuch as has been defcribed in the preceding
chapter, is quite infufficient for retaining or for
feparating thefe diverfified products, and we are
obliged to have recourfe, for this purpofe, to
methods of a more complicated nature.

The apparatus I am about to defcribe is calculated for the moft complicated diftillations,
and may be fimplified according to circumftances. It confifts of a tubulated glafs retort A,
Pl. IV. Fig. 1. having its beak fitted to a tubulated balloon or recipient BC ; to the upper orifice D of the balloon a bent tube DE*fg* is adjufted, which, at its other extremity *g*, is plunged into the liquor contained in the bottle L,
with three necks *xxx*. Three other fimilar
bottles are connected with this firft one, by
means of three fimilar bent tubes difpofed in
the fame manner ; and the fartheft neck of the
laft bottle is connected with a jar in a pneumato-chemical apparatus, by means of a bent
tube.

tube *. A determinate weight of diftilled water is ufually put into the firft bottle, and the other three have each a folution of cauftic potafh in water. The weight of all thefe bottles, and of the water and alkaline folution they contain, muft be accurately afcertained. Every thing being thus difpofed, the junctures between the retort and recipient, and of the tube D of the latter, muft be luted with fat lute, covered over with flips of linen, fpread with lime and white of egg; all the other junctures are to be fecured by a lute made of wax and rofin melted together.

When all thefe difpofitions are completed, and when, by means of heat applied to the retort A, the fubftance it contains becomes decompofed, it is evident that the leaft volatile products muft condenfe or fublime in the beak or neck of the retort itfelf, where moft of the concrete fubftances will fix themfelves. The more volatile fubftances, as the lighter oils, ammoniac, and feveral others, will condenfe in the recipient GC, whilft the gaffes, which are not fufceptible of condenfation by cold, will pafs on by the tubes, and boil up through the liquors in the feveral bottles. Such as are abforbable

by

* The reprefentation of this apparatus, Pl. IV. Fig. 1. will convey a much better idea of its difpofition than can poffibly be given by the moft laboured defcription.—E.

by water will remain in the firſt bottle, and thoſe which cauſtic alkali can abſorb will remain in the others; whilſt ſuch gaſſes as are not ſuſceptible of abſorption, either by water or alkalies, will eſcape by the tube RM, at the end of which they may be received into jars in a pneumato-chemical apparatus. The charcoal and fixed earth, &c. which form the ſubſtance or reſiduum, anciently called *caput mortuum*, remain behind in the retort.

In this manner of operating, we have always a very material proof of the accuracy of the analyſis, as the whole weights of the products taken together, after the proceſs is finiſhed, muſt be exactly equal to the weight of the original ſubſtance ſubmitted to diſtillation. Hence, for inſtance, if we have operated upon eight ounces of ſtarch or gum arabic, the weight of the charry reſiduum in the retort, together with that of all the products gathered in its neck and the balloon, and of all the gas received into the jars by the tube RM added to the additional weight acquired by the bottles, muſt, when taken together, be exactly eight ounces. If the product be leſs or more, it proceeds from error, and the experiment muſt be repeated until a ſatisfactory reſult be procured, which ought not to differ more than ſix or eight grains in the pound from the weight of the ſubſtance ſubmitted to experiment.

3 D   In

In experiments of this kind, I for a long time met with an almoſt inſurmountable difficulty, which muſt at laſt have obliged me to defiſt altogether, but for a very ſimple method of avoiding it, pointed out to me by Mr Haſſenfratz. The ſmalleſt diminution in the heat of the furnace, and many other circumſtances inſeparable from this kind of experiments, cauſe frequent reabſorptions of gas ; the water in the ciſtern of the pneumato chemical apparatus ruſhes into the laſt bottle through the tube RM, the ſame circumſtance happens from one bottle into another, and the fluid is often forced even into the recipient C. This accident is prevented by uſing bottles having three necks, as repreſented in the plate, into one of which, in each bottle, a capillary glaſs-tube St, st, st, st, is adapted, ſo as to have its lower extremity t immerſed in the liquor. If any abſorption takes place, either in the retort, or in any of the bottles, a ſufficient quantity of external air enters, by means of theſe tubes, to fill up the void ; and we get rid of the inconvenience at the price of having a ſmall mixture of common air with the products of the experiment, which is thereby prevented from failing altogether. Though theſe tubes admit the external air, they cannot permit any of the gaſſeous ſubſtances to eſcape, as they are always ſhut below by the water of the bottles.

It

It is evident that, in the courfe of experiments with this apparatus, the liquor of the bottles muft rife in thefe tubes in proportion to the
preffure fuftained by the gas or air contained in
the bottles ; and this preffure is determined by
the height and gravity of the column of fluid
contained in all the fubfequent bottles. If we
fuppofe that each bottle contains three inches
of fluid, and that there are three inches of water in the ciftern of the connected apparatus a-
bove the orifice of the tube RM, and allowing
the gravity of the fluids to be only equal to that
of water, it follows that the air in the firft bottle muft fuftain a preffure equal to twelve inches of water; the water muft therefore rife
twelve inches in the tube S, connected with the
firft bottle, nine inches in that belonging to the
fecond, fix inches in the third, and three in the
laft ; wherefore thefe tubes muft be made
fomewhat more than twelve, nine, fix, and three
inches long refpectively, allowance being made
for ofcillatory motions, which often take place
in the liquids. It is fometimes neceffary to introduce a fimilar tube between the retort and
recipient ; and, as the tube is not immerfed in
fluid at its lower extremity, until fome has collected in the progrefs of the diftillation, its upper end muft be fhut at firft with a little lute, fo
as to be opened according to neceffity, or after
                                                    there

there is fufficient liquid in the recipient to fecure its lower extremity.

This apparatus cannot be ufed in very accurate experiments, when the fubftances intended
to be operated upon have a very rapid action
upon each other, or when one of them can
only be introduced in fmall fucceffive portions,
as in fuch as produce violent effervefcence when
mixed together. In fuch cafes, we employ a
tubulated retort A, Pl. VII. Fig. 1. into which
one of the fubftances is introduced, preferring
always the folid body, if any fuch is to be treated, we then lute to the opening of the retort a
bent tube BCDA, terminating at its upper extremity B in a funnel, and at its other end A in
a capillary opening. The fluid material of the
experiment is poured into the retort by means
of this funnel, which muft be made of fuch a
length, from B to C, that the column of liquid
introduced may counterbalance the refiftance
produced by the liquors contained in all the
bottles, Pl. IV. Fig. 1.

Thofe who have not been accuftomed to ufe
the above defcribed diftilling apparatus may
perhaps be ftartled at the great number of openings which require luting, and the time neceffary for making all the previous preparations in
experiments of this kind. It is very true that,
if we take into account all the neceffary weigh-
ings of materials and products, both before and

after

after the experiments, thefe preparatory and fucceeding fteps require much more time and attention than the experiment itfelf. But, when the experiment fucceeds properly, we are well rewarded for all the time and trouble beftowed, as by one procefs carried on in this accurate manner much more juft and extenfive knowledge is acquired of the nature of the vegetable or animal fubftance thus fubmitted to inveftigation, than by many weeks affiduous labour in the ordinary method of proceeding.

When in want of bottles with three orifices, thofe with two may be ufed ; it is even poffible to introduce all the three tubes at one opening, fo as to employ ordinary wide-mouthed bottles, provided the opening be fufficiently large. In this cafe we muft carefully fit the bottles with corks very accurately cut, and boiled in a mixture of oil, wax, and turpentine. Thefe corks are pierced with the neceffary holes for receiving the tubes by means of a round file, as in Pl. IV. Fig. 8.

SECT.

S E C T.   II.

Of Metallic Diffolutions.

I have already pointed out the difference between folution of falts in water and metallic
diffolutions. The former requires no particular
veffels, whereas the latter requires very complicated veffels of late invention, that we may not
lofe any of the products of the experiment, and
may thereby procure truly conclufive refults of
the phenomena which occur. The metals, in
general, diffolve in acids with effervefcence,
which is only a motion excited in the folvent by
the difengagement of a great number of bubbles
of air or aëriform fluid, which proceed from the
furface of the metal, and break at the furface of
the liquid.

Mr Cavendifh and Dr Prieftley were the firft
inventors of a proper apparatus for collecting
thefe elaftic fluids. That of Dr Prieftley is extremely fimple, and confifts of a bottle A,
Pl. VII. Fig. 2. with its cork B, through which
paffes the bent glafs tube BC, which is engaged
under a jar filled with water in the pneumato-
chemical apparatus, or fimply in a bafon full of
water. The metal is firft introduced into the
bottle,

bottle, the acid is then poured over it, and the bottle is inftantly clofed with its cork and tube, as reprefented in the plate. But this apparatus has its inconveniencies. When the acid is much concentrated, or the metal much divided, the effervefcence begins before we have time to cork the bottle properly, and fome gas efcapes, by which we are prevented from afcertaining the quantity difengaged with rigorous exactnefs. In the next place, when we are obliged to employ heat, or when heat is produced by the procefs, a part of the acid diftills, and mixes with the water of the pneumato-chemical apparatus, by which means we are deceived in our calculation of the quantity of acid decompofed. Befides thefe, the water in the ciftern of the apparatus abforbs all the gas produced which is fufceptible of abforption, and renders it impoffible to collect thefe without lofs.

To remedy thefe inconveniencies, I at firft ufed a bottle with two necks, Pl. VII. Fig. 3. into one of which the glafs funnel BC is luted fo as to prevent any air efcaping ; a glafs rod DE is fitted with emery to the funnel, fo as to ferve the purpofe of a ftopper. When it is ufed, the matter to be diffolved is firft introduced into the bottle, and the acid is then permitted to pafs in as flowly as we pleafe, by raifing the glafs rod gently as often as is neceffary until faturation is produced.

Another

Another method has been fince employed, which ferves the fame purpofe, and is preferable to the laft defcribed in fome inftances. This confifts in adapting to one of the mouths of the bottle A, Pl. VII. Fig. 4. a bent tube DEFG, having a capillary opening at D, and ending in a funnel at G. This tube is fecurely luted to the mouth C of the bottle. When any liquid is poured into the funnel, it falls down to F; and, if a fufficient quantity be added, it paffes by the curvature E, and falls flowly into the bottle, fo long as frefh liquor is fupplied at the funnel. The liquor can never be forced out of the tube, and no gas can efcape through it, becaufe the weight of the liquid ferves the purpofe of an accurate cork.

To prevent any diftillation of acid, efpecially in diffolutions accompanied with heat, this tube is adapted to the retort A, Pl. VII. Fig. 1. and a fmall tubulated recipient, M, is applied, in which any liquor which may diftill is condenfed. On purpofe to feparate any gas that is abforbable by water, we add the double necked bottle L, half filled with a folution of cauftic potafh; the alkali abforbs any carbonic acid gas, and ufually only one or two other gaffes pafs into the jar of the connected pneumato-chemical apparatus through the tube NO. In the firft chapter of this third part we have directed how thefe are to be feparated and examined.

If

If one bottle of alkaline folution be not thought fufficient, two, three, or more, may be added.

S E C T.   III.

Apparatus neceffary in Experiments upon Vinous and Putrefactive Fermentations.

For thefe operations a peculiar apparatus, efpecially intended for this kind of experiment, is requifite. The one I am about to defcribe is finally adopted, as the beft calculated for the purpofe, after numerous corrections and improvements. It confifts of a large matrafs, **A,** Pl. X. fig. 1. holding about twelve pints, with a cap of brafs *a b*, ftrongly cemented to its mouth, and into which is fcrewed a bent tube *c d*, furnifhed with a ftop-cock *e*. To this tube is joined the glafs recipient B, having three openings, one of which communicates with the bottle C, placed below it. To the pofterior opening of this recipient is fitted a glafs tube *g h i*, cemented at *g* and *i* to collets of brafs, and intended to contain a very deliquefcent concrete neutral falt, fuch as nitrat or muriat of lime, acetite of potafh, &c. This tube communicates with two bottles D and E, filled to *x* and *y* with a folution of cauftic potafh.

3 E   All

All the parts of this machine are joined together by accurate ſcrews, and the touching parts have greaſed leather interpoſed, to prevent any paſſage of air. Each piece is likewiſe furniſhed with two ſtop-cocks, by which its two extremities may be cloſed, ſo that we can weigh each ſeparately at any period of the operation.

The fermentable matter, ſuch as ſugar, with a proper quantity of yeaſt, and diluted with water, is put into the matraſs. Sometimes, when the fermentation is too rapid, a conſiderable quantity of froth is produced, which not only fills the neck of the matraſs, but paſſes into the recipient, and from thence runs down into the bottle C. On purpoſe to collect this ſcum and muſt, and to prevent it from reaching the tube filled with deliqueſcent ſalts, the recipient and connected bottle are made of conſiderable capacity.

In the vinous fermentation, only carbonic acid gas is diſengaged, carrying with it a ſmall proportion of water in ſolution. A great part of this water is depoſited in paſſing through the tube *g h i*, which is filled with a deliqueſcent ſalt in groſs powder, and the quantity is aſcertained by the augmentation of the weight of the ſalt. The carbonic acid gas bubbles up through the alkaline ſolution in the bottle D, to which it is conveyed by the tube *k l m*. Any ſmall portion which may not be abſorbed by this

firſt

firſt bottle is ſecured by the ſolution in the ſecond bottle E, ſo that nothing, in general,
paſſes into the jar F, except the common air
contained in the veſſels at the commencement
of the experiment.

The ſame apparatus anſwers extremely well
for experiments upon the putrefactive fermentation ; but, in this caſe, a conſiderable quantity of hydrogen gas is diſengaged through the
tube *q r s t u*, by which it is conveyed into the
jar F ; and, as this diſengagement is very rapid,
eſpecially in ſummer, the jar muſt be frequently
changed. Theſe putrefactive fermentations require conſtant attendance from the above circumſtance, whereas the vinous fermentation
hardly needs any. By means of this apparatus
we can aſcertain, with great preciſion, the
weights of the ſubſtances ſubmitted to fermentation, and of the liquid and aëriform products
which are diſengaged. What has been already
ſaid in Part I. Chap. XIII. upon the products
of the vinous fermentation, may be conſulted.

S E C T.

S E C T. IV.

Apparatus for the Decompofition of Water.

Having already given an account, in the firft part of this work, of the experiments relative to the decompofition of water, 1 fhall avoid any unneceffary repetitions, and only give a few fummary obfervations upon the fubject in this fection. The principal fubftances which have the power of decompofing water are iron and charcoal; for which purpofe, they require to be made red hot, otherwife the water is only reduced into vapours, and condenfes afterwards by refrigeration, without fuftaining the fmalleft alteration. In a red heat, on the contrary, iron or charcoal carry off the oxygen from its union with hydrogen; in the firft cafe, black oxyd of iron is produced, and the hydrogen is difengaged pure in form of gas; in the other cafe, carbonic acid gas is formed, which difengages, mixed with the hydrogen gas; and this latter is commonly carbonated, or holds charcoal in folution.

A mufket barrel, without its breach pin, anfwers exceedingly well for the decompofition of water, by means of iron, and one fhould be

chofen

chofen of confiderable length, and pretty ftrong. When too fhort, fo as to run the rifk of heating the lute too much, a tube of copper is to be ftrongly foldered to one end. The barrel is placed in a long furnace, CDEF, Pl. VII. Fig. 11. fo as to have a few degrees of inclination from E to F; a glafs retort A, is luted to the upper extremity E, which contains water, and is placed upon the furnace VVXX. The lower extremity F is luted to a worm SS, which is connected with the tubulated bottle H, in which any water diftilled without decompofition, during the operation, collects, and the difengaged gas is carried by the tube KK to jars in a pneumato-chemical apparatus. Inftead of the retort a funnel may be employed, having its lower part fhut by a ftop-cock, through which the water is allowed to drop gradually into the gun-barrel. Immediately upon getting into contact with the heated part of the iron, the water is converted into fteam, and the experiment proceeds in the fame manner as if it were furnifhed in vapours from the retort.

In the experiment made by Mr Meufnier and me before a committee of the Academy, we ufed every precaution to obtain the greateft poffible precifion in the refult of our experiment, having even exhaufted all the veffels employed before we began, fo that the hydrogen gas obtained might be free from any mixture
of

of azotic gas. The refults of that experiment will hereafter be given at large in a particular memoir.

In numerous experiments, we are obliged to ufe tubes of glafs, porcelain, or copper, inftead of gun-barrels ; but glafs has the difadvantage of being eafily melted and flattened, if the heat be in the fmalleft degree raifed too high ; and porcelain is moftly full of fmall minute pores, through which the gas efcapes, efpecially when compreffed by a column of water. For thefe reafons I procured a tube of brafs, which Mr de la Briche got caft and bored out of the folid for me at Strafburg, under his own infpection. This tube is extremely convenient for decompofing alkohol, which refolves into charcoal, carbonic acid gas, and hydrogen gas ; it may likewife be ufed with the fame advantage for decompofing water by means of charcoal, and in a great number of experiments of this nature.

C H A P.

C H A P.   VII.

Of the Compofition and Application of Lutes.

THE neceffity of properly fecuring the
junctures of chemical veffels to prevent
the efcape of any of the products of experiments,
muft be fufficiently apparent ; for this purpofe
lutes are employed, which ought to be of fuch a
nature as to be equally impenetrable to the moft
fubtile fubftances, as glafs itfelf, through which
only caloric can efcape.

This firft object of lutes is very well accomplifhed by bees wax, melted with about an
eighth part of turpentine. This lute is very
eafily managed, fticks very clofely to glafs, and
is very difficultly penetrable; it may be rendered
more confiftent, and lefs or more hard or pliable,
by adding different kinds of refinous matters.
Though this fpecies of lute anfwers extremely
well for retaining gaffes and vapours, there are
many chemical experiments which produce confiderable heat, by which this lute becomes liquified, and confequently the expanfive vapours
muft very readily force through and efcape.

For

For fuch cafes, the following fat lute is the
beft hitherto difcovered, though not without its
difadvantages, which fhall be pointed out.
Take very pure and dry unbaked clay, reduced
to a very fine powder, put this into a brafs mortar, and beat it for feveral hours with a heavy
iron peftle, dropping in flowly fome boiled lintfeed oil; this is oil which has been oxygenated,
and has acquired a drying quality, by being
boiled with litharge. This lute is more tena-
cious, and applies better, if amber varnifh be u-
fed inftead of the above oil. To make this varnifh, melt fome yellow amber in an iron laddle,
by which operation it lofes a part of its fuccinic
acid, and effential oil, and mix it with lintfeed
oil. Though the lute prepared with this varnifh is better than that made with boiled oil,
yet, as its additional expence is hardly compenfated by its fuperior quality, it is feldom
ufed.

The above fat lute is capable of fuftaining a
very violent degree of heat, is impenetrable by
acids and fpiritous liquors, and adheres exceedingly well to metals, ftone ware, or glafs, providing they have been previoufly rendered perfectly dry. But if, unfortunately, any of the
liquor in the courfe of an experiment gets
through, either between the glafs and the lute,
or between the layers of the lute itfelf, fo as to
moiften the part, it is extremely difficult to clofe
the

the opening. This is the chief inconvenience which attends the ufe of fat lute, and perhaps the only one it is fubject to. As it is apt to foften by heat, we muft furround all the junctures with flips of wet bladder applied over the luting, and fixed on by pack-thread tied round both above and below the joint ; the bladder, and confequently the lute below, muft be farther fecured by a number of turns of pack-thread all over it. By thefe precautions, we are free from every danger of accident; and the junctures fecured in this manner may be confidered, in experiments, as hermetically fealed.

It frequently happens that the figure of the junctures prevents the application of ligatures, which is the cafe with the three-necked bottles formerly defcribed ; and it even requires great addrefs to apply the twine without fhaking the apparatus ; fo that, where a number of junctures require luting, we are apt to difplace feveral while fecuring one. In thefe cafes, we may fubftitute flips of linen, fpread with white of egg and lime mixed together, inftead of the wet bladder. Thefe are applied while ftill moift, and very fpeedily dry and acquire confiderable hardnefs. Strong glue diffolved in water may anfwer inftead of white of egg. Thefe fillets are ufefully applied likewife over junctures luted together with wax and rofin.

3 F   Before

Before applying a lute, all the junctures of the veſſels muſt be accurately and firmly fitted to each other, ſo as not to admit of being moved. If the beak of a retort is to be luted to the neck of a recipient, they ought to fit pretty accurately ; otherwiſe we muſt fix them, by introducing ſhort pieces of ſoft wood or of cork. If the diſproportion between the two be very conſiderable, we muſt employ a cork which fits the neck of the recipient, having a circular hole of proper dimenſions to admit the beak of the retort. The ſame precaution is neceſſary in adapting bent tubes to the necks of bottles in the apparatus repreſented Pl. IV. Fig. 1. and others of a ſimilar nature. Each mouth of each bottle muſt be fitted with a cork, having a hole made with a round file of a proper ſize for containing the tube. And, when one mouth is intended to admit two or more tubes, which frequently happens when we have not a ſufficient number of bottles with two or three necks, we muſt uſe a cork with two or three holes, Pl. IV. Fig. 8.

When the whole apparatus is thus ſolidly joined, ſo that no part can play upon another, we begin to lute. The lute is ſoftened by kneading and rolling it between the fingers, with the aſſiſtance of heat, if neceſſary. It is rolled into little cylindrical pieces, and applied to the junctures, taking great care to make it

apply

apply clofe, and adhere firmly, in every part ; a
fecond roll is applied over the firft, fo as to pafs
it on each fide, and fo on till each juncture be
fufficiently covered ; after this, the flips of bladder, or of linen, as above directed, muft be
carefully applied over all. Though this operation may appear extremely fimple, yet it requires
peculiar delicacy and management ; great care
muft be taken not to difturb one juncture whilft
luting another, and more efpecially when applying the fillets and ligatures.

Before beginning any experiment, the clofe-
nefs of the luting ought always to be previoufly
tried, either by flightly heating the retort A,
Pl. IV. Fig. 1, or by blowing in a little air by
fome of the perpendicular tubes $S s s s$; the alteration of preffure caufes a change in the level
of the liquid in thefe tubes. If the apparatus
be accurately luted, this alteration of level will
be permanent ; whereas, if there be the fmalleft
opening in any of the junctures, the liquid will
very foon recover its former level. It muft always be remembered, that the whole fuccefs of
experiments in modern chemiftry depends upon
the exactnefs of this operation, which therefore
requires the utmoft patience, and moft attentive
accuracy.

It would be of infinite fervice to enable chemifts, efpecially thofe who are engaged in pneumatic procefles, to difpenfe with the ufe of lutes,

or

or at leaſt to diminiſh the number neceſſary in
complicated inſtruments. I once thought of
having my apparatus conſtructed ſo as to unite
in all its parts by fitting with emery, in the way
of bottles with criſtal ſtoppers ; but the execu-
tion of this plan was extremely difficult. I have
ſince thought it preferable to ſubſtitute columns
of a few lines of mercury in place of lutes,
and have got an apparatus conſtructed upon
this principle, which appears capable of very
convenient application in a great number of
circumſtances.

It conſiſts of a double necked bottle A, Pl.
XII. Fig. 12. ; the interior neck *bc* communicates with the inſide of the bottle, and the ex-
terior neck or rim *de* leaves an interval between
the two necks, forming a deep gutter intended
to contain the mercury. The cap or lid of
glaſs B enters this gutter, and is properly fitted
to it, having notches in its lower edge for the
paſſage of the tubes which convey the gas.
Theſe tubes, inſtead of entering directly into
the bottles as in the ordinary apparatus, have a
double bend for making them enter the gutter,
as repreſented in Fig. 13. and for making them
fit the notches of the cap B ; they riſe again
from the gutter to enter the inſide of the bottle
over the border of the inner mouth. When the
tubes are diſpoſed in their proper places, and
the cap firmly fitted on, the gutter is filled with
mer-

mercury, by which means the bottle is completely excluded from any communication, excepting through the tubes. This apparatus may
be very convenient in many operations in which
the fubftances employed have no action upon
Mercury. Pl. XII. Fig. 14. reprefents an apparatus upon this principle properly fitted together.

Mr Seguin, to whofe active and intelligent
affiftance I have been very frequently much indebted, has befpoken for me, at the glafs-hou-
fes, fome retorts hermetically united to their
recipients, by which luting will be altogether
unneceffary.

C H A P.

C H A P. VIII.

Of Operations upon Combuſtion and Deflagration.

S E C T. I.

Of Combuſtion in general.

COMBUSTION, according to what has been already ſaid in the Firſt Part of this Work, is the decompoſition of oxygen gas produced by a combuſtible body. The oxygen which forms the baſe of this gas is abſorbed by, and enters into, combination with the burning body, while the caloric and light are ſet free. Every combuſtion, therefore, neceſſarily ſuppoſes oxygenation; whereas, on the contrary, every oxygenation does not neceſſarily imply concomitant combuſtion; becauſe combuſtion, properly ſo called, cannot take place without diſengagement of caloric and light. Before combuſtion can take place, it is neceſſary that the baſe of oxygen gas ſhould have greater affinity to the combuſtible body than it has to caloric;

loric ; and this elective attraction, to ufe Bergman's expreffion, can only take place at a certain degree of temperature, which is different
for each combuftible fubftance ; hence the neceffity of giving a firft motion or beginning to
every combuftion by the approach of a heated
body. This neceffity of heating any body we
mean to burn depends upon certain confiderations, which have not hitherto been attended to
by any natural philofopher, for which reafon I
fhall enlarge a little upon the fubject in this
place.

Nature is at prefent in a ftate of equilibrium,
which cannot have been attained until all the
fpontaneous combuftions or oxygenations poffible in the ordinary degrees of temperature had
taken place. Hence, no new combuftions or
oxygenations can happen without deftroying
this equilibrium, and raifing the combuftible
fubftances to a fuperior degree of temperature.
To illuftrate this abftract view of the matter by
example : Let us fuppofe the ufual temperature
of the earth a little changed, and that it is raifed
only to the degree of boiling water ; it is evident, that, in this cafe, phofphorus, which is
combuftible in a confiderably lower degree of
temperature, would no longer exift in nature in
its pure and fimple ftate, but would always be
procured in its acid or oxygenated ftate, and its
radical would become one of the fubftances unknown

known to chemiftry. By gradually increafing
the temperature of the earth the fame circumftance would fucceffively happen to all the bodies capable of combuftion; and, at laft, every
poffible combuftion having taken place, there
would no longer exift any combuftible body
whatever, as every fubftance fufceptible of that
operation would be oxygenated, and confequently incombuftible.

There cannot therefore exift, fo far as relates
to us, any combuftible body, except fuch as are
incombuftible in the ordinary temperatures of
the earth; or, what is the fame thing, in other
words, that it is effential to the nature of every
combuftible body not to poffefs the property of
combuftion, unlefs heated, or raifed to the degree
of temperature at which its combuftion naturally takes place. When this degree is once produced, combuftion commences, and the caloric
which is difengaged by the decompofition of
the oxygen gas keeps up the temperature neceffary for continuing combuftion. When this
is not the cafe, that is, when the difengaged caloric is infufficient for keeping up the neceffary
temperature, the combuftion ceafes: This circumftance is expreffed in common language
by faying, that a body burns ill, or with difficulty.

Although combuftion poffeffes fome circumftances in common with diftillation, efpecially
with

with the compound kind of that operation, they differ in a very material point. In diſtillation there is a ſeparation of one part of the elements of the ſubſtance from each other, and a combination of theſe, in a new order, occaſioned by the affinities which take place in the increaſed temperature produced during diſtillation : This likewiſe happens in combuſtion, but with this farther circumſtance, that a new element, not originally in the body, is brought into action; oxygen is added to the ſubſtance ſubmitted to the operation, and caloric is diſengaged.

The neceſſity of employing oxygen in the ſtate of gas in all experiments with combuſtion, and the rigorous determination of the quantities employed, render this kind of operations peculiarly troubleſome. As almoſt all the products of combuſtion are diſengaged in the ſtate of gas, it is ſtill more difficult to retain them than even thoſe furniſhed during compound diſtillation; hence this precaution was entirely neglected by the ancient chemiſts; and this ſet of experiments excluſively belong to modern chemiſtry.

Having thus pointed out, in a general way, the objects to be had in view in experiments upon combuſtion, I proceed, in the following ſections of this chapter, to deſcribe the different inſtruments I have uſed with this view. The following arrangement is formed, not upon the

3 G nature

nature of the combuftible bodies, but upon that
of the inftruments neceffary for combuftion.

S E C T. II.

Of the Combuftion of Phofphorus.

In thefe combuftions we begin by filling a jar,
capable at leaft of holding fix pints, with oxygen gas in the water apparatus, Pl. V. Fig. 1.;
when it is perfectly full, fo that the gas begins
to flow out below, the jar, A, is carried to the
mercury apparatus, Pl. IV. Fig. 3. We then
dry the furface of the mercury, both within and
without the jar, by means of blotting-paper, taking care to keep the paper for fome time entirely immerfed in the mercury before it is introduced under the jar, left we let in any common air, which fticks very obftinately to the
furface of the paper. The body to be fubmitted to combuftion, being firft very accurately
weighed in nice fcales, is placed in a fmall flat
fhallow difh, D, of iron or porcelain; this is
covered by the larger cup P, which ferves the
office of a diving bell, and the whole is paffed
through the mercury into the jar, after which
the larger cup is retired. The difficulty of paffing the materials of combuftion in this manner
through

through the mercury may be avoided by raiſing one of the ſides of the jar, A, for a moment, and ſlipping in the little cup, D, with the com-buſtible body as quickly as poſſible. In this manner of operating, a ſmall quantity of com-mon air gets into the jar, but it is ſo very in-conſiderable as not to injure either the progreſs or accuracy of the experiment in any ſenſible degree.

When the cup, D, is introduced under the jar, we ſuck out a part of the oxygen gas, ſo as to raiſe the mercury to EF, as formerly directed, Part I. Chap. V. otherwiſe, when the combuſ-tible body is ſet on fire, the gas becoming di-lated would be in part forced out, and we ſhould no longer be able to make any accurate calcu-lation of the quantities before and after the ex-periment. A very convenient mode of draw-ing out the air is by means of an air-pump ſy-ringe adapted to the ſyphon, GHI, by which the mercury may be raiſed to any degree under twenty-eight inches. Very inflammable bodies, as phoſphorus, are ſet on fire by means of the crooked iron wire, MN, Pl. IV. Fig. 16. made red hot, and paſſed quickly through the mercu-ry. Such as are leſs eaſily ſet on fire have a ſmall portion of tinder, upon which a minute particle of phoſphorus is fixed, laid upon them before uſing the red hot iron.

In

In the firft moment of combuftion the air, being heated, rarifies, and the mercury defcends; but when, as in combuftions of phofphorus and iron, no elaftic fluid is formed, abforption becomes prefently very fenfible, and the mercury rifes high into the jar. Great attention muft be ufed not to burn too large a quantity of any fubftance in a given quantity of gas, otherwife, towards the end of the experiment, the cup would approach fo near the top of the jar as to endanger breaking it by the great heat produced, and the fudden refrigeration from the cold mercury. For the methods of meafuring the volume of the gaffes, and for correcting the meafures according to the heighth of the barometer and thermometer, &c. fee Chap. II. Sect. V. and VI. of this part.

The above procefs anfwers very well for burning all the concrete fubftances, and even for the fixed oils: Thefe laft are burnt in lamps under the jar, and are readily fet on fire by means of tinder, phofphorus, and hot iron. But it is dangerous for fubftances fufceptible of evaporating in a moderate heat, fuch as ether, alkohol, and the effential oils; thefe fubftances diffolve in confiderable quantity in oxygen gas; and, when fet on fire, a dangerous and fudden explofion takes place, which carries up the jar to a great height, and dafhes it in a thoufand pieces. From two fuch explofions fome of the
members

members of the Academy and myfelf efcaped
very narrowly. Befides, though this manner of
operating is fufficient for determining pretty accurately the quantity of oxygen gas abforbed,
and of carbonic acid produced, as water is likewife formed in all experiments upon vegetable
and animal matters which contain an excefs of
hydrogen, this apparatus can neither collect it
nor determine its quantity. The experiment
with phofphorus is even incomplete in this way,
as it is impoffible to demonftrate that the weight
of the phofphoric acid produced is equal to the
fum of the weights of the phofphorus burnt and
oxygen gas abforbed during the procefs. I have
been therefore obliged to vary the inftruments
according to circumftances, and to employ feveral of different kinds, which I fhall defcribe
in their order, beginning with that ufed for
burning phofphorus.

Take a large balloon, A, Pl. IV. Fig. 4. of
criftal or white glafs, with an opening, EF, about
two inches and a half, or three inches, diameter, to which a cap of brafs is accurately fitted
with emery, and which has two holes for the
paffage of the tubes *xxx*, *yyy*. Before fhutting
the balloon with its cover, place within it the
ftand, BC, fupporting the cup of porcelain, D,
which contains the phofphorus. Then lute on
the cap with fat lute, and allow it to dry for
fome days, and weigh the whole accurately;
after

after this exhauft the balloon by means of an
air-pump connected with the tube *x x x*, and fill
it with oxygen gas by the tube *y y y*, from the
gazometer, Pl. VIII. Fig. 1. defcribed Chap. II.
Sect. II. of this part. The phofphorus is then
fet on fire by means of a burning-glafs, and is
allowed to burn till the cloud of concrete phofphoric acid ftops the combuftion, oxygen gas
being continually fupplied from the gazometer.
When the apparatus has cooled, it is weighed
and unluted; the tare of the inftrument being
allowed, the weight is that of the phofphoric
acid contained. It is proper, for greater accuracy, to examine the air or gas contained in the
balloon after combuftion, as it may happen to
be fomewhat heavier or lighter than common
air; and this difference of weight muft be taken
into account in the calculations upon the refults
of the experiment.

S E C T.   III.

Of the Combuftion of Charcoal.

The apparatus I have employed for this procefs confifts of a fmall conical furnace of hammered copper, reprefented in perfpective, Pl. XII.
Fig. 9. and internally difplayed Fig. 11. It is
divided

divided into the furnace, ABC, where the charcoal is burnt, the grate, *d e*, and the afh-hole,
F; the tube, GH, in the middle of the dome
of the furnace ferves to introduce the charcoal,
and as a chimney for carrying off the air which
has ferved for combuftion. Through the tube,
l m n, which communicates with the gazometer,
the hydrogen gas, or air, intended for fupporting the combuftion, is conveyed into the afh-
hole, F, whence it is forced, by the application
of preffure to the gazometer, to pafs through the
grate, *d e*, and to blow upon the burning charcoal placed immediately above.

Oxygen gas, which forms $\frac{28}{100}$ of atmofpheric
air, is changed into carbonic acid gas during
combuftion with charcoal, whilft the azotic gas
of the air is not altered at all. Hence, after
the combuftion of charcoal in atmofpheric air,
a mixture of carbonic acid gas and azotic gas
muft remain; to allow this mixture to pafs off,
the tube, *o p*, is adapted to the chimney, GH,
by means of a fcrew at G, and conveys the gas
into bottles half filled with folution of cauftic
potafh. The carbonic acid gas is abforbed by
the alkali, and the azotic gas is conveyed into
a fecond gazometer, where its quantity is afcertained.

The weight of the furnace, ABC, is firft accurately determined, then introduce the tube
RS, of known weight, by the chimney, GH,

till

till its lower end S, refts upon the grate, *d e,* which it occupies entirely; in the next place, fill the furnace with charcoal, and weigh the whole again, to know the exact quantity of charcoal fubmitted to experiment. The furnace is now put in its place, the tube, *l m n,* is fcrewed to that which communicates with the gazometer, and the tube, *o p,* to that which communicates with the bottles of alkaline folution. Every thing being in readinefs, the ftop-cock of the gazometer is opened, a fmall piece of burning charcoal is thrown into the tube, RS, which is inftantly withdrawn, and the tube, *o p,* is fcrewed to the chimney, GH. The little piece of charcoal falls upon the grate, and in this manner gets below the whole charcoal, and is kept on fire by the ftream of air from the ga-zometer. To be certain that the combuftion is begun, and goes on properly, the tube, *q r s,* is fixed to the furnace, having a piece of glafs ce-mented to its upper extremity, *s,* through which we can fee if the charcoal be on fire.

I neglected to obferve above, that the furnace, and its appendages, are plunged in water in the ciftern, TVXY, Fig. 11. Pl. XII. to which ice may be added to moderate the heat, if neceffary; though the heat is by no means very confide-rable, as there is no air but what comes from the gazometer, and no more of the charcoal
burns

burns at one time than what is immediately over the grate.

As one piece of charcoal is confumed another falls down into its place, in confequence of the declivity of the fides of the furnace; this gets into the ftream of air from the grate, *d e*, and is burnt; and fo on, fucceffively, till the whole charcoal is confumed. The air which has ferved the purpofe of the combuftion paffes through the mafs of charcoal, and is forced by the preffure of the gazometer to efcape through the tube, *o p*, and to pafs through the bottles of alkaline folution.

This experiment furnifhes all the neceffary data for a complete analyfis of atmofpheric air and of charcoal. We know the weight of charcoal confumed; the gazometer gives us the meafure of the air employed; the quantity and quality of gas remaining after combuftion may be determined, as it is received, either in another gazometer, or in jars, in a pneumato-chemical apparatus; the weight of afhes remaining in the afh-hole is readily afcertained; and, finally, the additional weight acquired by the bottles of alkaline folution gives the exact quantity of carbonic acid formed during the procefs. By this experiment we may likewife determine, with fufficient accuracy, the proportions in which charcoal and oxygen enter into the compofition of carbonic acid.

3 H   In

In a future memoir I fhall give an account
to the Academy of a feries of experiments I
have undertaken, with this inftrument, upon all
the vegetable and animal charcoals. By fome
very flight alterations, this machine may be
made to anfwer for obferving the principal phenomena of refpiration.

S E C T.   IV.

Of the Combuftion of Oils.

Oils are more compound in their nature than
charcoal, being formed by the combination of
at leaft two elements, charcoal and hydrogen;
of courfe, after their combuftion in common
air, water, carbonic acid gas, and azotic gas,
remain.   Hence the apparatus employed for
their combuftion requires to be adapted for
collecting thefe three products, and is confequently more complicated than the charcoal
furnace.

The apparatus I employ for this purpofe is
compofed of a large jar or pitcher A, Pl. XII.
Fig. 4. furrounded at its upper edge by a rim of
iron properly cemented at DE, and receding
from the jar at BC, fo as to leave a furrow or
gutter *xx*, between it and the outfide of the jar,
                                              fomewhat

somewhat more than two inches deep. The cover or lid of the jar, Fig. 5. is likewife furrounded by an iron rim *f g*, which adjufts into the gutter *x x*, Fig. 4. which being filled with mercury, has the effect of clofing the jar hermetically in an inftant, without ufing any lute; and, as the gutter will hold about two inches of mercury, the air in the jar may be made to fuftain the preffure of more than two feet of water, without danger of its efcaping.

The lid has four holes, T *h i k*, for the paffage of an equal number of tubes. The opening T is furnifhed with a leather box, through which paffes the rod, Fig. 3. intended for raifing and lowering the wick of the lamp, as will be afterwards directed. The three other holes are intended for the paffage of three feveral tubes, one of which conveys the oil to the lamp, a fecond conveys air for keeping up the combuftion, and the third carries off the air, after it has ferved for combuftion. The lamp in which the oil is burnt is reprefented Fig. 2 ; *a* is the refervoir of oil, having a funnel by which it is filled ; *b c d e f g h* is a fyphon 'which conveys the oil to the lamp 11 ; 7, 8, 9, 10, is the tube which conveys the air for combuftion from the gazometer to the fame lamp. The tube *b c* is formed externally, at its lower end *b*, into a male fcrew, which turns in a female fcrew in the lid of the refervoir of oil *a*; fo that, by turning
the

the refervoir one way or the other, it is made
to rife or fall, by which the oil is kept at the
neceffary level.

When the fyphon is to be filled, and the
communication formed between the refervoir of
oil and the lamp, the ftop-cock *c* is fhut, and
that at *e* opened, oil is poured in by the opening *f* at the top of the fyphon, till it rifes within three or four lines of the upper edge of the
lamp, the ftop-cock *k* is then fhut, and that at
c opened; the oil is then poured in at *f*, till the
branch *b c d* of the fyphon is filled, and then the
ftop-cock *e* is clofed. The two branches of the
fyphon being now completely filled, a communication is fully eftablifhed between the refervoir and the lamp.

In Pl. XII. Fig. 1. all the parts of the lamp
11, Fig. 2. are reprefented magnified, to fhow
them diftinctly. The tube *i k* carries the oil
from the refervoir to the cavity *a a a a*, which
contains the wick; the tube 9, 10, brings the
air from the gazometer for keeping up the combuftion; this air fpreads through the cavity
d d d d, and, by means of the paffages *c c c c* and
b b b b, is diftributed on each fide of the wick,
after the principles of the lamps conftructed by
Argand, Quinquet, and Lange.

To render the whole of this complicated apparatus more eafily underftood, and that its defcription may make all others of the fame kind

more

more readily followed, it is reprefented, completely conneĉted together for ufe, in Pl. XI.
The gazometer P furnifhes air for the combuftion by the tube and ftop-cock 1, 2 ; the tube
2, 3, communicates with a fecond gazometer,
which is filled whilft the firft one is emptying during the procefs, that there may be no interrup-
tion to the combuftion ; 4, 5, is a tube of glafs
filled with deliquefcent falts, for drying the air
as much as poffible in its paffage ; and the
weight of this tube and its contained falts, at
the beginning of the experiment, being known,
it is eafy to determine the quantity of water abforbed by them from the air.   From this deliquefcent tube the air is conduĉted through the
pipe 5, 6, 7, 8, 9, 10, to the lamp 11, where it
fpreads on both fides of the wick, as before defcribed, and feeds the flame.   One part of this
air, which ferves to keep up the combuftion of
the oil, forms carbonic acid gas and water, by
oxygenating its elements.   Part of this water
condenfes upon the fides of the pitcher A, and
another part is held in folution in the air by
means of caloric furnifhed by the combuftion.
This air is forced by the compreffion of the gazometer to pafs through the tube 12, 13, 14,
15, into the bottle 16, and the worm 17, 18,
where the water is fully condenfed from the refrigeration of the air ; and, if any water ftill remains

mains in folution, it is abforbed by deliquefcent
falts contained in the tube 19, 20.

All thefe precautions are folely intended for
collecting and determining the quantity of water formed during the experiment; the carbonic acid and azotic gas remains to be afcertained. The former is abforbed by cauftic alkaline
folution in the bottles 22 and 25. I have only
reprefented two of thefe in the figure, but nine
at leaft are requifite ; and the laft of the feries
may be half filled with lime-water, which is the
moft certain reagent for indicating the prefence
of carbonic acid ; if the lime-water is not rendered turbid, we may be certain that no fenfible
quantity of that acid remains in the air.

The reft of the air which has ferved for combuftion, and which chiefly confifts of azotic gas,
though ftill mixed with a confiderable portion
of oxygen gas, which has efcaped unchanged
from the combuftion, is carried through a third
tube 28, 29, of deliquefcent falts, to deprive it
of any moifture it may have acquired in the bottles of alkaline folution and lime-water, and
from thence by the tube 29, 30, into a gazometer, where its quantity is afcertained. Small
effays are then taken from it, which are expofed
to a folution of fulphuret of potafh, to afcertain
the proportions of oxygen and azotic gas it contains.

In

In the combuftion of oils the wick becomes
charred at laft, and obftructs the rife of the oil;
befides, if we raife the wick above a certain
height, more oil rifes through its capillary tubes
than the ftream of air is capable of confuming,
and fmoke is produced. Hence it is neceffary
to be able to lengthen or fhorten the wick without opening the apparatus; this is accomplifhed by means of the rod 31, 32, 33, 34, which
paffes through a leather-box, and is connected
with the fupport of the wick; and that the motion of this rod, and confequently of the wick,
may be regulated with the utmoft fmoothnefs
and facility; it is moved at pleafure by a pin-
nion which plays in a toothed rack. The rod,
with its appendages, are reprefented Pl. XII.
Fig. 3. It appeared to me, that the combuftion
would be affifted by furrounding the flame of
the lamp with a fmall glafs jar open at both
ends, as reprefented in its place in Pl. XI.

I fhall not enter into a more detailed defcription of the conftruction of this apparatus, which
is ftill capable of being altered and modified in
many refpects, but fhall only add, that when it
is to be ufed in experiment, the lamp and refervoir with the contained oil muft be accurately
weighed, after which it is placed as before directed, and lighted; having then formed the
connection between the air in the gazometer
and the lamp, the external jar A, Pl. XI. is fixed

ed over all, and fecured by means of the board BC and two rods of iron which conneƈt this board with the lid, and are fcrewed to it. A fmall quantity of oil is burnt while the jar is adjufting to the lid, and the produƈt of that combuftion is loft; there is likewife a fmall portion of air from the gazometer loft at the fame time. Both of thefe are of very inconfiderable confequence in extenfive experiments, and they are even capable of being va¹ued in our calculation of the refults.

In a particular memoir, I fhall give an account to the Academy of the difficulties infeparable from this kind of experiments: Thefe are fo infurmountable and troublefome, that I have not hitherto been able to obtain any rigorous determination of the quantities of the produƈts. I have fufficient proof, however, that the fixed oils are entirely refolved during combuftion into water and carbonic acid gas, and confequently that they are compofed of hydrogen and charcoal; but I have no certain knowledge refpeƈting the proportions of thefe ingredients.

S E C T.

SECT. V.

Of the Combuſtion of Alkohol.

The combuſtion of alkohol may be very readily performed in the apparatus already deſcribed
for the combuſtion of charcoal and phoſphorus.
A lamp filled with alkohol is placed under the
jar A, Pl. IV. Fig. 3. a ſmall morſel of phoſphorus is placed upon the wick of the lamp,
which is ſet on fire by means of the hot iron, as
before directed. This proceſs is, however, liable to conſiderable inconveniency ; it is dangerous to make uſe of oxygen gas at the beginning of the experiment for fear of deflagration,
which is even liable to happen when common
air is employed. An inſtance of this had very
near proved fatal to myſelf, in preſence of ſome
members of the Academy. Inſtead of preparing
the experiment, as uſual, at the time it was to
be performed, I had diſpoſed every thing in order the evening before; the atmoſpheric air of the
jar had thereby ſufficient time to diſſolve a good
deal of the alkohol ; and this evaporation had
even been conſiderably promoted by the height
of the column of mercury, which I had raiſed
to EF, Pl. IV. Fig. 3. The moment I attempt-

3 I ed

ed to fet the little morfel of phofphorus on fire
by means of the red hot iron, a violent explofion took place, which threw the jar with great
violence againft the floor of the laboratory, and
dafhed it in a thoufand pieces.

Hence we can only operate upon very fmall
quantities, fuch as ten or twelve grains of alkohol, in this manner ; and the errors which may
be committed in experiments upon fuch fmall
quantities prevents our placing any confidence
in their refults. I endeavoured to prolong the
combuftion, in the experiments contained in the
Memoirs of the Academy for 1784, p. 593.
by lighting the alkohol firft in common air, and
furnifhing oxygen gas afterwards to the jar, in
proportion as it confumed ; but the carbonic
acid gas produced by the procefs became a great
hinderance to the combuftion, the more fo that
alkohol is but difficultly combuftible, efpecially
in worfe than common air ; fo that even in this
way very fmall quantities only could be burnt.

Perhaps this combuftion might fucceed better
in the oil apparatus, Pl. XI. ; but I have not
hitherto ventured to try it. The jar A in which
the combuftion is performed is near 1400 cubical inches in dimenfion ; and, were an explofion to take place in fuch a veffel, its confequences would be very terrible, and very difficult to guard againft. I have not, however,
defpaired of making the attempt.

From

From all thefe difficulties, I have been hitherto obliged to confine myfelf to experiments upon very fmall quantities of alkohol, or at leaft
to combuftions made in open veffels, fuch as
that reprefented in Pl. IX. Fig. 5. which will
be defcribed in Section VII. of this chapter. If
I am ever able to remove thefe difficulties, I
fhall refume this inveftigation.

S E C T.   VI.

Of the Combuftion of Ether.

Tho' the combuftion of ether in clofe veffels
does not prefent the fame difficulties as that of
alkohol, yet it involves fome of a different kind,
not more eafily overcome, and which ftill prevent the progrefs of my experiments. I endeavoured to profit by the property which ether
poffeffes of diffolving in atmofpheric air, and
rendering it inflammable without explofion.
For this purpofe, I conftructed the refervoir of
ether *a b c d*, Plate XII. Fig. 8. to which air is
brought from the gazometer by the tube 1, 2,
3, 4. This air fpreads, in the firft place, in the
double lid *ac* of the refervoir, from which it
paffes through feven tubes *ef*, *gh*, *ik*, &c. which
defcend to the bottom of the ether, and it is
forced

forced by the preſſure of the gazometer to boil
up through the ether in the reſervoir. We
may replace the ether in this firſt reſervoir, in
proportion as it is diſſolved and carried off by
the air, by means of the ſupplementary reſervoir E, connected by a braſs tube fifteen or
eighteen inches long, and ſhut by a ſtop-cock.
This length of the connecting tube is to enable
the deſcending ether to overcome the reſiſtance
occaſioned by the preſſure of the air from the
gazometer.

The air, thus loaded with vapours of ether,
is conducted by the tube 5, 6, 7, 8, 9, to the
jar A, into which it is allowed to eſcape through
a capillary opening, at the extremity of which
it is ſet on fire. The air, when it has ſerved
the purpoſe of combuſtion, paſſes through the
bottle 16, Pl. XI. the worm 17, 18, and the
deliqueſcent tube 19, 20, after which it paſſes
through the alkaline bottles ; in theſe its carbonic acid gas is abſorbed, the water formed
during the experiment having been previouſly
depoſited in the former parts of the apparatus.

When I cauſed conſtruct this apparatus, I
ſuppoſed that the combination of atmoſpheric air
and ether formed in the reſervoir *a b c d*, Pl. XII.
Fig. 8. was in proper proportion for ſupporting
combuſtion ; but in this I was miſtaken ; for
there is a very conſiderable quantity of exceſs of
ether ; ſo that an additional quantity of atmoſpheric

fpheric air is neceffary to enable it to burn fully.
Hence a lamp conftructed upon thefe principles
will burn in common air, which furnifhes the
quantity of oxygen neceffary for combuftion,
but will not burn in clofe veffels in which the
air is not renewed. From this circumftance,
my ether lamp went out foon after being lighted and fhut up in the jar A, Pl. XII. Fig. 8.
To remedy this defect, I endeavoured to bring
atmofpheric air to the lamp by ths lateral tube
10, 11, 12, 13, 14, 15, which I diftributed
circularly round the flame; but the flame is fo
exceedingly rare, that it is blown out by the
gentleft poffible ftream of air, fo that I have not
hitherto fucceeded in burning ether. I do not,
however, defpair of being able to accomplifh it
by means of fome changes I am about to have
made upon this apparatus.

S E C T.   VII.

Of the Combuftion of Hydrogen Gas, and the For-
mation of Water.

In the formation of water, two fubftances,
hydrogen and oxygen, which are both in the
aëriform ftate before combuftion, are tranf-
formed into liquid or water by the operation.
This

This experiment would be very eafy, and would require very fimple inftruments, if it were poffible to procure the two gaffes perfectly pure, fo that they might burn without any refiduum. We might, in that cafe, operate in very fmall veffels, and, by continually furnifhing the two gaffes in proper proportions, might continue the combuftion indefinitely. But, hitherto, chemifts have only employed oxygen gas, mixed with azotic gas ; from which circumftance, they have only been able to keep up the combuftion of hydrogen gas for a very limited time in clofe veffels, becaufe, as the refiduum of azotic gas is continually increafing, the air becomes at laft fo much contaminated, that the flame weakens and goes out. This inconvenience is fo much the greater in proportion as the oxygen gas employed is lefs pure. From this circumftance, we muft either be fatisfied with operating upon fmall quantities, or muft exhauft the veffels at intervals, to get rid of the refiduum of azotic gas ; but, in this cafe, a portion of the water formed during the experiment is evaporated by the exhauftion ; and the refulting error is the more dangerous to the accuracy of the procefs, that we have no certain means of valuing it.

Thefe confiderations make me defirous to repeat the principal experiments of pneumatic chemiftry with oxygen gas entirely free from

any

any admixture of azotic gas; and this may be procured from oxygenated muriat of potafh. The oxygen gas extracted from this falt does not appear to contain azote, unlefs accidentally, fo that, by proper precautions, it may be obtained perfectly pure. In the mean time, the apparatus employed by Mr Meufnier and me for the combuftion of hydrogen gas, which is defcribed in the experiment for recompofition of water, Part I. Chap. VIII. and need not be here repeated, will anfwer the purpofe; when pure gaffes are procured, this apparatus will require no alterations, except that the capacity of the veffels may then be diminifhed. See Pl. IV. Fig. 5.

The combuftion, when once begun, continues for a confiderable time, but weakens gradually, in proportion as the quantity of azotic gas remaining from the combuftion incieafes, till at laft the azotic gas is in fuch over proportion that the combuftion can no longer be fupported, and the flame goes out. This fpontaneous extinction muft be prevented, becaufe, as the hydrogen gas is preffed upon in its refervoir, by an inch and a half of water, whilft the oxygen gas fuffers a preffure only of three lines, a mixture of the two would take place in the balloon, which would at laft be forced by the fuperior preffure into the refervoir of oxygen gas. Wherefore the combuftion muft be ftopped.

ped, by fhutting the ftop-cock of the tube *d*D*d*
whenever the flame grows very feeble ; for
which purpofe it muft be attentively watch-
ed.

There is another apparatus for combuftion,
which, though we cannot with it perform experiments with the fame fcrupulous exactnefs as
with the preceding inftruments, gives very ftriking refults that are extremely proper to be
fhewn in courfes of philofophical chemiftry.   It
confifts of a worm EF, Pl. IX. Fig. 5. contained
in a metallic cooller ABCD.   To the upper
part of this worm E, the chimney GH is fixed,
which is compofed of two tubes, the inner of
which is a continuation of the worm, and the
outer one is a cafe of tin-plate, which furrounds
it at about an inch diftance, and the interval is
filled up with fand.   At the inferior extremity
K of the inner tube, a glafs tube is fixed, to
which we adopt the Argand lamp LM for burning alkohol, &c.

Things being thus difpofed, and the lamp
being filled with a determinate quantity of alkohol, it is fet on fire ; the water which is formed
during the combuftion rifes in the chimney KE,
and being condenfed in the worm, runs out at
its extremity F into the bottle P.   The double
tube of the chimney, filled with fand in the in-
terftice, is to prevent the tube from cooling in
its upper part, and condenfing the water ; o-
therwife,

therwife, it would fall back in the tube, and we
fhould not be able to afcertain its quantity, and
befides it might fall in drops upon the wick,
and extinguifh the flame. The intention of this
conftruction, is to keep the chimney always hot,
and the worm always cool, that the water may
be preferved in the ftate of vapour whilft ri-
fing, and may be condenfed immediately upon
getting into the defcending part of the apparatus. By this inftrument, which was contrived
by Mr Meufnier, and which is defcribed by me
in the Memoirs of the Academy for 1784, p.
593. we may, with attention to keep the worm
always cold, collect nearly feventeen ounces of
water from the combuftion of fixteen ounces of
alkohol.

S E C T.   VIII.

Of the Oxydation of Metals.

The term *oxydation* or *calcination* is chiefly u-
fed to fignify the procefs by which metals expofed to a certain degree of heat are converted
into oxyds, by abforbing oxygen from the air.
This combination takes place in confequence of
oxygen poffeffing a greater affinity to metals, at
a certain temperature, than to caloric, which

3 K   becomes

becomes difengaged in its free ſtate ; but, as
this difengagement, when made in common air,
is flow and progreſſive, it is ſcarcely evident to
the ſenſes. It is quite otherwiſe, however, when
oxydation takes place in oxygen gas ; for, being
produced with much greater rapidity, it is generally accompanied with heat and light, ſo as
evidently to ſhow that metallic ſubſtances are
real combuſtible bodies.

All the metals have not the ſame degree of
affinity to oxygen. Gold, ſilver, and platina, for
inſtance, are incapable of taking it away from
its combination with caloric, even in the greateſt
known heat ; whereas the other metals abſorb it
in a larger or ſmaller quantity, until the affinities of the metal to oxygen, and of the latter to
caloric, are in exaƈt equilibrium. Indeed, this
ſtate of equilibrium of affinities may be aſſumed
as a general law of nature in all combinations.

In all operations of this nature, the oxydation
of metals is accelerated by giving free acceſs
to the air ; it is ſometimes much aſſiſted by
joining the aƈtion of a bellows, which direƈts a
ſtream of air over the ſurface of the metal.
This proceſs becomes greatly more rapid if a
ſtream of oxygen gas be uſed, which is readily
done by means of the gazometer formerly deſcribed. The metal, in this caſe, throws out a
brilliant flame, and the oxydation is very quickly

ly accomplifhed ; but this method can only be ufed in very confined experiments, on account of the expence of procuring oxygen gas. In the effay of ores, and in all the common operations of the laboratory, the calcination or oxydation of metals is ufually performed in a difh of baked clay, Pl. IV. Fig. 6. commonly called a *roafting teft*, placed in a ftrong furnace. The fubftances to be oxydated are frequently ftirred, on purpofe to prefent frefh furfaces to the air.

Whenever this operation is performed upon ametal which is not volatile, and from which nothing flies off into the furrounding air during the procefs, the metal acquires additional weight ; but the caufe of this increafed weight during oxydation could never have been difco‧vered by means of experiments performed in free air ; and it is only fince thefe operations have been performed in clofe veffels, and in determinate quantities of air, that any juft conjectures have been formed concerning the caufe of this phenomenon. The firft method for this purpofe is due to Dr Prieftley, who expofes the metal to be calcined in a porcelain cup N, Pl. IV. Fig. 11. placed upon the ftand IK, under a jar A, in the bafon BCDE, full of water ; the water is made to rife up to GH, by fucking out the air with a fyphon, and the focus of a burning glafs is made to fall upon the metal. In a few minutes the oxydation takes place,

a

a part of the oxygen contained in the air combines with the metal, and a proportional diminution of the volume of air is produced ; what
remains is nothing more than azotic gas, ftill
however mixed with a fmall quantity of oxygen
gas.   I have given an account of a feries of experiments made with this apparatus in my Phyfical and Chemical Effays, firft publifhed in
1773.   Mercury may be ufed inftead of water
in this experiment, whereby the refults are rendered ftill more conclufive.

Another procefs for this purpofe was invented
by Mr Boyle, and of which I gave an account
in the Memoirs of the Academy for 1774, p.
351.   The metal is introduced into a retort,
Pl. III. Fig. 20. the beak of which is hermetically fealed ; the metal is then oxydated by
means of heat applied with great precaution.
The weight of the veffel, and its contained fubftances, is not at all changed by this procefs,
until the extremity of the neck of the retort is
broken ; but, when that is done, the external
air rufhes in with a hiffing noife.   This operation is attended with danger, unlefs a part of
the air is driven out of the retort, by means of
heat, before it is hermetically fealed, as otherwife the retort would be apt to burft by the di-
lation of the air when placed in the furnace.
The quantity of air driven out may be received
under a jar in the pneumato-chemical apparatus,

tus, by which its quantity, and that of the air remaining in the retort, is afcertained. I have not multiplied my experiments upon oxydation of metals fo much as I could have wifhed ; neither have I obtained fatisfactory refults with any metal except tin. It is much to be wifhed that fome perfon would undertake a feries of experiments upon oxydation of metals in the feveral gaffes ; the fubject is important, and would fully repay any trouble which this kind of experiment might occafion.

As all the oxyds of mercury are capable of revivifying without addition, and reftore the oxygen gas they had before abforbed, this feemed to be the moft proper metal for becoming the fubject of conclufive experiments upon oxydation. I formerly endeavoured to accomplifh the oxydation of mercury in clofe veffels, by filling a retort, containing a fmall quantity of mercury, with oxygen gas, and adapting a bladder half full of the fame gas to its beak ; See Pl. IV. Fig. 12. Afterwards, by heating the mercury in the retort for a very long time, I fucceeded in oxydating a very fmall portion, fo as to form a little red oxyd floating upon the furface of the running mercury ; but the quantity was fo fmall, that the fmalleft error committed in the determination of the quantities of oxygen gas before and after the operation muft have thrown very great uncertainty upon the
refults

refults of the experiment. I was, befides, dif-
fatisfied with this procefs, and not without
caufe, left any air might have efcaped through
the pores of the bladder, more efpecially as it
becomes fhrivelled by the heat of the furnace,
unlefs covered over with cloths kept conftantly
wet.

This experiment is performed with more certainty in the apparatus defcribed in the Memoirs of the Academy for 1775, p. 580. This
confifts of a retort, A, Pl. IV. Fig. 2. having a
crooked glafs tube BCDE of ten or twelve lines
internal diameter, melted on to its beak, and
which is engaged under the bell glafs FG,
ftanding with its mouth downwards, in a bafon
filled with water or mercury. The retort is
placed upon the bars of the furnace MMNN,
Pl. IV. Fig. 2. or in a fand bath, and by means
of this apparatus we may, in the courfe of feveral days, oxydate a fmall quantity of mercury
in common air ; the red oxyd floats upon the
furface, from which it may be collected and revivified, fo as to compare the quantity of oxygen gas obtained in revivification with the abforption which took place during oxydation.
This kind of experiment can only be performed
upon a fmall fcale, fo that no very certain conclufions can be drawn from them *.

The

* See an account of this experiment, Part. I. Chap,
iii.—A.

The combuftion of iron in oxygen gas being
a true oxydation of that metal, ought to be
mentioned in this place. The apparatus employed by Mr Ingenhoufz for this operation is
reprefented in Pl. IV. Fig. 17.; but, having already defcribed it fufficiently in Chap. III. I fhall
refer the reader to what is faid of it in that
place. Iron may likewife be oxydated by combuftion in veffels filled with oxygen gas, in the
way already directed for phofphorus and charcoal. This apparatus is reprefented Pl. IV.
Fig. 3. and defcribed in the fifth chapter of the
firft part of this work. We learn from Mr Ingenhoufz, that all the metals, except gold, filver, and mercury, may be burnt or oxydated
in the fame manner, by reducing them into very
fine wire, or very thin plates cut into narrow
flips ; thefe are twifted round with iron-wire,
which communicates the property of burning
to the other metals.

Mercury is even difficultly oxydated in free
air. In chemical laboratories, this procefs is
ufually carried on in a matrafs A, Pl. IV. Fig.
having a very flat body, and a very long neck
BC, which veffel is commonly called *Boyle's
hell.* A quantity of mercury is introduced fufficient to cover the bottom, and it is placed in
a fand-bath, which keeps up a conftant heat
approaching to that of boiling mercury. By
continuing this operation with five or fix fimilar matraffes during feveral months, and re-
newing

newing the mercury from time to time, a few
ounces of red oxyd are at laſt obtained. The
great ſlowneſs and inconvenience of this apparatus ariſes from the air not being ſufficiently
renewed; but if, on the other hand, too free a
circulation were given to the external air, it
would carry off the mercury in ſolution in the
ſtate of vapour, ſo that in a few days none
would remain in the veſſel.

As, of all the experiments upon the oxydation of metals, thoſe with mercury are the moſt
concluſive, it were much to be wiſhed that a
ſimple apparatus could be contrived by which
this oxydation and its reſults might be demon-
ſtrated in public courſes of chemiſtry. This
might, in my opinion, be accompliſhed by methods ſimilar to thoſe I have already deſcribed
for the combuſtion of charcoal and the oils;
but, from other purſuits, I have not been able
hitherto to reſume this kind of experiment.

The oxyd of mercury revives without addition, by being heated to a ſlightly red heat. In
this degree of temperature, oxygen has greater
affinity to caloric than to mercury, and forms
oxygen gas. This is always mixed with a ſmall
portion of azotic gas, which indicates that the
mercury abſorbs a ſmall portion of this latter
gas during oxydation. It almoſt always contains a little carbonic acid gas, which muſt undoubtedly be attributed to the foulneſſes of the
oxyd;

oxyd; thefe are charred by the heat, and convert a part of the oxygen gas into carbonic acid.

If chemifts were reduced to the neceffity of procuring all the oxygen gas employed in their experiments from mercury oxydated by heat without addition, or, as it is called, *calcined* or *precipitated* per fe, the exceffive dearnefs of that preparation would render experiments, even upon a moderate fcale, quite impracticable. But mercury may likewife be oxydated by means of nitric acid; and in this way we procure a red oxyd, even more pure than that produced by calcination. I have fometimes prepared this oxyd by diffolving mercury in nitric acid, evaporating to drynefs, and calcining the falt, either in a retort, or in capfules formed of pieces of broken matraffes and retorts, in the manner formerly defcribed; but I have never fucceeded in making it equally beautiful with what is fold by the druggifts, and which is, I believe, brought from Holland. In choofing this, we ought to prefer what is in folid lumps compofed of foft adhering fcales, as when in powder it is fometimes adulterated with red oxyd of lead.

To obtain oxygen gas from the red oxyd of mercury, I ufually employ a porcelain retort, having a long glafs tube adapted to its beak, which is engaged under jars in the water pneu-

3 L   mato-

mato-chemical apparatus, and I place a bottle
in the water, at the end of the tube, for receiving the mercury, in proportion as it revives
and diſtils over.   As the oxygen gas never appears till the retort becomes red, it ſeems to
prove the principle eſtabliſhed by Mr Berthollet, that an obſcure heat can never form oxygen
gas, and that light is one of its conſtituent elements.   We muſt rejeft the firſt portion of gas
which comes over, as being mixed with common air, from what was contained in the retort at the beginning of the experiment ; but,
even with this precaution, the oxygen gas procured is uſually contaminated with a tenth part
of azotic gas, and with a very ſmall portion of
carbonic acid gas.   This latter is readily got
rid of, by making the gas paſs through a ſolution of cauſtic alkali ; but we know of no method for ſeparating the azotic gas ; its proportions may however be aſcertained, by leaving
a known quantity of the oxygen gas contaminated with it for a fortnight, in contaft with
ſulphuret of ſoda or potaſh, which abſorbs the
oxygen gas ſo as to convert the ſulphur into
ſulphuric acid, and leaves the azotic gas remaining pure.

We may likewiſe procure oxygen gas from
black oxyd of manganeſe or nitrat of potaſh,
by expoſing them to a red heat in the apparatus already deſcribed for operating upon red

oxyd

oxyd of mercury; only, as it requires fuch a heat as is at leaft capable of foftening glafs, we muft employ retorts of ftone or of porcelain. But the pureft and beft oxygen gas is what is difengaged from oxygenated muriat of potafh by fimple heat. This operation is performed in a glafs retort, and the gas obtained is perfectly pure, provided that the firft portions, which are mixed with the common air of the veffels, be rejected.

C H A P

C H A P.   IX.

Of Deflagration.

I HAVE already fhown, Part I. Chap. IX.
that oxygen does not always part with the
whole of the caloric it contained in the ftate of
gas when it enters into combination with other
bodies. It carries almoft the whole of its caloric alongft with it in entering into the combinations which form nitric acid and oxygenated
muriatic acid; fo that in nitrats, and more efpecially in oxygenated muriats, the oxygen is, in
a certain degree, in the ftate of oxygen gas,
condenfed, and reduced to the fmalleft volume
it is capable of occupying.

In thefe combinations, the caloric exerts a
conftant action upon the oxygen to bring it
back to the ftate of gas; hence the oxygen adheres but very flightly, and the fmalleft additional force is capable of fetting it free; and,
when fuch force is applied, it often recovers the
ftate of gas inftantaneoufly. This rapid paffage
from the folid to the aëriform ftate is called
detonation, or fulmination, becaufe it is ufually
accompanied with noife and explofion. Deflagrations are commonly produced by means of
combinations of charcoal either with nitre or
oxygenated

oxygenated muriat of potafh; fometimes, to affift the inflammation, fulphur is added; and,
upon the juft proportion of thefe ingredients,
and the proper manipulation of the mixture,
depends the art of making gun-powder.

As oxygen is changed, by deflagration with
charcoal, into carbonic acid, inftead of oxygen
gas, carbonic acid gas is difengaged, at leaft
when the mixture has been made in juft proportions. In deflagration with nitre, azotic
gas is likewife difengaged, becaufe azote is one
of the conftituent elements of nitric acid.

The fudden and inftantaneous difengagement and expanfion of thefe gaffes is not, however, fufficient for explaining all the phenomena of deflagration; becaufe, if this were the fole
operating power, gun powder would always be
fo much the ftronger in proportion as the quantity of gas difengaged in a given time was the
more confiderable, which does not always ac-
cord with experiment. I have tried fome kinds
which produced almoft double the effect of ordinary gun powder, although they gave out a
fixth part lefs of gas during deflagration. It
would appear that the quantity of caloric difengaged at the moment of detonation contributes
confiderably to the expanfive effects produced;
for, although caloric penetrates freely through
the pores of every body in nature, it can only
do fo progreffively, and in a given time; hence,

when

when the quantity difengaged at once is too large to get through the pores of the furround-ing bodies, it muft necefarily act in the fame way with ordinary elaftic fluids, and overturn every thing that oppofes its paffage. This muft, at leaft in part, take place when gun-powder is fet on fire in a cannon; as, although the metal is permeable to caloric, the quantity difengaged at once is too large to find its way through the pores of the metal, it muft therefore make an effort to efcape on every fide; and, as the re-fiftance all around, excepting towards the muz-zle, is too great to be overcome, this effort is employed for expelling the bullet.

The caloric produces a fecond effect, by means of the repulfive force exerted between its particles; it caufes the gaffes, difengaged at the moment of deflagration, to expand with a degree of force proportioned to the temperature produced.

It is very probable that water is decompofed during the deflagration of gun-powder, and that part of the oxygen furnifhed to the nafcent car-bonic acid gas is produced from it. If fo, a confiderable quantity of hydrogen gas muft be difengaged in the inftant of deflagration, which expands, and contributes to the force of the ex-plofion. It may readily be conceived how great-ly this circumftance muft increafe the effect of powder, if we confider that a pint of hydrogen

gas

gas weighs only one grain and two thirds; hence a very fmall quantity in weight muſt oc-cupy a very large fpace, and it muſt exert a prodigious expanfive force in paffing from the liquid to the aëriform ſtate of exiſtence.

In the laſt place, as a portion of undecom-pofed water is reduced to vapour during the deflagration of gun-powder, and as water, in the ſtate of gas, occupies feventeen or eighteen hundred times more fpace than in its liquid ſtate, this circumſtance muſt likewife contribute largely to the exploſive force of the powder.

I have already made a confiderable feries of experiments upon the nature of the elaſtic fluids difengaged during the deflagration of nitre with charcoal and fulphur; and have made fome, likewife, with the oxygenated muriat of potafh. This method of inveſtigation leads to tollerably accurate concluſions with refpect to the conſti-tuent elements of thefe falts. Some of the prin-cipal refults of thefe experiments, and of the confequences drawn from them refpecting the analyfis of nitric acid, are reported in the col-lection of memoirs prefented to the Academy by foreign philofophers, vol. xi. p. 625. Since then I have procured more convenient inſtru-ments, and I intend to repeat thefe experiments upon a larger fcale, by which I ſhall procure more accurate precifion in their refults; the following, however, is the procefs I have hither-

to employed. I would very earneftly advife fuch
as intend to repeat fome of thefe experiments,
to be very much upon their guard in operating
upon any mixture which contains nitre, charcoal, and fulphur, and more efpecially with thofe
in which oxygenated muriat of potafh is mixed
with thefe two materials.

I make ufe of piftol barrels, about fix inches
long, and of five or fix lines diameter, having
the touch-hole fpiked up with an iron nail
ftrongly driven in, and broken in the hole, and
a little tin-fmith's folder run in to prevent any
poffible iffue for the air. Thefe are charged
with a mixture of known quantities of nitre and
charcoal, or any other mixture capable of deflagration, reduced to an impalpable powder,
and formed into a pafte with a moderate quantity of water. Every portion of the materials
introduced muft be rammed down with a ram-
mer nearly of the fame caliber with the barrel,
four or five lines at the muzzle muft be left
empty, and about two inches of quick match
are added at the end of the charge. The only
difficulty in this experiment, efpecially when fulphur is contained in the mixture, is to difcover
the proper degree of moiftening; for, if the
pafte be too much wetted, it will not take fire,
and if too dry, the deflagration is apt to become
too rapid, and even dangerous.

When

When the experiment is not intended to be rigoroufly exact, we fet fire to the match, and, when it is juft about to communicate with the charge, we plunge the piftol below a large bell-glafs full of water, in the pneumato chemical apparatus. The deflagration begins, and continues in the water, and gas is difengaged with lefs or more rapidity, in proportion as the mixture is more or lefs dry. So long as the deflagration continues, the muzzle of the piftol muft be kept fomewhat inclined downwards, to prevent the water from getting into its barrel. In this manner I have fometimes collected the gas produced from the deflagration of an ounce and half, or two ounces, of nitre.

In this manner of operating it is impoffible to determine the quantity of carbonic acid gas difengaged, becaufe a part of it is abforbed by the water while paffing through it; but, when the carbonic acid is abforbed, the azotic gas remains; and, if it be agitated for a few minutes in cauftic alkaline folution, we obtain it pure, and can eafily determine its volume and weight. We may even, in this way, acquire a tollerably exact knowledge of the quantity of carbonic acid by repeating the experiment a great many times, and varying the proportions of charcoal, till we find the exact quantity requifite to deflagrate the whole nitre employed. Hence, by means of the weight of charcoal employed, we

3 M   determine

determine the weight of oxygen neceſſary for ſaturation, and deduce the quantity of oxygen contained in a given weight of nitre.

I have uſed another proceſs, by which the reſults of this experiment are conſiderably more accurate, which conſiſts in receiving the diſengaged gaſſes in bell-glaſſes filled with mercury. The mercurial apparatus I employ is large enough to contain jars of from twelve to fifteen pints in capacity, which are not very readily managed when full of mercury, and even require to be filled by a particular method. When the jar is placed in the ciſtern of mercury, a glaſs ſyphon is introduced, connected with a ſmall air-pump, by means of which the air is exhauſted, and the mercury riſes ſo as to fill the jar. After this, the gas of the deflagration is made to paſs into the jar in the ſame manner as directed when water is employed.

I muſt again repeat, that this ſpecies of experiment requires to be performed with the greateſt poſſible precautions. I have ſometimes ſeen, when the diſengagement of gas proceeded with too great rapidity, jars filled with more than an hundred and fifty pounds of mercury driven off by the force of the exploſion, and broken to pieces, while the mercury was ſcattered about in great quantities.

When the experiment has ſucceeded, and the gas is collected under the jar, its quantity in general,

general, and the nature and quantities of the feveral fpecies of gaffes of which the mixture is
compofed, are accurately afcertained by the methods already pointed out in the fecond chapter
of this part of my work. I have been prevented from putting the laft hand to the experiments I had begun upon deflagration, from their
connection with the objects I am at prefent engaged in; and I am in hopes they will throw
confiderable light upon the operations belonging to the manufacture of gun-powder.

C H A P.

C H A P. X.

Of the Instruments necessary for Operating upon Bodies in very high Temperatures.

S E C T. I.

Of Fusion.

WE have already seen, that, by aqueous solution, in which the particles of bodies are separated from each other, neither the solvent nor the body held in solution are at all decomposed; so that, whenever the cause of separation ceases, the particles reunite, and the saline substance recovers precisely the same appearance and properties it possessed before solution. Real solutions are produced by fire, or by introducing and accumulating a great quantity of caloric between the particles of bodies; and this species of solution in caloric is usually called *fusion.*

This operation is commonly performed in vessels called crucibles, which must necessarily be be lefs fufible than the bodies they are intended to contain. Hence, in all ages, chemifts have been extremely folicitous to procure crucibles of very refractory materials, or fuch as are capable of refifting a very high degree of heat. The beft are made of very pure clay or of porcelain earth ; whereas fuch as are made of clay mixed with calcareous or filicious earth are very fufible. All the crucibles made in the neighbourhood of Paris are of this kind, and confequently unfit for moft chemical experiments. The Heffian crucibles are tolerably good ; but the beft are made of Limoges earth, which feems abfolutely infufible. We have, in France, a great many clays very fit for making crucibles ; fuch, for inftance, is the kind ufed for making melting pots at the glafs-manufactory of St Gobin.

Crucibles are made of various forms, according to the operations they are intended to perform. Several of the moft common kinds are reprefented Pl. VII. Fig. 7. 8. 9. and 10. the one reprefented at Fig. 9. is almoft fhut at its mouth.

Though fufion may often take place without changing the nature of the fufed body, this operation is frequently employed as a chemical means of decompofing and recompounding bodies. In this way all the metals are extracted from their ores ; and, by this procefs, they are revivified,
moulded,

moulded, and alloyed with each other. By this procefs fand and alkali are combined to form glafs, and by it likewife paftes, or coloured ftones, enamels, &c. are formed.

The action of violent fire was much more frequently employed by the ancient chemifts than it is in modern experiments. Since greater precifion has been employed in philofophical refearches, the *humid* has been preferred to the *dry* method of procefs, and fufion is feldom had recourfe to until all the other means of analyfis have failed.

S E C T.　II.

Of Furnaces.

Thefe are inftruments of moft univerfal ufe in chemiftry ; and, as the fuccefs of a great number of experiments depends upon their being well or ill conftructed, it is of great importance that a laboratory be well provided in this refpect. A furnace is a kind of hollow cylindrical tower, fometimes widened above, Pl. XIII. Fig. 1. ABCD, which muft have at leaft two lateral openings ; one in its upper part F, which is the door of the fire-place, and one below, G, leading to the afh-hole. Between thefe the furnace

nace is divided by a horizontal grate, intended for fupporting the fewel, the fituation of which is marked in the figure by the line HI. Though this be the leaft complicated of all the chemical furnaces, yet it is applicable to a great number of purpofes. By it lead, tin, bifmuth, and, in general, every fubftance which does not require a very ftrong fire, may be melted in crucibles ; it will ferve for metallic oxydations, for evaporatory veffels, and for fand-baths, as in Pl. III. Fig. 1. and 2. To render it proper for thefe purpofes, feveral notches, *m m m m*, Pl. XIII. Fig. 1. are made in its upper edge, as otherwife any pan which might be placed over the fire would ftop the paffage of the air, and prevent the fewel from burning. This furnace can only produce a moderate degree of heat, becaufe the quantity of charcoal it is capable of confuming is limited by the quantity of air which is allowed to pafs through the opening G of the afh-hole. Its power might be confiderably augmented by enlarging this opening, but then the great ftream of air which is convenient for fome operations might be hurtful in others ; wherefore we muft have furnaces of different forms, conftructed for different purpofes, in our laboratories : There ought efpecially to be feveral of the kind now defcribed of different fizes.

The reverberatory furnace, Pl. XIII. Fig. 2. is perhaps more neceffary. This, like the common

mon furnace, is compofed of the afh-hole HIKL, the fire-place KLMN, the laboratory MNOP, and the dome RRSS, with its funnel or chimney TTVV ; and to this laft feveral additional tubes may be adapted, according to the nature of the different experiments. The retort A is placed in the divifion called the laboratory, and fupported by two bars of iron which run acrofs the furnace, and its beak comes out at a round hole in the fide of the furnace, one half of which is cut in the piece called the laboratory, and the other in the dome. In moft of the ready made reverberatory furnaces which are fold by the potters at Paris, the openings both above and below are too fmall : Thefe do not allow a fufficient volume of air to pafs through ; hence, as the quantity of charcoal confumed, or, what is much the fame thing, the quantity of caloric difengaged, is nearly in proportion to the quantity of air which paffes through the furnace, thefe furnaces do not produce a fufficient effect in a great number of experiments. To remedy this defect, there ought to be two openings GG to the afh-hole ; one of thefe is fhut up when only a moderate fire is required ; and both are kept open when the ftrongeft power of the furnace is to be exerted. The opening of the dome SS ought likewife to be confiderably larger than is ufually made.

It

It is of great importance not to employ retorts of too large fize in proportion to the furnace, as a fufficient fpace ought always to be allowed for the paffage of the air between the
fides of the furnace and the veffel. The retort
A in the figure is too fmall for the fize of the
furnace, yet I find it more eafy to point out the
error than to correct it. The intention of the
dome is to oblige the flame and heat to furround
and ftrike back or reverberate upon every part
of the retort, whence the furnace gets the name
of reverberatory. Without this circumftance
the retort would only be heated in its bottom,
the vapours raifed from the contained fubftance
would condenfe in the upper part, and a continual cohabitation would take place without any
thing paffing over into the receiver, but, by
means of the dome, the retort is equally heated
in every part, and the vapours being forced out,
can only condenfe in the neck of the retort, or
in the recipient.

To prevent the bottom of the retort from being either heated or coolled too fuddenly, it is
fometimes placed in a fmall fand-bath of baked
clay, ftanding upon the crofs bars of the furnace. Likewife, in many operations, the retorts
are coated over with lutes, fome of which are
intended to preferve them from the too fudden
influence of heat or of cold, while others are for
fuftaining the glafs, or forming a kind of fecond

3 N   retort,

retort, which fupports the glafs one during operations wherein the ftrength of the fire might
foften it. The former is made of brick-clay
with a little cow's hair beat up alongft with it,
into a pafte or mortar, and fpread over the glafs
or ftone retorts. The latter is made of pure
clay and pounded ftone-ware mixed together,
and ufed in the fame manner. This dries and
hardens by the fire, fo as to form a true fupple-
mentary retort capable of retaining the materials, if the glafs retort below fhould crack or
foften. But, in experiments which are intended for collecting gafles, this lute, being porous,
is of no manner of ufe.

In a great many experiments wherein very
violent fire is not required, the reverberatory
furnace may be ufed as a melting one, by leaving out the piece called the laboratory, and
placing the dome immediately upon the fire-
place, as reprefented Pl. XIII. Fig. 3. The furnace reprefented in Fig. 4. is very convenient
for fufions; it is compofed of the fire-place and
afh-hole ABD, without a door, and having a
hole E, which receives the muzzle of a pair of
bellows ftrongly luted on, and the dome ABGH,
which ought to be rather lower than is reprefented in the figure. This furnace is not capable of producing a very ftrong heat, but is
fufficient for ordinary operations, and may be
readily moved to any part of the laboratory
where

where it is wanted. Though thefe particular furnaces are very convenient, every laboratory muft be provided with a forge furnace, having a good pair of bellows, or, what is more necef-fary, a powerful melting furnace. I fhall de-fcribe the one I ufe, with the principles upon which it is conftructed.

The air circulates in a furnace in confequence of being heated in its paffage through the burn-ing coals; it dilates, and, becoming lighter than the furrounding air, is forced to rife upwards by the preffure of the lateral columns of air, and is replaced by frefh air from all fides, efpe-cially from below. This circulation of air even takes place when coals are burnt in a common chaffing difh; but we can readily conceive, that, in a furnace open on all fides, the mafs of air which paffes, all other circumftances being equal, cannot be fo great as when it is obliged to pafs through a furnace in the fhape of a hol-low tower, like moft of the chemical furnaces, and confequently, that the combuftion muft be more rapid in a furnace of this latter con-ftruction. Suppofe, for inftance, the furnace ABCDEF open above, and filled with burning coals, the force with which the air paffes through the coals will be in proportion to the difference between the fpecific gravity of two columns equal to AC, the one of cold air without, and the other of heated air within the furnace.

There

There muſt be ſome heated air above the opening AB, and the ſuperior levity of this ought
likewiſe to be taken into conſideration ; but, as
this portion is continually coolled and carried
off by the external air, it cannot produce any
great effect.

But, if we add to this furnace a large hollow
tube GHAB of the ſame diameter, which pre-
ſerves the air which has been heated by the
burning coals from being coolled and diſperſed
by the ſurrounding air, the difference of ſpecific
gravity which cauſes the circulation will then be
between two columns equal to GC.   Hence, if
GC be three times the length of AC, the circulation will have treble force.   This is upon
the ſuppoſition that the air in GHCD is as
much heated as what is contained in ABCD,
which is not ſtrictly the caſe, becauſe the heat
muſt decreaſe between AB and GH ; but, as
the air in GHAB is much warmer than the external air, it follows, that the addition of the
tube muſt increaſe the rapidity of the ſtream of
air, that a larger quantity muſt paſs through
the coals, and conſequently that a greater degree of combuſtion muſt take place.

We muſt not, however, conclude from theſe
principles, that the length of this tube ought to
be indefinitely prolonged ; for, ſince the heat of
the air gradually diminiſhes in paſſing from AB
to GH, even from the contact of the ſides of the
tube,

tube, if the tube were prolonged to a certain degree, we would at laſt come to a point where the ſpecific gravity of the included air would be equal to the air without; and, in this caſe, as the cool air would no longer tend to riſe upwards, it would become a gravitating maſs, reſiſting the aſcenſion of the air below. Beſides, as this air, which has ſerved for combuſtion, is neceſſarily mixed with carbonic acid gas, which is conſiderably heavier than common air, if the tube were made long enough, the air might at laſt approach ſo near to the temperature of the external air as even to gravitate downwards; hence we muſt conclude, that the length of the tube added to a furnace muſt have ſome limit beyond which it weakens, inſtead of ſtrengthening the force of the fire.

From theſe reflections it follows, that the firſt foot of tube added to a furnace produces more effect than the ſixth, and the ſixth more than the tenth; but we have no data to aſcertain at what height we ought to ſtop. This limit of uſeful addition is ſo much the farther in proportion as the materials of the tube are weaker conductors of heat, becauſe the air will thereby be ſo much leſs coolled; hence baked earth is much to be preferred to plate iron. It would be even of conſequence to make the tube double, and to fill the interval with rammed charcoal, which is one of the worſt conductors of heat known;

known; by this the refrigeration of the air will be retarded, and the rapidity of the ftream of air confequently increafed; and, by this means, the tube may be made fo much the longer.

As the fire-place is the hotteft part of a furnace, and the part where the air is moft dilated in its paffage, this part ought to be made with a confiderable widening or belly. This is the more neceffary, as it is intended to contain the charcoal and crucible, as well as for the paffage of the air which fupports, or rather produces the combuftion; hence we only allow the inter-ftices between the coals for the paffage of the air.

From thefe principles my melting furnace is conftructed, which I believe is at leaft equal in power to any hitherto made, though I by no means pretend that it poffeffes the greateft poffible intenfity that can be produced in chemical furnaces. The augmentation of the volume of air produced during its paffage through a melting furnace not being hitherto afcertained from experiment, we are ftill unacquainted with the proportions which fhould exift between the inferior and fuperior apertures, and the abfolute fize of which thefe openings fhould be made is ftill lefs underftood; hence data are wanting by which to proceed upon principle, and we can only accomplifh the end in view by repeated trials.

This

This furnace, which, according to the above
ftated rules, is in form of an eliptical fpheroid,
is reprefented Pl. XIII. Fig. 6. ABCD; it is cut
off at the two ends by two plains, which pafs,
perpendicular to the axis, through the foci of
the elipfe. From this fhape it is capable of containing a confiderable quantity of charcoal,
while it leaves fufficient fpace in the intervals
for the paffage of the air. That no obftacle
may oppofe the free accefs of external air, it is
perfectly open below, after the model of Mr
Macquer's melting furnace, and ftands upon an
iron tripod. The grate is made of flat bars fet
on edge, and with confiderable interftices. To
the upper part is added a chimney, or tube, of
baked earth, ABFG, about eighteen feet long,
and almoft half the diameter of the furnace.
Though this furnace produces a greater heat
than any hitherto employed by chemifts, it is
ftill fufceptible of being confiderably increafed
in power by the means already mentioned, the
principal of which is to render the tube as bad
a conductor of heat as poffible, by making it
double, and filling the interval with rammed
charcoal.

When it is required to know if lead contains
any mixture of gold or filver, it is heated in a
ftrong fire in capfules of calcined bones, which
are called cuppels. The lead is oxydated, becomes vitrified, and finks into the fubftance of
the

the cuppel, while the gold or filver, being incapable of oxydation, remain pure. As lead
will not oxydate without free accefs of air, this
operation cannot be performed in a crucible
placed in the middle of the burning coals of a
furnace, becaufe the internal air, being moftly
already reduced by the combuftion into azotic
and carbonic acid gas, is no longer fit for the
oxydation of metals. It was therefore neceffary
to contrive a particular apparatus, in which the
metal fhould be at the fame time expofed to the
influence of violent heat, and defended from
contact with air rendered incombuftible by its
paffage through burning coals. The furnace
intended for anfwering this double purpofe is
called the cuppelling or effay furnace. It is
ufually made of a fquare form, as reprefented
Pl. XIII. Fig. 8. and 10. having an afh-hole
AABB, a fire-place BBCC, a laboratory CCDD,
and a dome DDEE. The muffle or fmall oven
of baked earth GH, Fig. 9. being placed in the
laboratory of the furnace upon crofs bars of iron,
is adjufted to the opening GG, and luted with
clay foftened in water. The cuppels are placed
in this oven or muffle, and charcoal is conveyed into the furnace through the openings of the
dome and fire-place. The external air enters
through the openings of the afh-hole for fupporting the combuftion, and efcapes by the fuperior opening or chimney at EE ; and air is
admitted

admitted through the door of the muffle GG
for oxydating the contained metal.

Very little reflection is sufficient to discover
the erroneous principles upon which this furnace is constructed. When the opening GG is
shut, the oxydation is produced slowly, and with
difficulty, for want of air to carry it on ; and,
when this hole is open, the stream of cold air
which is then admitted fixes the metal, and ob-
structs the procefs. These inconveniencies may
be easily remedied, by constructing the muffle
and furnace in such a manner that a stream of
fresh external air should always play upon the
surface of the metal, and this air should be
made to pass through a pipe of clay kept continually red hot by the fire of the furnace. By
this means the inside of the muffle will never be
coolled, and procefses will be finished in a few
minutes, which at present require a confiderable
space of time.

Mr Sage remedies these inconveniencies in a
different manner ; he places the cuppel containing lead, alloyed with gold or filver, amongst
the charcoal of an ordinary furnace, and covered by a small porcelain muffle ; when the whole
is fufficiently heated, he directs the blast of a
common pair of hand-bellows upon the furface
of the metal, and completes the cuppellation in
this way with great ease and exactnefs.

3 O   S E C T.

S E C T.   III.

Of increasing the Action of Fire, by using Oxygen Gas instead of Atmospheric Air.

By means of large burning glasses, such as those of Tchirnausen and Mr de Trudaine, a degree of heat is obtained somewhat greater than has hitherto been produced in chemical furnaces, or even in the ovens of furnaces used for baking hard porcelain. But these instruments are extremely expensive, and do not even produce heat sufficient to melt crude platina; so that their advantages are by no means sufficient to compensate for the difficulty of procuring, and even of using them. Concave mirrors produce somewhat more effect than burning glasses of the same diameter, as is proved by the experiments of Messrs Macquer and Beaumé with the speculum of the Abbé Bouriot; but, as the direction of the reflected rays is necessarily from below upwards, the substance to be operated upon must be placed in the air without any support, which renders most chemical experiments impossible to be performed with this instrument.

For

For thefe reafons, I firft endeavoured to employ oxygen gas for combuftion, by filling large
bladders with it, and making it pafs through a
tube capable of being fhut by a ftop cock ; and
in this way I fucceeded in caufing it to fupport
the combuftion of lighted charcoal. The intenfity of the heat produced, even in my firft
attempt, was fo great as readily to melt a fmall
quantity of crude platina. To the fuccefs of
this attempt is owing the idea of the gazometer, defcribed p. 308. *et feq.* which I fubftituted
inftead of the bladders ; and, as we can give
the oxygen gas any neceffary degree of preffure,
we can with this inftrument keep up a continued ftream, and give it even a very confiderable force.

The only apparatus neceffary for experiments
of this kind confifts of a fmall table ABCD,
Pl XII. Fig. 15. with a hole F, through which
paffes a tube of copper or filver, ending in a
very fmall opening at G, and capable of being
opened or fhut by the ftop-cock H. This tube
is continued below the table at *l m n o*, and is
connected with the interior cavity of the gazometer. When we mean to operate, a hole of a few
lines deep muft be made with a chizel in a piece
of charcoal, into which the fubftance to be treated is laid ; the charcoal is fet on fire by means
of a candle and blow-pipe, after which it is expofed

pofed to a rapid ftream of oxygen gas from the extremity G of the tube FG.

This manner of operating can only be ufed with fuch bodies as can be placed, without inconvenience, in contact with charcoal, fuch as metals, fimple earths, &c. But, for bodies whofe elements have affinity to charcoal, and which are confequently decompofed by that fubftance, fuch as fulphats, phofphats, and moft of the neutral falts, metallic glaffes, enamels, &c. we muft ufe a lamp, and make the ftream of oxygen gas pafs through its flame. For this purpofe, we ufe the elbowed blow-pipe ST, inftead of the bent one FG, employed with charcoal. The heat produced in this fecond manner is by no means fo intenfe as in the former way, and is very difficultly made to melt platina. In this manner of operating with the lamp, the fubftances are placed in cuppels of calcined bones, or little cups of porcelain, or even in metallic difhes. If thefe laft are fufficiently large, they do not melt, becaufe, metals being good conductors of heat, the caloric fpreads rapidly through the whole mafs, fo that none of its parts are very much heated.

In the Memoirs of the Academy for 1782, p. 476. and for 1783, p. 573. the feries of experiments I have made with this apparatus may be feen at large. The following are fome of the principal refults.

1. Rock

1. Rock criftal, or pure filicious earth, is in-fufible, but becomes capable of being foftened or fufed when mixed with other fubftances.

2. Lime, magnefia, and barytes, are infu-fible, either when alone, or when combined together ; but, efpecially lime, they affift the fufion of every other body.

3. Argill, or pure bafe of alum, is completely fufible *per fe* into a very hard opake vitreous fubftance, which fcratches glafs like the preci-ous ftones.

4. All the compound earths and ftones are readily fufed into a brownifh glafs.

5. All the faline fubftances, even fixed alkali, are volatilized in a few feconds.

6. Gold, filver, and probably platina, are flowly volatilized without any particular pheno-menon.

7. All other metallic fubftances, except mer-cury, become oxydated, though placed upon charcoal, and burn with different coloured flames, and at laft diffipate altogether.

8. The metallic oxyds likewife all burn with flames. This feems to form a diftinctive character for thefe fubftances, and even leads me to believe, as was fufpected by Bergman, that barytes is a metallic oxyd, though we have not hitherto been able to obtain the metal in its pure or reguline ftate.

9. Some

9. Some of the precious ſtones, as rubies, are capable of being ſoftened and ſoldered together, without injuring their colour, or even diminiſhing their weights. The hyacinth, tho' almoſt equally fixed with the ruby, loſes its colour very readily. The Saxon and Braſilian topaz, and the Braſilian ruby, loſe their colour very quickly, and loſe about a fifth of their weight, leaving a white earth, reſembling white quartz, or unglazed china. The emerald, chryſolite, and garnet, are almoſt inſtantly melted into an opake and coloured glaſs.

10. The diamond preſents a property peculiar to itſelf; it burns in the ſame manner with combuſtible bodies, and is entirely diſſipated.

There is yet another manner of employing oxygen gas for conſiderably increaſing the force of fire, by uſing it to blow a furnace. Mr Achard firſt conceived this idea; but the proceſs he employed, by which he thought to dephlogiſticate, as it is called, atmoſpheric air, or to deprive it of azotic gas, is abſolutely unſatisfactory. I propoſe to conſtruct a very ſimple furnace, for this purpoſe, of very refractory earth, ſimilar to the one repreſented Pl. XIII. Fig. 4. but ſmaller in all its dimenſions. It is to have two openings, as at E, through one of which the nozle of a pair of bellows is to paſs, by which the heat is to be raiſed as high as poſſible with common air; after which, the

ſtream

ftream of common air from the bellows being fuddenly ftopt, oxygen gas is to be admitted by a tube, at the other opening, communicating with a gazometer having the preffure of four or five inches of water. I can in this manner unite the oxygen gas from feveral gazometers, fo as to make eight or nine cubical feet of gas pafs through the furnace; and in this way I expect to produce a heat greatly more intenfe than any hitherto known. The upper orifice of the furnace muft be carefully made of confiderable dimenfions, that the caloric produced may have free iffue, left the too fudden expanfion of that highly elaftic fluid fhould produce a dangerous explofion.

F I N I S.

APPENDIX.

No. I.

TABLE *for Converting Lines, or Twelfth Parts of an Inch, and Fractions of Lines, into Decimal Fractions of the Inch.*

Twelfth Parts of a Line.	Decimal Fractions.	Lines.	Decimal Fractions.
1	0.00694	1	0.08333
2	0.01389	2	0.16667
3	0.02083	3	0.25000
4	0.02778	4	0.33333
5	0.03472	5	0.41667
6	0.04167	6	0.50000
7	0.04861	7	0.58333
8	0.05556	8	0.66667
9	0.06250	9	0.75000
10	0.06944	10	0.83333
11	0.07639	11	0.91667
12	0.08333	12	1.00000

No.

No. II.

Table *for Converting the Obferved Heighths of Water in the Jars of the Pneumato-Chemical Apparatus, expreffed in Inches and Decimals, into Correfponding Heighths of Mercury.*

Water.	Mercury.	Water.	Mercury.
.1	.00737	4.	.29480
.2	.01474	5.	.36851
.3	.02201	6.	.44221
.4	.02948	7.	.51591
.5	.03685	8.	.58961
.6	.04422	9.	.66332
.7	.05159	10.	.73702
.8	.05896	11.	.81072
.9	.06633	12.	.88442
1.	.07370	13.	.96812
2.	.14740	14.	1.04182
3.	.22010	15.	1.11525

No,

No. III.

TABLE *for Converting the Ounce Measures used by Dr Priestly into French and English Cubical Inches.*

Ounce measures.	French cubical inches.	English cubical inches.
1	1.567	1.898
2	3.134	3.796
3	4.701	5.694
4	6.268	7.592
5	7.835	9.490
6	9.402	11.388
7	10.969	13.286
8	12.536	15.184
9	14.103	17.082
10	15.670	18.980
20	31.340	37.960
30	47.010	56.940
40	62.680	75.920
50	78.350	94.900
60	94.020	113.880
70	109.690	132.860
80	125.360	151.840
90	141.030	170.820
100	156.700	189.800
1000	1567.000	1898.000

No.

No IV. Additional.

Table *for Reducing the Degrees of Reaumeur's Thermometer into its corresponding Degrees of Fahrenheit's Scale.*

R.	F.	R.	F.	R.	F.	R.	F.
0=32		21=	79.25	41=	124.25	61=	169.25
1=34.25		22=	81.5	42=	126.5	62=	171.5
2=36.5		23=	83.75	43=	128.75	63=	173.75
3=38.75		24=	86	44=	131	64=	176.
4=41		25=	88.25	45=	133.25	65=	178.25
5=43.25		26=	90.5	46=	135.5	66=	180.5
6=45.5		27=	92.75	47=	137.75	67=	182.75
7=47.75		28=	95	48=	140	68=	185
8=50		29=	97.25	49=	142.25	69=	187.25
9=52.25		30=	99.5	50=	144.5	70=	189.5
10=54.5		31=	101.75	51=	146.75	71=	191.75
11=56.75		32=	104	52=	149	72=	194.
12=59		33=	106.25	53=	151.25	73=	196 25
13=61.25		34=	108.5	54=	153.5	74=	198.5
14=63.5		35=	110.75	55=	155.75	75=	200.75
15=65.75		36=	113	56=	158	76=	203
16=68		37=	115.25	57=	160.25	77=	205.25
17=70.25		38=	117.5	58=	162.5	78=	207.5
18=72.5		39=	119.75	59=	164.75	79=	209.75
19=74.75		40=	122	60=	167	80=	212
20=77							

Note—Any degree, either higher or lower, than what is contained in the above Table, may be at any time converted, by remembering that one degree of Reaumeur's scale is equal to 2.25° of Fahrenheit ; or it may be done without the Table by the following formula, $\frac{R \times 9}{4} + 32 = F$; that is, multiply the degree of Reaumeur by 9, divide the product by 4, to the quotient add 32, and the sum is the degree of Fahrenheit. --E.

No. V. Additional.

RULES *for converting French Weights and Meafures into correfpondent Englifh Denominations* *.

§ 1. *Weights.*

The Paris pound, poids de mark of Charlemagne, contains 9216 Paris grains; it is divided into 16 ounces, each ounce into 8 gros, and each gros into 72 grains. It is equal to 7561 Englifh Troy grains.

The Englifh Troy pound of 12 ounces contains 5760 Englifh Troy grains, and is equal to 7021 Paris grains.

The Englifh averdupois pound of 16 ounces contains 7000 Englifh Troy grains, and is equal to 8538 Paris grains.

To reduce Paris *grs.* to Englifh Troy
grs. divide by . . .
To reduce Englifh Troy *grs.* to Pa- } 1.2189
ris *grs.* multiply by . . .

To reduce Paris ounces to Englifh
Troy, divide by . . .
To reduce Englifh Troy ounces to } 1.015734
Paris, multiply by . .

Or

---

* For the materials of this Article the Tranflator is indebted to Profeffor Robertfon.

Or the converfion may be made by means
of the following Tables.

I. *To reduce French to Englifh Troy Weight.*

The Paris pound = 7561
The ounce   =   472.5625 } Englifh.
The gros   =   59.0703 } Troy.
The grain   =   .8194 } Grains.

II. *To Reduce Englifh Troy to Paris Weight.*

The Englifh Troy pound } = 7021.
of 12 ounces
The Troy ounce   = 585.0830 | Paris
The dram of 60 grs.   = 73.1353 }
The penny weight, or } = 29.2540 | grains.
denier, of 24 grs. }
The fcruple, of 20 grs. = 24.3784 }

III. *To Reduce Englifh Averdupois to Paris Weight.*

The averdupois pound of }
16 ounces, or 7000 } = 8538.   } Paris
Troy grains. }   } grains.
The ounce   ·   = 533.6250 }

§ 2.

§ 2. *Long and Cubical Meafures.*

To reduce Paris feet or inches into
   Englifh, multiply by   -
Englifh feet or inches into Paris, } 1.065977
   divide by - - -

To reduce Paris cubic feet or inches to Englifh, multiply by -
Englifh cubic feet or inches to Pa- } 1.211278
   ris, divide by - - -

Or by means of the following tables :

IV. *To Reduce Paris Long Meafure to Englifh.*

The Paris royal foot of
   12 inches - - } =12.7977] Englifh
The inch - - = 1.0659
The line, or $\frac{1}{12}$ of an inch = .0888] inches.
The $\frac{1}{12}$ of a line - = .0074

V. *To Reduce Englifh Long Meafure to French,*

The Englifh foot =11.2596
The inch - = .9383
The $\frac{1}{8}$ of an inch = .1173 } Paris inches.
The $\frac{1}{10}$ - = .0938
The line, or $\frac{1}{12}$ = .0782

VI.

VI. *To Reduce French Cube Measure to English.*

The Paris
cube foot = 1.211278 } English cubical feet, or } 2093.088384 } inches.
The cubic
inch   = .000700 }   1.211278 }

VII. *To Reduce English Cube Measure to French.*

The English cube foot,
or 1728 cubical inches } = 1427.4864 } French
The cubical inch   =   .8260 } cubical
The cube tenth   =   .0008 } inches.

§ 3.   *Measure of Capacity.*

The Paris pint contains 58.145 * English cubical inches, and the English wine pint contains
28.85 cubical inches; or, the Paris pint contains
2.01508

* It is said, *Belidor Archit. Hydrog.* to contain 31 *oz.*
64 *grs.* of water, which makes it 58.075 English inches; but, as there is confiderable uncertainty in the de-
terminations of the weight of the French cubical meafure of water, owing to the uncertainty of the ftandards
made ufe of, it is better to abide by Mr Everard's
meafure, which was with the Exchequer ftandards, and
by the proportions of the English and French foot, as
eftablished by the French Academy and Royal Society.

2.01508 Englifh pints, and the Englifh pint contains .49617 Paris pints; hence,

To reduce the Paris pint to the Eng-
  lifh, multiply by ⋅ -
To reduce the Englifh pint to the } 2.01508.
  Paris, divide by - -

3 Q   No.

No. VI.

TABLE *of the Weights of the different Gasses, at*
28 French inches, or 29.84 English inches ba-
rometrical pressure, and at 10° (54.5°) of tem-
perature, expressed in English measure and En-
glish Troy weight.

Names of the Gasses.	Weight of a cubical inch.	Weight of a cubical foot.		
*	qrs.	oz.	dr.	qrs.
Atmospheric air	.32112	1	1	15
Azotic gas	.30064	1	0	39.5
Oxygen gas	.34211	1	1	51
Hydrogen gas	.02394	0	0	41.26
Carbonic acid gas	.44108	1	4	41
**				
Nitrous gas	.37000	1	2	39
Ammoniacal gas	.18515	0	5	19.73
Sulphurous acid gas	.71580	2	4	38

No.

* These five were ascertained by Mr Lavoisier himself.—E.

** The last three are inserted by Mr Lavoisier upon
the authority of Mr Kirwan.—E.

No. VII.

TABLES *of the Specific Gravities of different bodies.*

§ 1. *Metallic Subſtances.*

G O L D.

Pure gold of 24 carats melted but not hammered . . .	19.2581
The ſame hammered . .	19.3617
Gold of the Pariſian ſtandard, 22 carats fine, not hammered * .	17.4863
The ſame hammered . .	17.5894
Gold of the ſtandard of French coin, 21$\frac{2}{3}\frac{2}{4}$ carats fine, not hammered	17.4022
The ſame coined . .	17.6474
Gold of the French trinket ſtandard, 20 carats fine, not hammered .	15.7090
The ſame hammered . .	15.7746

S I L V E R.

Pure or virgin ſilver, 24 deniers, not hammered . . .	10.4743
The ſame hammered . .	10.5107
Silver of the Paris ſtandard, 11 deniers 10 grains fine, not hammered †	10.1752
The ſame hammered .	10.3765

Silver,

* The ſame with Sterling.
† This is 10 *grs.* finer than Sterling.

Silver, ftandard of French coin, 10 de-
niers 21 grains fine, not hammered   10.0476
The fame coined   .   .   10.4077

PLATINA.

Crude platina in grains   .   .   15.6017
The fame, after being treated with mu-
riatic acid   .   .   .   16.7521
Purified platina, not hammered   .   19.5000
The fame hammered   ᴄ   .   20.3366
The fame drawn into wire   .   21.0417
The fame paffed through rollers   22.0690

COPPER AND BRASS.

Copper, not hammered   .   .   7.7880
The fame wire drawn   .   .   8.8785
Brafs, not hammered   .   .   8.3958
The fame wire drawn   .   .   8.5441

IRON AND STEEL.

Caft iron   .   .   .   7.2070
Bar iron, either fcrewed or not   .   7.7880
Steel neither tempered nor fcrewed   7.8331
Steel fcrewed but not tempered   .   7.8404
Steel tempered and fcrewed   .   7.8180
Steel tempered and not fcrewed   .   7.8163

TIN.

T I N.

Pure tin from Cornwall melted and not screwed	7.2914
The same screwed	7.2994
Malacca tin, not screwed	7.2963
The same screwed	7.3065
Molten lead	11.3523
Molten zinc	7.1908
Molten bismuth	9.8227
Molten cobalt	7.8119
Molten arsenic	5.7633
Molten nickel	7.8070
Molten antimony	6.7021
Crude antimony	4.0643
Glass of antimony	4.9464
Molybdena	4.7385
Tungstein	6.0665
Mercury	13.5681

§ 2. *Precious Stones.*

White Oriental diamond	3.5212
Rose-coloured Oriental ditto	3.5310
Oriental ruby	4.2833
Spinell ditto	3.7600
Ballas ditto	3.6458
Brasillian ditto	3.5311
Oriental topas	4.0106

Ditto

Ditto Piftachio ditto	4.0615
Brafillian ditto	3.5365
Saxon topas	3.5640
Ditto white ditto	3.5535
Oriental faphir	3.9941
Ditto white ditto	3.9911
Saphir of Puy	4.0769
Ditto of Brafil	3.1307
Girafol	4.0000
Ceylon jargon	4.4161
Hyacinth	3.6873
Vermillion	4.2299
Bohemian garnet	4.1888
Dodecahedral ditto	4.0627
Syrian ditto	4.0000
Volcanic ditto, with 24 fides	2.4684
Peruvian emerald	2.7755
Cryfolite of the jewellers	2.7821
Ditto of Brafil	2.6923
Beryl, or Oriental aqua marine	3.5489
Occidental aqua marine	2.7227

§ 3. *Silicious Stones.*

Pure rock criftal of Madagafcar	2.6530
Ditto of Brafil	2.6526
Ditto of Europe, or gelatinous	2.6548
Criftallized quartz	2.6546
Amorphous ditto	2.6471

Oriental

Oriental agate . . .	2.5901
Agate onyx . . .	2.6375
Tranfparent calcedony . .	2.6640
Carnelian . . .	2.6137
Sardonyx . . .	2.6025
Prafe	2.5805
Onyx pebble . . .	2.6644
Pebble of Rennes . .	2.6538
White jade . - .	2.9502
Green jade . . .	2.9660
Red jafper . . .	2.6612
Brown ditto . .	2.6911
Yellow ditto . .	2.7101
Violet ditto . . .	2.7111
Gray ditto . . .	2.7640
Jafponyx . . .	2.8160
Black prifmatic hexahedral fchorl .	3.3852
Black fpary ditto . . .	3.3852
Black amorphous fchorl, called antique bafaltes . . .	2.9225
Paving ftone . . .	2.4158
Grind ftone . . .	2.1429
Cutler's ftone . . .	2.1113
Fountainbleau ftone . .	2.5616
Scyth ftone of Auvergne .	2.5638
Ditto of Lorrain . .	2.5298
Mill ftone . . .	2.4835
White flint . . .	2.5941
Blackifh ditto . . .	2.5817

§ 4.

§ 4. *Various Stones, &c.*

Opake green Italian ferpentine, or gabro of the Florentines . .	2.4295
Coarfe Briancon chalk . .	2.7274
Spanifh chalk . . .	2.7902
Foliated lapis ollaris of Dauphiny .	2.7687
Ditto ditto from Sweden . .	2.8531
Mufcovy talc . . .	2.7917
Black mica . . .	2.9004
Common fchiftus or flate .	2.6718
New flate . . .	2.8535
White rafor hone . .	2.8763
Black and white hone . .	3.1311
Rhombic or Iceland criftal .	2.7151
Pyramidal calcareous fpar . .	2.7141
Oriental or white antique alabafter	2.7302
Green Campan marble . .	2.7417
Red Campan marble . .	2.7242
White Carara marble . .	2.7168
White Parian marble . .	2.8376
Various kinds of calcareous ftones ⎱ ufed in France for building. ⎰	from 1.3864 to 2.3902
Heavy fpar . . .	4.4300
White fluor . . .	3.1555
Red ditto . . .	3.1911
Green ditto . . .	3.1817
Blue ditto . . .	3.1688
Violet ditto . . .	3.1757

Red

Red fcintilant zeolite from Edelfors	2.4868
White fcintilant zeolite . .	2.0739
Criftallized zeolite . .	2.0833
Black pitch ftone . .	2.0499
Yellow pitch ftone . .	2.0860
Red ditto . . .	2.6695
Blackifh ditto . . .	2.3191
Red porphyry . . .	2.7651
Ditto of Dauphiny . .	2.7033
Green ferpentine . .	2.8960
Black ditto of Dauphiny, called variolite	2.9339
Green ditto from Dauphiny .	2.9883
Ophites . . .	2.9722
Granitello . . .	3.0626
Red Egyptian granite . .	2.6541
Beautiful red granite . .	2.7609
Granite of Girardmas . .	2.7163
Pumice ftone . . .	.9145
Lapis obfidianus . .	2.3480
Pierre de Volvic . .	2.3205
Touch ftone . . .	2.4153
Bafaltes from Giants Caufeway .	2.8642
Ditto prifmatic from Auvergne .	2.4153
Glafs gall . . .	2.8548
Bottle glafs . . .	2.7325
Green glafs . . .	2.6423
White glafs . . .	2.8922
St Gobin criftal . .	2.4882
Flint glafs . . .	3.3293
Borax glafs . . .	2.6070

3 R   Seves

Seves porcelain	2.1457
Limoges ditto	2.3410
China ditto	2.3847
Native fulphur	2.0332
Melted fulphur	1.9907
Hard peat	1.3290
Ambergreafe	.9263
Yellow tranfparent amber.	1.0780

§ 5. *Liquids.*

Diftilled water	1.0000
Rain water	1.0000
Filtered water of the Seine	1.00015
Arcueil water	1.00046
Avray water	1.00043
Sea water	1.0263
Water of the Dead Sea	1.2403
Burgundy wine	.9915
Bourdeaux ditto	.9939
Malmfey Madeira	1.0382
Red beer	1.0338
White ditto	1.0231
Cyder	1.0181
Highly rectified alkohol	.8293
Common fpirits of wine	.8371

Alkohol

Alkohol	15 pts. water	1 part.	.8527
	14	2	.8674
	13	3	.8815
	12	4	.8947
	11	5	.9075
	10	6	.9199
	9	7	.9317
	8	8	.9427
	7	9	.9519
	6	10	.9594
	5	11	.9674
	4	12	.6733
	3	13	.9791
	2	14	.9852
	1	15	.9919

Sulphuric ether   •   •   •   .7394
Nitric ether   •   •   •   •   .9088
Muriatic ether   •   •   •   .7298
Acetic ether   •   •   •   .8664
Sulphuric acid   •   •   •   •   1.8409
Nitric ditto   •   •   •   1.2715
Muriatic ditto   •   •   •   •   1.1940
Red acetous ditto   •   •   •   1.0251
White acetous ditto   •   •   •   1.0135
Diſtilled ditto ditto   •   •   1.0095
Acetic ditto   •   •   •   •   1.0626
Formic ditto   •   •   •   •   .9942
Solution of cauſtic ammoniac, or vola-
til alkali fluor   •   •   .8970

Eſſential

Effential or volatile oil of turpentine	.8697
Liquid turpentine	.9910
Volatile oil of lavender	.8938
Volatile oil of cloves	1.0363
Volatile oil of cinnamon	1.0439
Oil of olives	.9153
Oil of fweet almonds	.9170
Lintfeed oil	.9403
Oil of poppy feed	.9288
Oil of beech maft	.9176
Whale oil	.9233
Womans milk	1.0203
Mares milk	1.0346
Afs milk	1.0355
Goats milk	1.0341
Ewe milk	1.0409
Cows milk	1.0324
Cow whey	1.0193
Human urine	1.0106

§ 6. *Refins and Gums*

Common yellow or white rofin	1.0727
Arcanfon	1.0857
Galipot *	1.0819
Baras *	1.0441
Sandarac	

* Refinous juices extracted in France from the Pine.
Vide Bomare's **Dict.**

Sandarac	1.0920
Maftic	1.0742
Storax	1.1098
Opake copal	1.1398
Tranfparent ditto	1.0452
Madagafcar ditto	1.0600
Chinefe ditto	1.0628
Elemi	1.0182
Oriental anime	1.0284
Occidental ditto	1.0426
Labdanum	1.1862
Ditto *in tortis*	2.4933
Refin of guaiac	1.2289
Ditto of jallap	1.2185
Dragons blood	1.2045
Gum lac	1.1390
Tacamahaca	1.0463
Benzoin	1.0924
Alouchi *	1.0604
Caragna †	1.1244
Elaftic gum	.9335
Camphor	.9887
Gum ammoniac	1.2071
Sagapenum	1.2008
	Ivy

* Odoriferous gum from the tree which produces the Cortex Winteranus. *Bomare.*

† Refin of the tree called in Mexico Caragna, or Tree of Madnefs. *Ibid.*

Ivy gum * . . .	1.2948
Gamboge . . .	1.2216
Euphorbium . . .	1.1244
Olibanum . . .	1.1732
Myrrh	1.3600
Bdellium . . .	1.3717
Aleppo Scamony . .	1.2354
Smyrna ditto . . .	1.2743
Galbanum . . .	1.2120
Affafoetida . . .	1.3275
Sarcocolla	1.2684
Opoponax . . .	1.6226
Cherry tree gum . . .	1.4817
Gum Arabic . . .	1.4523
Tragacanth . . .	1.3161
Bafora gum	1.4346
Acajou gum † . . .	1.4456
Monbain gum ‡ . . .	1.4206
Infpiffated juice of liquorice . .	1.7228
————— Acacia . .	1.5153
————— Areca .	1.4573
Terra Japonica	1.3980
Hepatic aloes . . .	1.3586
Socotrine aloes . .	1.3795
Infpiffated juice of St John's wort .	1.5263
	Opium

* Extracted in Perfia and the warm countries from Hedera terreftris.—*Bomare.*

† From a Brafilian tree of this name.—*Ibid.*

‡ From a tree of this name.—*Ibid.*

Opium . . .	1.3366
Indigo . . .	.7690
Arnotto	.5956
Yellow wax . . .	.9648
White ditto . . .	.9686
Ouarouchi ditto * . . .	.8970
Cacao butter . . .	.8916
Spermaceti . . .	.9433
Beef fat	.9232
Veal fat	.9342
Mutton fat	.9235
Tallow	.9419
Hoggs fat . . .	.9368
Lard	.9478
Butter	.9423

§ 7. *Woods.*

Heart of oak 60 years old . .	1.1700
Cork	.2400
Elm trunk . . .	.6710
Aſh ditto . . .	.8450
Beech	.8520
Alder	.8000
Maple	.7550
Walnut	.6710
Willow	.5850
Linden . . .	.6040
	Male

* The produce of the Tallow Tree of Guayana. *Vid:*
Bomare's Dict.

Male fir . . .	.5500
Female ditto . . .	.4980
Poplar . . .	.3830
White Spanish ditto . .	.5294
Apple tree . . .	.7930
Pear tree . . .	.6610
Quince tree	.7050
Medlar - . . .	.9440
Plumb tree	.7850
Olive wood . . .	.9270
Cherry tree . . .	.7150
Filbert tree . . .	.6000
French box . . .	.9120
Dutch ditto . . .	1.3280
Dutch yew . . .	.7880
Spanish ditto . . .	.8070
Spanish cyprefs . .	.6440
American cedar . .	.5608
Pomgranate tree . . .	1.3540
Spanish mulberry tree . .	.8970
Lignum vitae . . .	1.3330
Orange tree . . .	.7050

No.

Note—The numbers in the above Table, if the Decimal point be carried thrèe figures farther to the right hand, nearly exprefs the abfolute weight of an Englifh cube foot of each fubftance in averdupois ounces. See No. VIII. of the Appendix.—E.

No. VIII. Additional.

Rules *for Calculating the Abfolute Gravity in Englifh Troy Weight of a Cubic Foot and Inch, Englifh Meafure, of any Subftance whofe Specific Gravity is known* *.

In 1696, Mr Everard, balance-maker to the Exchequer, weighed before the Commiffioners of the Houfe of Commons 2145.6 cubical inches, by the Exchequer ftandard foot, of diftilled water, at the temperature of 55° of Fahrenheit, and found it to weigh 1131 oz. 14 dts. Troy, of the Exchequer ftandard. The beam turned with 6 grs. when loaded with 30 pounds in each fcale. Hence, fuppofing the pound averdupois to weigh 7000 grs. Troy, a cubic foot of water weighs $62\frac{1}{2}$ pounds averdupois, or 1000 ounces averdupois, wanting 106 grains Troy. And hence, if the fpecific gravity of water be called 1000, the proportional fpecific gravities of all other bodies will nearly exprefs the number of averdupois ounces in a cubic foot. Or more accurately, fuppofing the fpecific gravity of water expreffed by 1. and of all other bodies in proportional numbers, as the

3 S   cubic

The whole of this and the following article was communicated to the Tranflator by Profeffor Robinfon. —E.

cubic foot of water weighs, at the above temperature, exactly 437489.4 grains Troy, and
the cubic inch of water 253.175 grains, the
abfolute weight of a cubical foot or inch of
any body in Troy grains may be found by mul-
tiplying their fpecific gravity by either of the
above numbers refpectively.

By Everard's experiment, and the proportions of the Englifh and French foot, as eftablifhed by the Royal Society and French Academy of Sciences, the following numbers are
afcertained.

Paris grains in a Paris cube foot of
   water   -   -   -   = 645511
Englifh grains in a Paris cube foot
   of water   -   -   -   = 529922
Paris grains in an Englifh cube foot
   of water   -   -   -   = 533247
Englifh grains in an Englifh cube
   foot of water   -   = 437489.4
Englifh grains in an Englifh cube
   inch of water   -   -   = 253.175

By an experiment of Picard with
   the meafure and weight of the
   Chatelet, the Paris cube foot of
   water contains of Paris grains   = 641326
By one of Du Hamel, made with
   great care   -   -   -   = 641376
By Homberg   -   -   -   = 641666

Thefe

These show some uncertainty in measures or in weights; but the above computation from Everard's experiment may be relied on, because the comparison of the foot of England with that of France was made by the joint labours of the Royal Society of London and the French Academy of Sciences: It agrees likewise very nearly with the weight assigned by Mr Lavoisier, 70 Paris pounds to the cubical foot of water.

No.

No. IX.

TABLES *for Converting Ounces, Drams, and Grains, Troy, into Decimals of the Troy Pound of* 12 *Ounces, and for Converting Decimals of the Pound Troy into Ounces,* &c.

I. *For Grains.*

Grains = Pound.		Grains = Pound.	
1	.0001736	100	.0173611
2	.0003472	200	.0374222
3	.0005208	300	.0520833
4	.0006944	400	.0694444
5	.0008681	500	.0868055
6	.0010417	600	.1041666
7	.0012153	700	.1215277
8	.0013889	800	.1388888
9	.0015625	900	.1562499
10	.0017361	1000	.1736110
20	.0034722	2000	.3472220
30	.0052083	3000	.5208330
40	,0069444	4000	.6944440
50	.0086806	5000	.8680550
60	.0104167	6000	1.0418660
70	.0121528	7000	1.2152770
80	.0138889	8000	1.3888880
90	.0156250	9000	1.5624990

II.

II. *For Drams.*

Drams =	Pound.
1	.0104167
2	.0208333
3	.0312500
4	.0416667
5	.0520833
6	.0625000
7	.0729167
8	.0833333

III. *For Ounces.*

Ounces =	Pounds.
1	.0833333
2	.1666667
3	.2500000
4	.3333333
5	.4166667
6	.5000000
7	.5833333
8	.6666667
9	.7500000
10	.8333333
11	.9166667
12	1.0000000

IV.

IV. *Decimals of the Pound into Ounces, &c.*

Tenth parts.				*Thousandths.*	
lib. =	*oz.*	*dr.*	*gr.*	*lib.* =	*grs.*
0.1	1	1	36	0.006	34.56
0.2	2	3	12	0.007	40.32
0.3	3	4	48	0.008	46.08
0.4	4	6	24	0.009	51.84
0.5	6	0	0	*Ten thousandth parts.*	
0.6	7	1	36	0.0001	0.576
0.7	8	3	12	0.0002	1.152
0.8	9	4	48	0.0003	1.728
0.9	10	6	24	0.0004	2.304
Hundredth parts.				0.0005	2.880
0.01	0	0	57.6	0.0006	3.456
0.02	0	1	55.2	0.0007	4.032
0.03	0	2	52.8	0.0008	4.608
0.04	0	3	50.4	0.0009	5.184
0.05	0	4	48.0	*Hundred thousandth*	
0.06	0	5	45.6	*parts.*	
0.07	0	6	43.2	0.00001	0.052
0.08	0	7	40.8	0.00002	0.115
0.09	0	3	38.4	0.00003	0.173
Thousandths.				0.00004	0.230
0.001	0	0	5.76	0.00005	0.288
0.002	0	0	11.52	0.00006	0.346
0.003	0	0	17.28	0.00007	0.403
0.004	0	0	23.04	0.00008	0.461
0.005	0	0	28.80	0.00009	0.518

No.

No. X.

TABLE *of the English Cubical Inches and Decimals corresponding to a determinate Troy Weight of Distilled Water at the Temperature of 55°, calculated from Everard's experiment.*

For Grains.		For Ounces.	
Grs.	Cubical inches.	Oz.	Cubical inches.
1 =	.0039	1 =	1.8927
2	.0078	2	3.7855
3	.0118	3	5.6782
4	.0157	4	7.5710
5	.0197	5	9.4631
6	.0236	6	11.3565
7	.0275	7	13.2493
8	.0315	8	15.1420
9	.0354	9	17.0748
10	.0394	10	18.9276
20	.0788	11	20.8204
30	.1182		
40	.1577		For Pounds.
50	.1971	Libs.	Cubical inches.

For Drams.			
		1 =	22.7131
		2	45.4263
Drams.	Cubical inches.	3	68.1394
1 =	.2365	4	90.8525
2	.4731	5	113.5657
3	.7094	6	136.2788
4	.9463	7	158.9919
5	1.1829	8	181.7051
6	1.4195	9	204.4183
7	1.6561	10	227.1314
		50	1135.6574
		100	2271.3148
		1000	22713.1488

THE END.

THE PLATES

Plate I

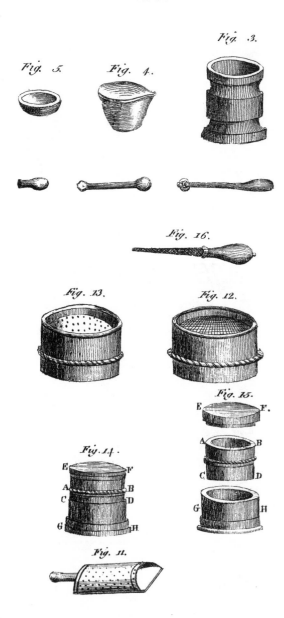

Plate I (continued)

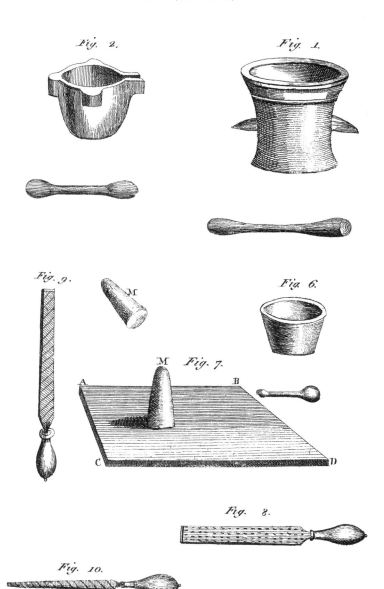

Fig. 2.

Fig. 1.

Fig. 9.

M

Fig. 6.

M   Fig. 7.

A   B

C   D

Fig. 8.

Fig. 10.

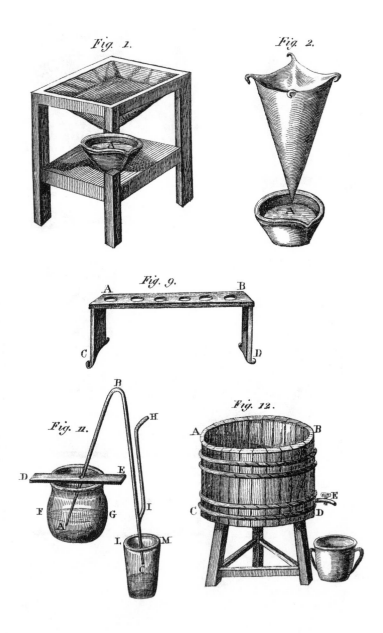

Plate II

Fig. 1.

Fig. 2.

Fig. 9.

Fig. 11.

Fig. 12.

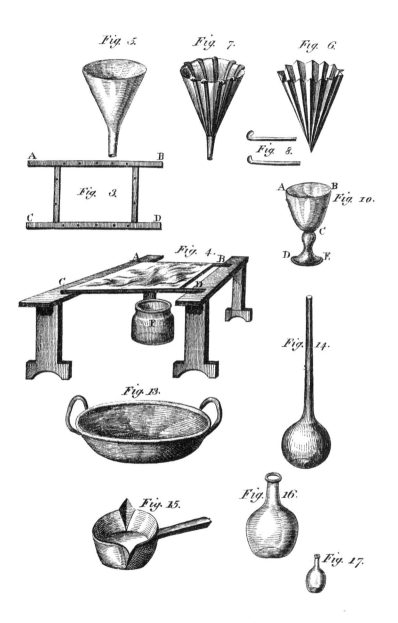

Plate II (continued)

Plate III

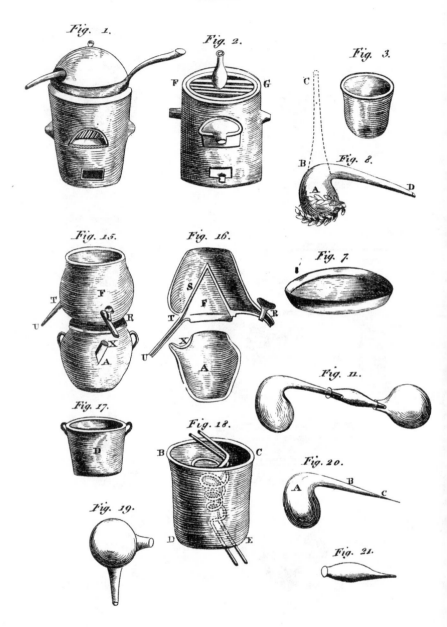

Fig. 1. Fig. 2. Fig. 3. Fig. 8. Fig. 15. Fig. 16. Fig. 7. Fig. 11. Fig. 17. Fig. 18. Fig. 20. Fig. 19. Fig. 21.

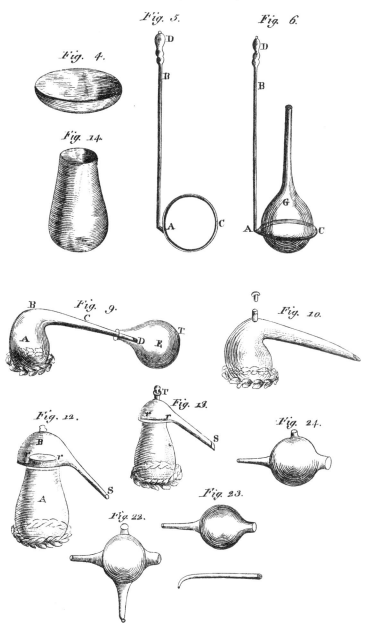

Plate III (*continued*)

Fig. 4.

Fig. 5.

Fig. 6.

Fig. 14.

Fig. 9.

Fig. 10.

Fig. 12.

Fig. 13.

Fig. 24.

Fig. 23.

Fig. 22.

Plate IV

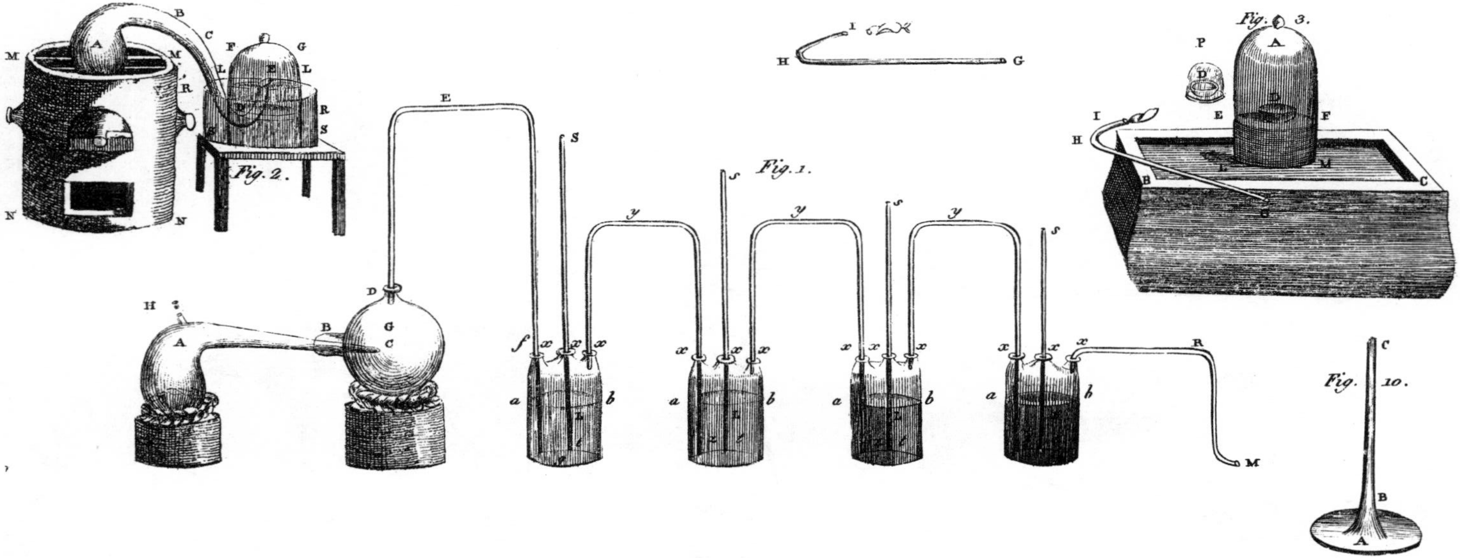

Fig. 3.

Fig. 10.

Fig. 1.

Fig. 2.

Plate IV (continued).

Fig. 11.

Fig. A. 17.

Fig. 9.

Fig. 15.

Fig. 16.

Fig. 14.

Fig. 13.

Fig. 7.

Fig. 12.

Fig. 5.

Fig. 4.

Fig. 8.

Fig. 6.

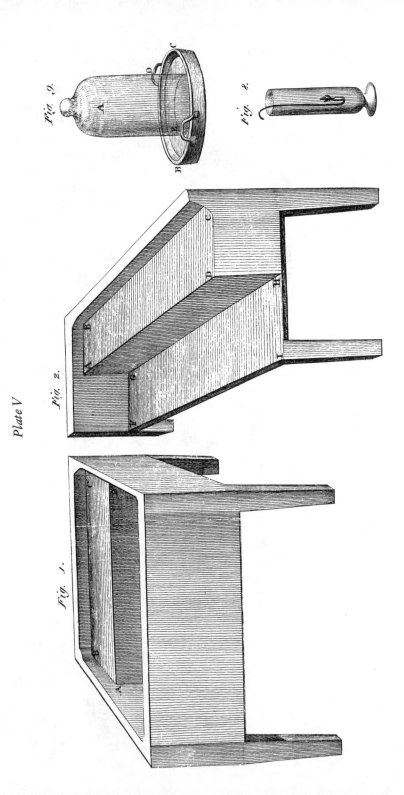

Plate V

Fig. 2.

Fig. 1.

Fig. 9.

Fig. 8.

Plate V (continued)

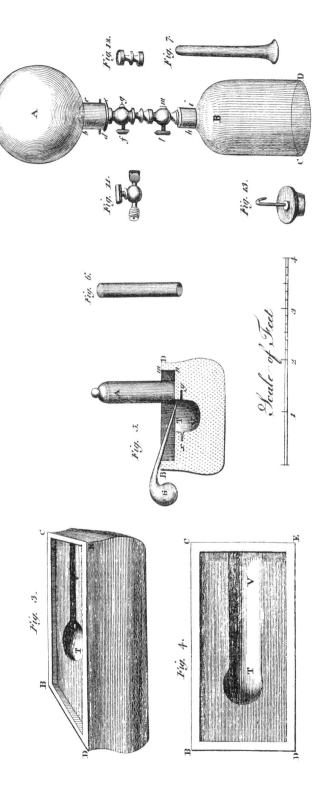

Scale of Feet

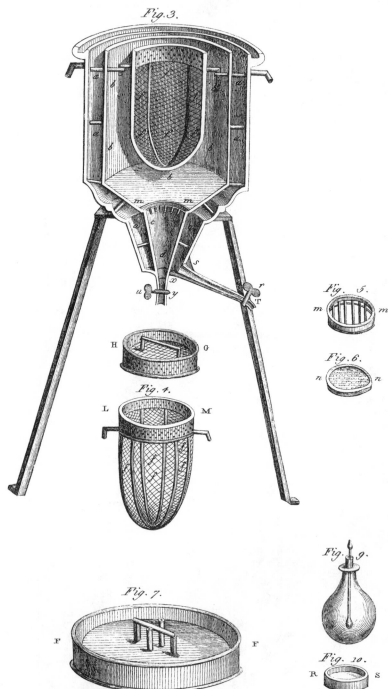

Plate VI

Fig. 3.

Fig. 5.

Fig. 6.

Fig. 4.

Fig. 7.

Fig. 9.

Fig. 10.

Plate VI (continued)

Fig. 1.

Scale of Feet.

Fig. 2.

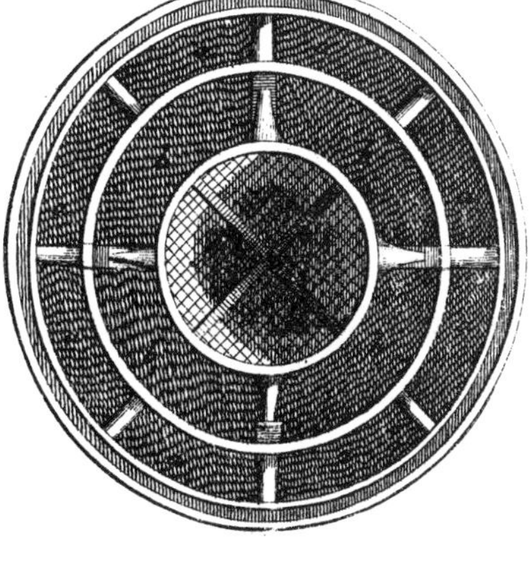

Fig. 8.

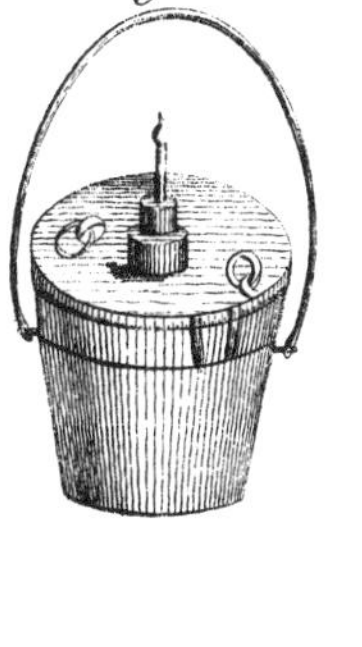

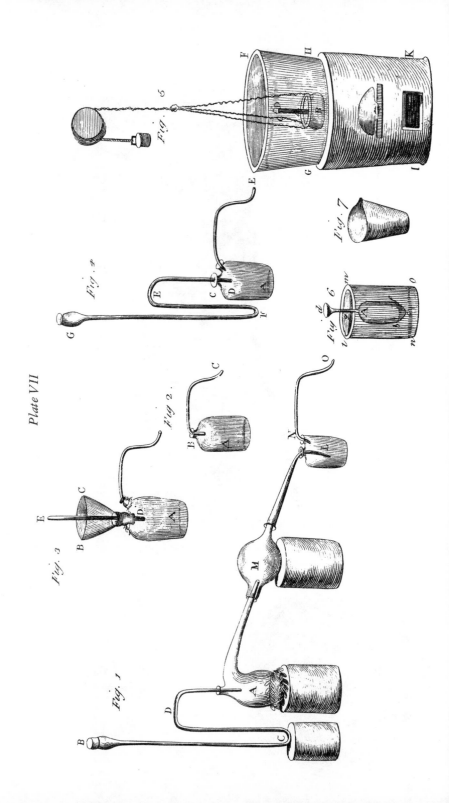

Plate VII

Fig. 1

Fig. 2

Fig. 3

Fig. 4

Fig. 5

Fig. 6

Fig. 7

Plate VII (continued)

Fig. 8

Fig. 11

Fig. 17

Fig. 9

Fig. 16

Fig. 10

Fig. 15

Fig. 13

Fig. 12

Fig. 14

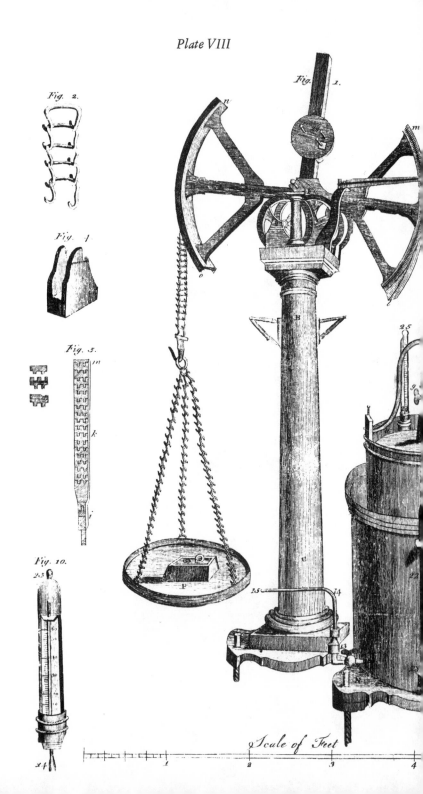

Plate VIII

Fig. 2.

Fig. 4.

Fig. 5.

Fig. 10.

Fig. 1.

Scale of Feet

Plate VIII (continued)

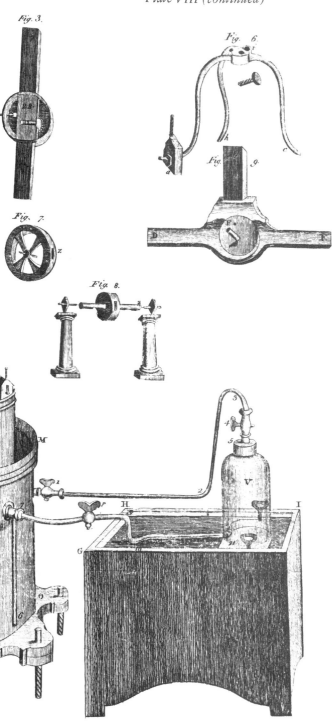

Fig. 3.

Fig. 6.

Fig. 7.

Fig. 8.

Fig. 9.

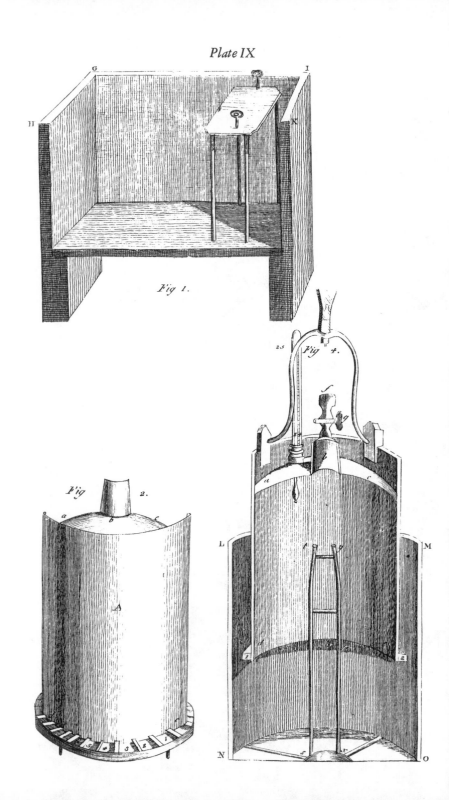

Plate IX

Fig 1.

Fig. 2.

Fig. 4.

Plate IX (continued)

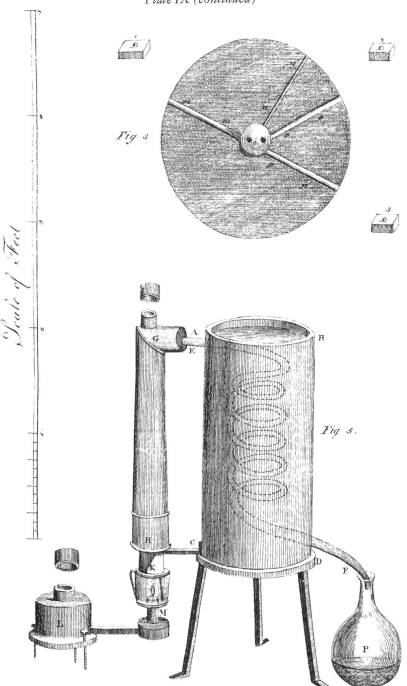

Fig 3

Fig 5.

Scale of Feet

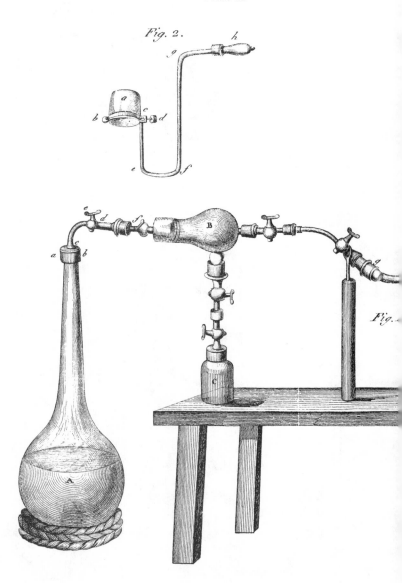

Plate X

Fig. 2.

Fig.

Plate X (continued)

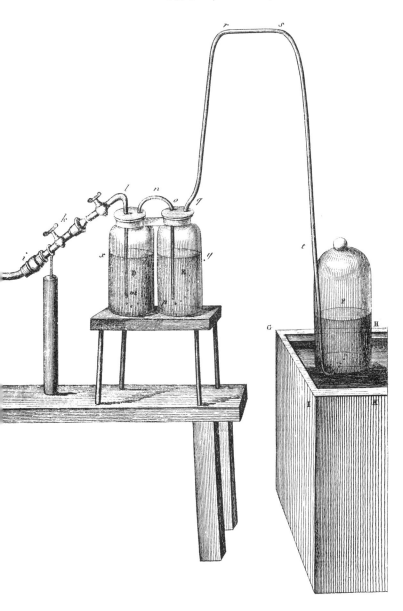

Plate XI

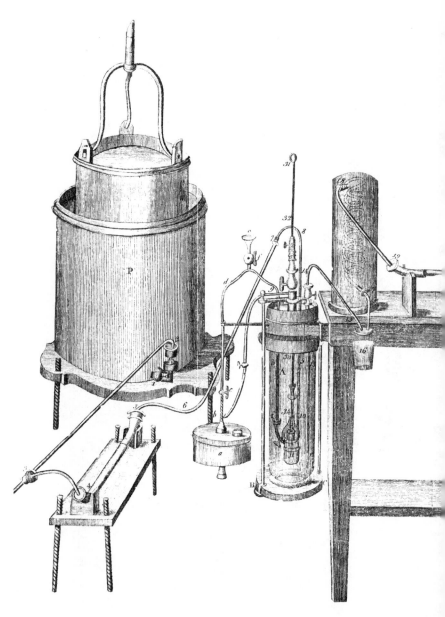

Plate XI *(continued)*

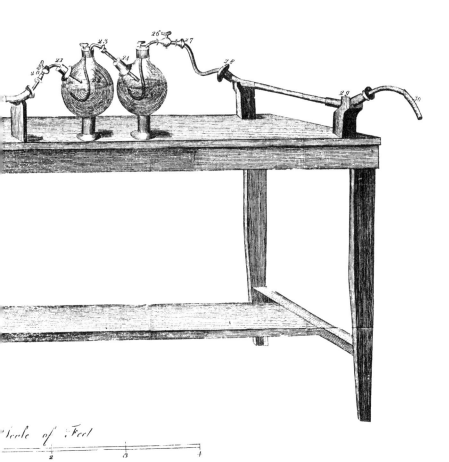

Plate XII

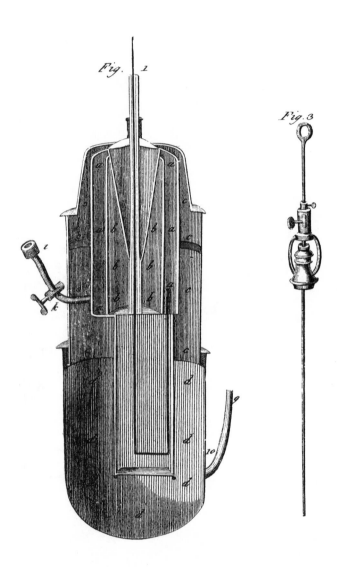

Plate XII (continued)

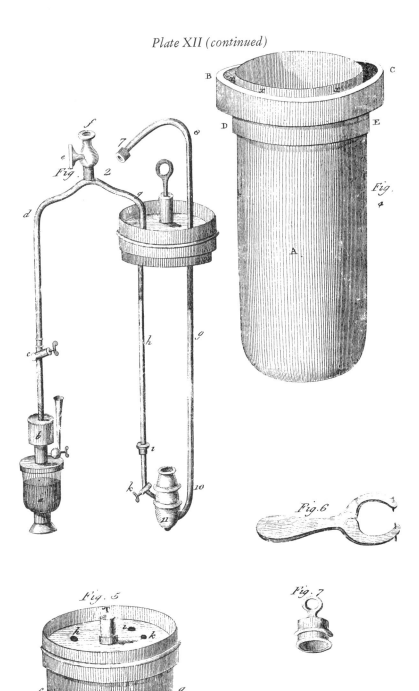

Fig.

Fig. 4

Fig. 6

Fig. 5

Fig. 7

Plate XII (continued)

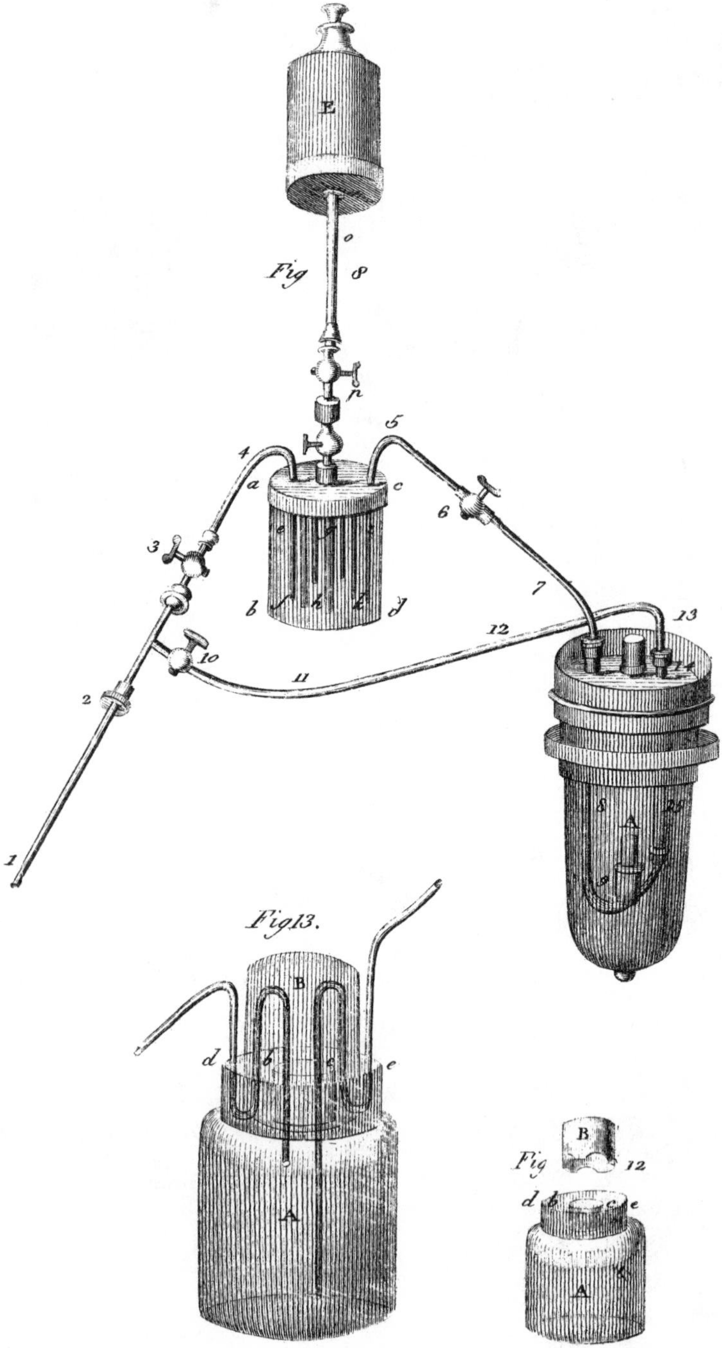

Fig 8

Fig13.

Fig 12

Plate XII (continued)

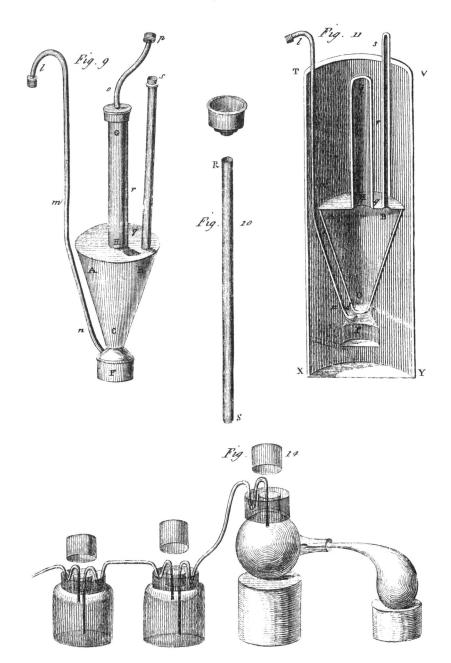

Plate XII (continued)

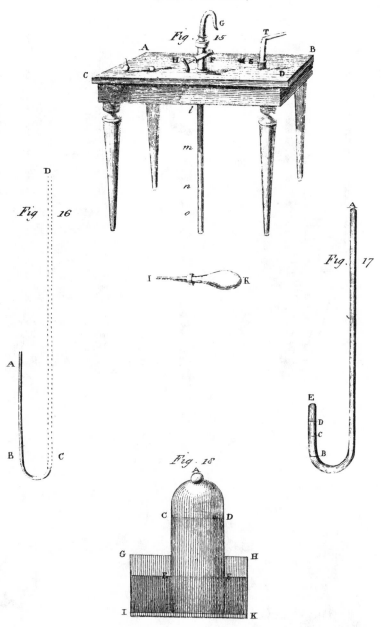

Fig. 15

Fig. 16

Fig. 17

Fig. 18

Plate XIII

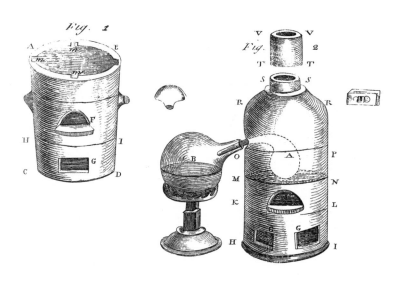

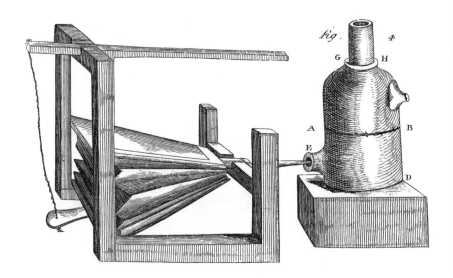

Plate XIII (continued)

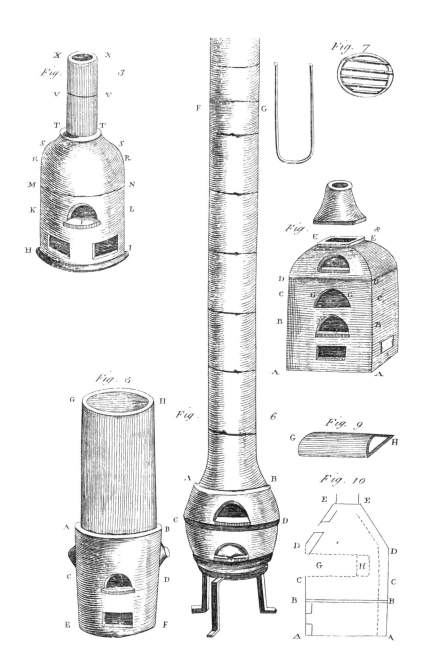

A CATALOG OF SELECTED

DOVER BOOKS
IN SCIENCE AND MATHEMATICS

A CATALOG OF SELECTED
DOVER BOOKS
IN SCIENCE AND MATHEMATICS

QUALITATIVE THEORY OF DIFFERENTIAL EQUATIONS, V.V. Nemytskii and V.V. Stepanov. Classic graduate-level text by two prominent Soviet mathematicians covers classical differential equations as well as topological dynamics and ergodic theory. Bibliographies. 523pp. 5⅜ x 8½. 65954-2 Pa. $14.95

MATRICES AND LINEAR ALGEBRA, Hans Schneider and George Phillip Barker. Basic textbook covers theory of matrices and its applications to systems of linear equations and related topics such as determinants, eigenvalues and differential equations. Numerous exercises. 432pp. 5⅜ x 8½. 66014-1 Pa. $12.95

QUANTUM THEORY, David Bohm. This advanced undergraduate-level text presents the quantum theory in terms of qualitative and imaginative concepts, followed by specific applications worked out in mathematical detail. Preface. Index. 655pp. 5⅜ x 8½. 65969-0 Pa. $15.95

ATOMIC PHYSICS (8th edition), Max Born. Nobel laureate's lucid treatment of kinetic theory of gases, elementary particles, nuclear atom, wave-corpuscles, atomic structure and spectral lines, much more. Over 40 appendices, bibliography. 495pp. 5⅜ x 8½. 65984-4 Pa. $13.95

ELECTRONIC STRUCTURE AND THE PROPERTIES OF SOLIDS: The Physics of the Chemical Bond, Walter A. Harrison. Innovative text offers basic understanding of the electronic structure of covalent and ionic solids, simple metals, transition metals and their compounds. Problems. 1980 edition. 582pp. 6⅛ x 9¼. 66021-4 Pa. $19.95

BOUNDARY VALUE PROBLEMS OF HEAT CONDUCTION, M. Necati Özisik. Systematic, comprehensive treatment of modern mathematical methods of solving problems in heat conduction and diffusion. Numerous examples and problems. Selected references. Appendices. 505pp. 5⅜ x 8½. 65990-9 Pa. $12.95

A SHORT HISTORY OF CHEMISTRY (3rd edition), J.R. Partington. Classic exposition explores origins of chemistry, alchemy, early medical chemistry, nature of atmosphere, theory of valency, laws and structure of atomic theory, much more. 428pp. 5⅜ x 8½. (Available in U.S. only) 65977-1 Pa. $12.95

A HISTORY OF ASTRONOMY, A. Pannekoek. Well-balanced, carefully reasoned study covers such topics as Ptolemaic theory, work of Copernicus, Kepler, Newton, Eddington's work on stars, much more. Illustrated. References. 521pp. 5⅜ x 8½. 65994-1 Pa. $15.95

PRINCIPLES OF METEOROLOGICAL ANALYSIS, Walter J. Saucier. Highly respected, abundantly illustrated classic reviews atmospheric variables, hydrostatics, static stability, various analyses (scalar, cross-section, isobaric, isentropic, more). For intermediate meteorology students. 454pp. 6⅛ x 9¼. 65979-8 Pa. $14.95

RELATIVITY, THERMODYNAMICS AND COSMOLOGY, Richard C. Tolman. Landmark study extends thermodynamics to special, general relativity; also applications of relativistic mechanics, thermodynamics to cosmological models. 501pp. 5⅜ x 8½. 65383-8 Pa. $15.95

APPLIED ANALYSIS, Cornelius Lanczos. Classic work on analysis and design of finite processes for approximating solution of analytical problems. Algebraic equations, matrices, harmonic analysis, quadrature methods, much more. 559pp. 5⅜ x 8½. 65656-X Pa. $16.95

INTRODUCTION TO ANALYSIS, Maxwell Rosenlicht. Unusually clear, accessible coverage of set theory, real number system, metric spaces, continuous functions, Riemann integration, multiple integrals, more. Wide range of problems. Undergraduate level. Bibliography. 254pp. 5⅜ x 8½. 65038-3 Pa. $9.95

INTRODUCTION TO QUANTUM MECHANICS With Applications to Chemistry, Linus Pauling & E. Bright Wilson, Jr. Classic undergraduate text by Nobel Prize winner applies quantum mechanics to chemical and physical problems. Numerous tables and figures enhance the text. Chapter bibliographies. Appendices. Index. 468pp. 5⅜ x 8½. 64871-0 Pa. $12.95

ASYMPTOTIC EXPANSIONS OF INTEGRALS, Norman Bleistein & Richard A. Handelsman. Best introduction to important field with applications in a variety of scientific disciplines. New preface. Problems. Diagrams. Tables. Bibliography. Index. 448pp. 5⅜ x 8½. 65082-0 Pa. $13.95

MATHEMATICS APPLIED TO CONTINUUM MECHANICS, Lee A. Segel. Analyzes models of fluid flow and solid deformation. For upper-level math, science and engineering students. 608pp. 5⅜ x 8½. 65369-2 Pa. $14.95

ELEMENTS OF REAL ANALYSIS, David A. Sprecher. Classic text covers fundamental concepts, real number system, point sets, functions of a real variable, Fourier series, much more. Over 500 exercises. 352pp. 5⅜ x 8½. 65385-4 Pa. $11.95

PHYSICAL PRINCIPLES OF THE QUANTUM THEORY, Werner Heisenberg. Nobel Laureate discusses quantum theory, uncertainty, wave mechanics, work of Dirac, Schroedinger, Compton, Wilson, Einstein, etc. 184pp. 5⅜ x 8½. 60113-7 Pa. $8.95

INTRODUCTORY REAL ANALYSIS, A.N. Kolmogorov, S.V. Fomin. Translated by Richard A. Silverman. Self-contained, evenly paced introduction to real and functional analysis. Some 350 problems. 403pp. 5⅜ x 8½. 61226-0 Pa. $11.95

PROBLEMS AND SOLUTIONS IN QUANTUM CHEMISTRY AND PHYSICS, Charles S. Johnson, Jr. and Lee G. Pedersen. Unusually varied problems, detailed solutions in coverage of quantum mechanics, wave mechanics, angular momentum, molecular spectroscopy, scattering theory, more. 280 problems plus 139 supplementary exercises. 430pp. 6½ x 9¼. 65236-X Pa. $14.95

ASYMPTOTIC METHODS IN ANALYSIS, N.G. de Bruijn. An inexpensive, comprehensive guide to asymptotic methods–the pioneering work that teaches by explaining worked examples in detail. Index. 224pp. 5⅜ x 8½. 64221-6 Pa. $7.95

OPTICAL RESONANCE AND TWO-LEVEL ATOMS, L. Allen and J. H. Eberly. Clear, comprehensive introduction to basic principles behind all quantum optical resonance phenomena. 53 illustrations. Preface. Index. 256pp. 5⅜ x 8½. 65533-4 Pa. $10.95

COMPLEX VARIABLES, Francis J. Flanigan. Unusual approach, delaying complex algebra till harmonic functions have been analyzed from real variable viewpoint. Includes problems with answers. 364pp. 5⅜ x 8½. 61388-7 Pa. $10.95

ATOMIC SPECTRA AND ATOMIC STRUCTURE, Gerhard Herzberg. One of best introductions; especially for specialist in other fields. Treatment is physical rather than mathematical. 80 illustrations. 257pp. 5⅜ x 8½. 60115-3 Pa. $7.95

APPLIED COMPLEX VARIABLES, John W. Dettman. Step-by-step coverage of fundamentals of analytic function theory–plus lucid exposition of five important applications: Potential Theory; Ordinary Differential Equations; Fourier Transforms; Laplace Transforms; Asymptotic Expansions. 66 figures. Exercises at chapter ends. 512pp. 5⅜ x 8½. 64670-X Pa. $14.95

ULTRASONIC ABSORPTION: An Introduction to the Theory of Sound Absorption and Dispersion in Gases, Liquids and Solids, A.B. Bhatia. Standard reference in the field provides a clear, systematically organized introductory review of fundamental concepts for advanced graduate students, research workers. Numerous diagrams. Bibliography. 440pp. 5⅜ x 8½. 64917-2 Pa. $11.95

UNBOUNDED LINEAR OPERATORS: Theory and Applications, Seymour Goldberg. Classic presents systematic treatment of the theory of unbounded linear operators in normed linear spaces with applications to differential equations. Bibliography. 199pp. 5⅜ x 8½. 64830-3 Pa. $7.95

LIGHT SCATTERING BY SMALL PARTICLES, H.C. van de Hulst. Comprehensive treatment including full range of useful approximation methods for researchers in chemistry, meteorology and astronomy. 44 illustrations. 470pp. 5⅜ x 8½. 64228-3 Pa. $12.95

CONFORMAL MAPPING ON RIEMANN SURFACES, Harvey Cohn. Lucid, insightful book presents ideal coverage of subject. 334 exercises make book perfect for self-study. 55 figures. 352pp. 5⅝ x 8¼. 64025-6 Pa. $11.95

OPTICKS, Sir Isaac Newton. Newton's own experiments with spectroscopy, colors, lenses, reflection, refraction, etc., in language the layman can follow. Foreword by Albert Einstein. 532pp. 5⅜ x 8½. 60205-2 Pa. $13.95

GENERALIZED INTEGRAL TRANSFORMATIONS, A.H. Zemanian. Graduate-level study of recent generalizations of the Laplace, Mellin, Hankel, K. Weierstrass, convolution and other simple transformations. Bibliography. 320pp. 5⅜ x 8½. 65375-7 Pa. $8.95

THE ELECTROMAGNETIC FIELD, Albert Shadowitz. Comprehensive undergraduate text covers basics of electric and magnetic fields, builds up to electromagnetic theory. Also related topics, including relativity. Over 900 problems. 768pp. 5⅝ x 8¼. 65660-8 Pa. $19.95

FOURIER SERIES, Georgi P. Tolstov. Translated by Richard A. Silverman. A valuable addition to the literature on the subject, moving clearly from subject to subject and theorem to theorem. 107 problems, answers. 336pp. 5⅜ x 8½. 63317-9 Pa. $11.95

THEORY OF ELECTROMAGNETIC WAVE PROPAGATION, Charles Herach Papas. Graduate-level study discusses the Maxwell field equations, radiation from wire antennas, the Doppler effect and more. xiii + 244pp. 5⅜ x 8½. 65678-0 Pa. $9.95

DISTRIBUTION THEORY AND TRANSFORM ANALYSIS: An Introduction to Generalized Functions, with Applications, A.H. Zemanian. Provides basics of distribution theory, describes generalized Fourier and Laplace transformations. Numerous problems. 384pp. 5⅜ x 8½. 65479-6 Pa. $13.95

THE PHYSICS OF WAVES, William C. Elmore and Mark A. Heald. Unique overview of classical wave theory. Acoustics, optics, electromagnetic radiation, more. Ideal as classroom text or for self-study. Problems. 477pp. 5⅜ x 8½. 64926-1 Pa. $14.95

CALCULUS OF VARIATIONS WITH APPLICATIONS, George M. Ewing. Applications-oriented introduction to variational theory develops insight and promotes understanding of specialized books, research papers. Suitable for advanced undergraduate/graduate students as primary, supplementary text. 352pp. 5⅜ x 8½. 64856-7 Pa. $9.95

A TREATISE ON ELECTRICITY AND MAGNETISM, James Clerk Maxwell. Important foundation work of modern physics. Brings to final form Maxwell's theory of electromagnetism and rigorously derives his general equations of field theory. 1,084pp. 5⅜ x 8½. 60636-8, 60637-6 Pa., Two-vol. set $27.90

AN INTRODUCTION TO THE CALCULUS OF VARIATIONS, Charles Fox. Graduate-level text covers variations of an integral, isoperimetrical problems, least action, special relativity, approximations, more. References. 279pp. 5⅜ x 8½. 65499-0 Pa. $8.95

HYDRODYNAMIC AND HYDROMAGNETIC STABILITY, S. Chandrasekhar. Lucid examination of the Rayleigh-Benard problem; clear coverage of the theory of instabilities causing convection. 704pp. 5⅝ x 8¼. 64071-X Pa. $17.95

CALCULUS OF VARIATIONS, Robert Weinstock. Basic introduction covering isoperimetric problems, theory of elasticity, quantum mechanics, electrostatics, etc. Exercises throughout. 326pp. 5⅜ x 8½. 63069-2 Pa. $9.95

DYNAMICS OF FLUIDS IN POROUS MEDIA, Jacob Bear. For advanced students of ground water hydrology, soil mechanics and physics, drainage and irrigation engineering and more. 335 illustrations. Exercises, with answers. 784pp. 6⅛ x 9¼. 65675-6 Pa. $19.95

DE RE METALLICA, Georgius Agricola. The famous Hoover translation of greatest treatise on technological chemistry, engineering, geology, mining of early modern times (1556). All 289 original woodcuts. 638pp. 6¾ x 11.   60006-8 Pa. $21.95

SOME THEORY OF SAMPLING, William Edwards Deming. Analysis of the problems, theory and design of sampling techniques for social scientists, industrial managers and others who find statistics increasingly important in their work. 61 tables. 90 figures. xvii + 602pp. 5⅜ x 8½.   64684-X Pa. $16.95

THE VARIOUS AND INGENIOUS MACHINES OF AGOSTINO RAMELLI: A Classic Sixteenth-Century Illustrated Treatise on Technology, Agostino Ramelli. One of the most widely known and copied works on machinery in the 16th century. 194 detailed plates of water pumps, grain mills, cranes, more. 608pp. 9 x 12.
28180-9 Pa. $24.95

LINEAR PROGRAMMING AND ECONOMIC ANALYSIS, Robert Dorfman, Paul A. Samuelson and Robert M. Solow. First comprehensive treatment of linear programming in standard economic analysis. Game theory, modern welfare economics, Leontief input-output, more. 525pp. 5⅜ x 8½.   65491-5 Pa. $17.95

ELEMENTARY DECISION THEORY, Herman Chernoff and Lincoln E. Moses. Clear introduction to statistics and statistical theory covers data processing, probability and random variables, testing hypotheses, much more. Exercises. 364pp. 5⅜ x 8½.   65218-1 Pa. $10.95

THE COMPLEAT STRATEGYST: Being a Primer on the Theory of Games of Strategy, J.D. Williams. Highly entertaining classic describes, with many illustrated examples, how to select best strategies in conflict situations. Prefaces. Appendices. 268pp. 5⅜ x 8½.   25101-2 Pa. $8.95

CONSTRUCTIONS AND COMBINATORIAL PROBLEMS IN DESIGN OF EXPERIMENTS, Damaraju Raghavarao. In-depth reference work examines orthogonal Latin squares, incomplete block designs, tactical configuration, partial geometry, much more. Abundant explanations, examples. 416pp. 5⅜ x 8¼.
65685-3 Pa. $10.95

THE ABSOLUTE DIFFERENTIAL CALCULUS (CALCULUS OF TENSORS), Tullio Levi-Civita. Great 20th-century mathematician's classic work on material necessary for mathematical grasp of theory of relativity. 452pp. 5⅜ x 8½.
63401-9 Pa. $11.95

VECTOR AND TENSOR ANALYSIS WITH APPLICATIONS, A.I. Borisenko and I.E. Tarapov. Concise introduction. Worked-out problems, solutions, exercises. 257pp. 5⅜ x 8¼.   63833-2 Pa. $9.95

THE FOUR-COLOR PROBLEM: Assaults and Conquest, Thomas L. Saaty and Paul G. Kainen. Engrossing, comprehensive account of the century-old combinatorial topological problem, its history and solution. Bibliographies. Index. 110 figures. 228pp. 5⅜ x 8½.   65092-8 Pa. $7.95

CATALYSIS IN CHEMISTRY AND ENZYMOLOGY, William P. Jencks. Exceptionally clear coverage of mechanisms for catalysis, forces in aqueous solution, carbonyl- and acyl-group reactions, practical kinetics, more. 864pp. 5⅜ x 8½.
65460-5 Pa. $19.95

PROBABILITY: An Introduction, Samuel Goldberg. Excellent basic text covers set theory, probability theory for finite sample spaces, binomial theorem, much more. 360 problems. Bibliographies. 322pp. 5⅜ x 8½. 65252-1 Pa. $10.95

LIGHTNING, Martin A. Uman. Revised, updated edition of classic work on the physics of lightning. Phenomena, terminology, measurement, photography, spectroscopy, thunder, more. Reviews recent research. Bibliography. Indices. 320pp. 5⅜ x 8¼. 64575-4 Pa. $8.95

PROBABILITY THEORY: A Concise Course, Y.A. Rozanov. Highly readable, self-contained introduction covers combination of events, dependent events, Bernoulli trials, etc. Translation by Richard Silverman. 148pp. 5⅜ x 8¼. 63544-9 Pa. $8.95

AN INTRODUCTION TO HAMILTONIAN OPTICS, H. A. Buchdahl. Detailed account of the Hamiltonian treatment of aberration theory in geometrical optics. Many classes of optical systems defined in terms of the symmetries they possess. Problems with detailed solutions. 1970 edition. xv + 360pp. 5⅜ x 8½.
67597-1 Pa. $10.95

STATISTICS MANUAL, Edwin L. Crow, et al. Comprehensive, practical collection of classical and modern methods prepared by U.S. Naval Ordnance Test Station. Stress on use. Basics of statistics assumed. 288pp. 5⅜ x 8½. 60599-X Pa. $8.95

DICTIONARY/OUTLINE OF BASIC STATISTICS, John E. Freund and Frank J. Williams. A clear concise dictionary of over 1,000 statistical terms and an outline of statistical formulas covering probability, nonparametric tests, much more. 208pp. 5⅜ x 8½. 66796-0 Pa. $7.95

STATISTICAL METHOD FROM THE VIEWPOINT OF QUALITY CONTROL, Walter A. Shewhart. Important text explains regulation of variables, uses of statistical control to achieve quality control in industry, agriculture, other areas. 192pp. 5⅜ x 8½. 65232-7 Pa. $8.95

METHODS OF THERMODYNAMICS, Howard Reiss. Outstanding text focuses on physical technique of thermodynamics, typical problem areas of understanding, and significance and use of thermodynamic potential. 1965 edition. 238pp. 5⅜ x 8½.
69445-3 Pa. $8.95

STATISTICAL ADJUSTMENT OF DATA, W. Edwards Deming. Introduction to basic concepts of statistics, curve fitting, least squares solution, conditions without parameter, conditions containing parameters. 26 exercises worked out. 271pp. 5⅜ x 8½.
64685-8 Pa. $9.95

TENSOR CALCULUS, J.L. Synge and A. Schild. Widely used introductory text covers spaces and tensors, basic operations in Riemannian space, non-Riemannian spaces, etc. 324pp. 5⅜ x 8¼. 63612-7 Pa. $11.95

A CONCISE HISTORY OF MATHEMATICS, Dirk J. Struik. The best brief history of mathematics. Stresses origins and covers every major figure from ancient Near East to 19th century. 41 illustrations. 195pp. 5⅜ x 8½. 60255-9 Pa. $8.95

A SHORT ACCOUNT OF THE HISTORY OF MATHEMATICS, W.W. Rouse Ball. One of clearest, most authoritative surveys from the Egyptians and Phoenicians through 19th-century figures such as Grassman, Galois, Riemann. Fourth edition. 522pp. 5⅜ x 8½. 20630-0 Pa. $13.95

HISTORY OF MATHEMATICS, David E. Smith. Nontechnical survey from ancient Greece and Orient to late 19th century; evolution of arithmetic, geometry, trigonometry, calculating devices, algebra, the calculus. 362 illustrations. 1,355pp. 5⅜ x 8½. 20429-4, 20430-8 Pa., Two-vol. set $27.90

THE GEOMETRY OF RENÉ DESCARTES, René Descartes. The great work founded analytical geometry. Original French text, Descartes' own diagrams, together with definitive Smith-Latham translation. 244pp. 5⅜ x 8½. 60068-8 Pa. $8.95

GAMES, GODS & GAMBLING: A History of Probability and Statistical Ideas, F. N. David. Episodes from the lives of Galileo, Fermat, Pascal, and others illustrate this fascinating account of the roots of mathematics. Features thought-provoking references to classics, archaeology, biography, poetry. 1962 edition. 304pp. 5⅜ x 8½. (USO) 40023-9 Pa. $9.95

THE HISTORY OF THE CALCULUS AND ITS CONCEPTUAL DEVELOPMENT, Carl B. Boyer. Origins in antiquity, medieval contributions, work of Newton, Leibniz, rigorous formulation. Treatment is verbal. 346pp. 5⅜ x 8½. 60509-4 Pa. $9.95

THE THIRTEEN BOOKS OF EUCLID'S ELEMENTS, translated with introduction and commentary by Sir Thomas L. Heath. Definitive edition. Textual and linguistic notes, mathematical analysis. 2,500 years of critical commentary. Not abridged. 1,414pp. 5⅜ x 8½. 60088-2, 60089-0, 60090-4 Pa., Three-vol. set $34.85

GAMES AND DECISIONS: Introduction and Critical Survey, R. Duncan Luce and Howard Raiffa. Superb nontechnical introduction to game theory, primarily applied to social sciences. Utility theory, zero-sum games, n-person games, decision-making, much more. Bibliography. 509pp. 5⅜ x 8½. 65943-7 Pa. $14.95

THE HISTORICAL ROOTS OF ELEMENTARY MATHEMATICS, Lucas N.H. Bunt, Phillip S. Jones, and Jack D. Bedient. Fundamental underpinnings of modern arithmetic, algebra, geometry and number systems derived from ancient civilizations. 320pp. 5⅜ x 8½. 25563-8 Pa. $9.95

CALCULUS REFRESHER FOR TECHNICAL PEOPLE, A. Albert Klaf. Covers important aspects of integral and differential calculus via 756 questions. 566 problems, most answered. 431pp. 5⅜ x 8½. 20370-0 Pa. $9.95

CHALLENGING MATHEMATICAL PROBLEMS WITH ELEMENTARY SOLUTIONS, A.M. Yaglom and I.M. Yaglom. Over 170 challenging problems on probability theory, combinatorial analysis, points and lines, topology, convex polygons, many other topics. Solutions. Total of 445pp. 5⅜ x 8½. Two-vol. set.

Vol. I: 65536-9 Pa. $8.95
Vol. II: 65537-7 Pa. $7.95

FIFTY CHALLENGING PROBLEMS IN PROBABILITY WITH SOLUTIONS, Frederick Mosteller. Remarkable puzzlers, graded in difficulty, illustrate elementary and advanced aspects of probability. Detailed solutions. 88pp. 5⅜ x 8½.

65355-2 Pa. $4.95

EXPERIMENTS IN TOPOLOGY, Stephen Barr. Classic, lively explanation of one of the byways of mathematics. Klein bottles, Moebius strips, projective planes, map coloring, problem of the Koenigsberg bridges, much more, described with clarity and wit. 43 figures. 210pp. 5⅜ x 8½.   25933-1 Pa. $8.95

RELATIVITY IN ILLUSTRATIONS, Jacob T. Schwartz. Clear nontechnical treatment makes relativity more accessible than ever before. Over 60 drawings illustrate concepts more clearly than text alone. Only high school geometry needed. Bibliography. 128pp. 6⅛ x 9¼.   25965-X Pa. $7.95

AN INTRODUCTION TO ORDINARY DIFFERENTIAL EQUATIONS, Earl A. Coddington. A thorough and systematic first course in elementary differential equations for undergraduates in mathematics and science, with many exercises and problems (with answers). Index. 304pp. 5⅜ x 8½.   65942-9 Pa. $9.95

FOURIER SERIES AND ORTHOGONAL FUNCTIONS, Harry F. Davis. An incisive text combining theory and practical example to introduce Fourier series, orthogonal functions and applications of the Fourier method to boundary-value problems. 570 exercises. Answers and notes. 416pp. 5⅜ x 8½.   65973-9 Pa. $13.95

AN INTRODUCTION TO ALGEBRAIC STRUCTURES, Joseph Landin. Superb self-contained text covers "abstract algebra": sets and numbers, theory of groups, theory of rings, much more. Numerous well-chosen examples, exercises. 247pp. 5⅜ x 8½.

65940-2 Pa. $8.95

STARS AND RELATIVITY, Ya. B. Zel'dovich and I. D. Novikov. Vol. 1 of *Relativistic Astrophysics* by famed Russian scientists. General relativity, properties of matter under astrophysical conditions, stars and stellar systems. Deep physical insights, clear presentation. 1971 edition. References. 544pp. 5⅜ x 8½.

69424-0 Pa. $14.95
